Liberalism 2.0 and the Rise of China

Innovation is crucial in this period of historic, global turbulence amidst seismic environmental, technological and (geo)political change. But what is innovation?

In *Liberalism 2.0 and the Rise of China*, Tyfield challenges the typical depiction of innovation as new technologies which 'solve' specific 'problems'. Innovation is presented as something much more complicated – a thoroughly social, cultural and political process with profound implications for the arrangement of power in society, and hence also a lens on emerging futures and how we can shape them. Indeed, exploring evidence from the key arena of low-carbon urban mobility innovation in the pivotal location of a rising China, this enlightening book describes the global systemic crisis of a neoliberal world order, manifest in Four Great Challenges, and the embryonic emergence of an alternative global power regime: a 'liberalism 2.0'.

Forecasting a digitally-based and complexity-adept revitalization of the classical liberalism of the nineteenth century, as well as new Dickensian inequalities and injustices that are reminiscent of the Victorian age, this title will be of interest to undergraduate and postgraduate students, as well as postdoctoral researchers, interested in fields such as Political Economy, contemporary China, Science and Technology, Sustainable Transitions and Mobilities.

David Tyfield is a Reader at the Lancaster Environment Centre, Lancaster University, a Director of the Joint Institute for the Environment, Guangzhou and Co-Director of the Centre for Mobilities Research.

Routledge Advances in Sociology

For a full list of titles in this series, please visit www.routledge.com/series/SE0511

Other books tell us what to think about China. This book shows us how to think with and through China. A stirring fusion of social analysis and Sinofuturism, David Tyfield's Liberalism 2.0 and the Rise of China lays out the logics of innovation through which the global system is being reinvented – as we speak.

Nigel Clark, *Chair of Social Sustainability, Lancaster Environment Centre, Lancaster University, UK*

Tyfield offers us a fascinating crystal ball into the future playing out of current crises of neoliberal global capitalism informed by his deep understanding of the dynamics of complex power/knowledge systems and the concept of innovation-as-politics. He combines this astute theoretical vision of the contradictions and 'monstrosities' of post-human technological change with an eye-opening empirical study of China's dynamic systems innovation, epitomized by turbulent struggles over transitions in electric auto-mobility and the disruptive emergence of mobility-as-a-service. You could not find a better starting place for insights into the future of urbanization in megacities, the failure of 'Googliberal' transformations of the global economy, and the tug-of-war of liberty-security logics that will shape the 21st century global economy.

Mimi Sheller, *Professor of Sociology and Director of the Center for Mobilities Research and Policy, Drexel University, USA*

A superb analysis of China's innovation system and the struggle for new mobility. This title is essential to understand the social shaping of technology and the fragmented but dynamic politics of innovation in China.

Boy Lüthje, *Visiting Professor School of Government, Sun Yat-sen University, Guangzhou, China*

Deeply questioning the global risks we face, Tyfield brilliantly shows the intrinsic limits built into the neoliberal model of innovation, epitomized by Silicon Valley, and reveals why and how new sources of power-knowledge system innovations are emerging in China. Anyone looking for a new technological and economic vision of ecological civilization should read this title.

Sang-Jin Han, *Professor Emeritus, Department of Sociology, Seoul National University, South Korea; Distinguished Visiting Professor, Department of Sociology, Beijing University, China*

Liberalism 2.0 and the Rise of China

Global Crisis, Innovation and Urban Mobility

David Tyfield

Routledge
Taylor & Francis Group

LONDON AND NEW YORK

First published 2018
by Routledge

2 Park Square, Milton Park, Abingdon, Oxfordshire OX14 4RN
711 Third Avenue, New York, NY 10017

Routledge is an imprint of the Taylor & Francis Group, an informa business

First issued in paperback 2018

British Library Cataloguing in Publication Data
A catalogue record for this book is available from the British Library

Library of Congress Cataloging in Publication Data
A catalog record for this book has been requested

ISBN: 978-1-138-83263-3 (hbk)
ISBN: 978-1-138-39304-2 (pbk)

Typeset in Galliard
by Wearset Ltd, Boldon, Tyne and Wear

For Deb

Contents

Outline of the book

Part I The problem: the global system crisis of neoliberalism

Chapter 1 Neoliberalism, knowledge and the global system

- Neoliberalism is the dominant regime of global system government today.
- Neoliberalism is a political project built upon epistemic market fundamentalism.
- This gives neoliberalism a specific set of dynamic relations to knowledge that can be both described in the abstract and traced in the concrete history of its evolution as a dominant regime to date.
- These relations to knowledge can be studied from the perspective of complex power/knowledge systems (CP/KS).

Chapter 2 Four Great Challenges

- There are four Great Challenges today confronting humanity: environment and Anthropocene; cosmopolitized globalism; post-human innovation; and the complex government of complex systems.
- All centrally involve innovation – and are driven by the specifically neoliberal model of innovation.
- Together they add up to the emergence of global risk-innovation-capitalism
- The existential challenge today is to forge new relations to knowledge adequate to the government of complex global systems.

Chapter 3 The genealogy of the emerging capitalist present

- Given problems – theoretical, empirical and political – with prognoses of a post-capitalism, utopia or dystopia, and robust evidence of its continuation, our default hypothesis for 'what next?' is a new regime of global capitalism.
- This hypothesis can be examined using a CP/KS analysis to construct a genealogy of the emerging present as a project of phronesis, or strategic situated practical wisdom.
- This analysis is illuminated by the abstract dynamic of liberal system growth, overthrow and resettlement: liberty-security, in innovation and new (discursive) power/knowledges.

Part II Where are we? Innovation in China

Chapter 4 Will China rule the world? The emergence of Chinese capitalism

- China manifests perhaps the most intense and globally significant forms of each of the Four Challenges (and associated liberty-security dynamics)
- At the heart of that dynamism is the contemporary climax of the historical encounter of (the CP/KSs of) global capitalism and China, as 'unstoppable force' (UF) and 'immoveable object' (IO) respectively.
- The key question of the moment is thus 'Will China rule the world?', but where each of these terms is themselves unsettled.
- Our default hypothesis may thus be rephrased as 'How are China and global capitalism shaping each other through processes of power/knowledge innovation?'

Chapter 5 The supply side: Debates and paradoxes regarding Chinese innovation upgrade

- China's innovation capacity is a crucial, politically fraught and confounding issue, containing multitudes and contradictions.
- Optimists and pessimists regarding China's imminent global leadership in innovation are both armed with convincing evidence – and both accounts are important half-truths.
- Chinese innovation capacity has improved in a characteristic non-linear fashion – of innovation and counter-innovation – that systematically creates unexpected innovation capacity that is 'not what the CCP wants'.
- To understand the whole, we must attend to this non-linear process of innovation-as-politics and its co-production with the intensifying dynamic of 'immoveable object' and 'unstoppable force'.

Chapter 6 The unexpected innovation hegemon

- The trajectory and conditions of Chinese innovation upgrade have produced a specific and surprising form of entrepreneurship: disruptive innovation.
- These firms, some now very large, technologically advanced and powerful, are shaping innovation-as-politics that is now pushing beyond merely economic/industrial disruption.
- Co-producing the heightening intensity of the IO vs. UF, these innovators are constructing the impetus, capacity, agency, process and conditions for a globally significant shift in dominant model of innovation-as-politics and global capitalism.

Chapter 7 The demand side: the emergence of risk/innovation-class in China

- The demand side of innovation is also crucial, if too often overlooked, in assessing the trajectories of Chinese innovation-as-politics (vis-à-vis the Four Challenges).

- In CP/KS terms, 'demand' means the coproduction of specific innovations and the changing socio-political profile of those demanding and enabled by them.
- In the shadow of the 'IO vs UF', the dynamics of liberty-security in Chinese socio-technical change are particularly intense.
- This is seeding the emergence of a new category of social stratification: risk/innovation-class.

Chapter 8 The emerging historic bloc: China's middle risk/innovation-class

- The rise of the Chinese (and global) 'middle class' is the socio-political story of the age.
- But what 'middle class' means is essentially contested.
- It is more illuminating to view 'middle class' as a productive political discourse for a socio-political group that is still in-formation as the system winners of an emerging CP/KS, the middle risk-class.
- This group are progressively enabling and being enabled by innovation-as-politics attentive to the Four Challenges
- The convergence of key socio-political dynamics through innovation demand is constructing the middle risk-class-for-itself as key agency of a new Chinese capitalism.

Part III Where are we going? Sharing and haggling the long complex journey to green urban mobility systems transition in China

Chapter 9 Electric vehicle innovation-as-politics in China

- Urban mobility innovation is a key case regarding the Four Challenges and the future of capitalism, with China at its apex.
- The official focus in China is on the electric car, generating familiar optimist and pessimist analyses.
- Meanwhile, the specifically Chinese disruptive innovation of electric two-wheelers and micro EVs is generally neglected by policy and analysis.
- Instantiating dynamics of liberty-security, Chinese e-mobility innovation presents an arena of contending 'monsters', none of which is currently 'domesticated' and capable of restabilizing urban mobility systems.

Chapter 10 Towards mobility-as-a-service

- Capping e-vehicle innovation and the crux of mobility innovation-as-politics today is the emerging imaginary of Mobility-as-as-Service (MaaS).
- This is dominated globally by late neoliberal ('Googliberal') ventures such as Uber.
- In China, though, it is Chinese disruptive innovators that are dominating MaaS.
- This trajectory of innovation-as-politics, co-produced with the exceptional dynamism of the IO vs UF and the emergence of China's middle risk-class, is generating a unique socio-political dynamism constructing a new Chinese, complexity-attentive capitalism.

Part IV What can be done? Conclusion

Chapter 11 Liberalism 2.0 and beyond

- The Chinese middle risk/innovation-class are emerging as historic bloc of a new complexity-attentive regime of global capitalism centred on (mega-city) China: liberalism 2.0.
- This liberal 2.0 regime will likely meaningfully address the Four Challenges, but in ways that systematically exclude a great many system losers alongside the system winners who are enabling and empowered by it, auguring a new 19th century.
- Working with this strategic tide and seeking to shape it to better futures, in the short-medium- and long-term, demands the new relation to knowledge of a phronetic virtue ethics of complexity, in practices that cultivate a strategic, situated and ethical wisdom.
- This underpins a left Liberalism 2.0 in the short-term towards a con-vivial neo-socialism in the long-term.
- A phronetic analysis and politics of innovation is crucial in this programme, incubating a phronetic civilization through virtuous innovation.

Figures

Tables

Acknowledgements

This book has been incubating for several years, even as it was finally written in an intense six-month period at the end of 2016. Along the way, though, it has inevitably benefitted immeasurably from numerous discussions. First and foremost, I must thank my partners in the ESRC project 'Low Carbon Innovation in China: Prospects, Politics and Practices', Adrian Ely, Sam Geall, Li Ping and Frauke Urban, and with a special thanks to Dennis Zuev for his excellent fieldwork, countless insightful discussions and tireless good humour. Through this project and much besides over the last decade, I must also single out my bottomless gratitude to John Urry, for everything he did to support me, and this work, before his untimely death. Similarly, I am forever grateful to Ulrich Beck for his support, insight and encouragement. Both of these intellectual giants are sorely missed. Other colleagues who have inspired and informed this work along the way include Anders Blok, Sabine Selchow, Ingrid Volkmer, Albert Gröber, Sang-Jin Han, Julia Guivant, John Schellnhuber, Monika Büscher, Mimi Sheller, Xu Honggang, Gu Shulin, Adrian Smith, Jesse Goldstein, Brian Garvey, Leo Freire de Mello, André Luiz Sica de Campos, Phil MacNaghten, Rebecca Lave, Sam Randalls, Charles Thorpe, Lixian Qian, Didier Soopramanien, Xu Changfu, Boy Lüthje, Julia Kirkegaard, Douglas Fuller, Paul Cambre and Sebastian Pfotenhauer. My apologies to anyone whom I may have missed. I am very grateful to Kevin Jones, Gan Zhang and Yongtao Li for their support, giving me the opportunity to live and work in China for two years – experience essential to this book. Nils Markusson, Kean Birch, Dean Curran and Nigel Clark all offered invaluable feedback on parts, or all, of the book, alongside further discussions, while Nigel and Leigh Martindale have stimulated countless fascinating conversations. All the errors that remain, though, of course are mine. Thanks also to Emily Briggs and Elena Chiu at Routledge for their patience and assistance. Finally, I must thank both my parents, Stuart and Linda, and my wife's, Graham and Maggie, for their numerous acts of great generosity and support, especially when living abroad and when back but juggling the writing up, childcare and illness. And, of course, my beloved daughters, Hannah and Adah, just for being in my life. It is to my wife, Deb, though that I dedicate this book. Without your sense of adventure, love and endless support, this work would never have been possible.

Introduction

The year is '47. A queue has gathered, stretching round the corner, for tickets to the latest blockbuster at the cinema. When the film starts, an image, destined to become iconic, flashes up on the silver screen before the delighted, expectant crowd. Some spontaneously cheer. Our hero rides into an abandoned town, peppered with remnants of a thriving commercial past, on his horse. Sand blows against his face and a pitiless sun beats down. He is here to save the day, to beat the corrupt, lawless bullies that have torn the community apart for their gain and by sowing discord through populist but undeliverable promises of simply rebuilding what was once there and flourishing.

Except this is not a Western, showing on American screens in 1947. It is 2047, when cinemas have achieved a late 'retro' boom. And this is not a mythical depiction of the nation-building of late nineteenth century America, but of the early twenty-first century ascendancy of China. Similarly, the hero is riding a peculiar kind of 'horse' – a (by then, antique) *Aima* (literally 'love(d) horse') electric bike. And the ruined urban landscape is that of twentieth century fossil fuel industrialism, the 'baddies' being the corporate and political (international) elites who have grown rich and powerful off that model of development and now stand in the way of changing it, even as it yields ever-decreasing social dividends and ever-increasing social problems. Of course, the sand and sun are the same. It's just that this is sand and sun unleashed with a new intensity by climate change, engulfing northern Chinese cities in growing deserts and extreme weather, while the sun is filtered through a choking, toxic haze. Finally, our hero is not in fact alone, but part of a crazy fleet of e-bikes – an entire workforce – weaving through a stationary traffic jam, onto pavements, jumping red lights, cursing at buses and lorries that cut them up.

What has the unlikely figure of the rider of an electric bike – a *Chinese* electric bike – got to do with 'saving the day' in the early twenty-first century? And how does he (or she – yes, sorry, removing the helmet, we find the hero is a woman) play such a pivotal role in the new mythology of a new age of Chinese greatness? The argument and narrative that follows offers an answer to precisely these questions.

Certainly, there is no shortage of evidence of the need for such a 'hero' today, actually situated in the early twenty-first century. It is easy – all too tempting – but probably fallacious to see the present as always the 'most important' and possibly 'most turbulent' phase in human history. Since the dawn of modernity, in particular, the demonstrable acceleration, expanded connectedness and socio-technical reach and power of human life offers some prima facie credibility to such alarmist inclinations. Nonetheless, we must be careful to avoid self-aggrandizing Jeremiad – and to balance any rigorous analysis of contemporary societal 'crisis' with an historical sensibility, a sense of geographical difference (e.g. crisis in the global North need not be 'global') and awareness of any secular tendencies that are improving lives, perhaps in historically unprecedented ways.

Even so, though, the present – i.e. the early twenty-first century – does indeed appear to be a moment of arguably unprecedented *global* socio-political turbulence. It is, I argue here, in fact a moment when humanity is confronting 4 Grand Challenges (see Chapter 2) – of global 'Anthropocenic' environment, cosmopolitized globalism, post-human innovation and complex government of complex systems – and only just beginning to acknowledge them and their (in turn) complex, 'wicked', knotted overlaps. Together these pose and instantiate a profound disorientation, calling for nothing less than new relations to knowledge. Naturally, this has stimulated a flood of work seeking to make sense of this predicament, to explain it and/or to resolve it, or at least various parts of it. Amongst the key buzzwords of the day, 'neoliberalism' has achieved particular (and legitimate) salience in these discussions. And, again, this book follows an ostensibly similar line of argument, opening with an explanation of just what neoliberalism is, and how the global turbulence is best understood precisely as a terminal crisis of that globally dominant and globalizing political economic regime (Chapter 1).

The main aim of this book, however, is to go far beyond an explanatory analysis of what is currently wrong in the world, alerting the reader to the enduring power and influence of the forces that have predominately shaped its perilous state. This is not just another critique of neoliberalism. Rather, our goal here is to explore what is being or may be constructed – and especially *how* – in the medium-term at the meso-level of globally dominant regime amidst the overlapping crises of neoliberalism that are the 4 Great Challenges.

The window we use on this crucial but heinously vexed question is perhaps the key mode of world- and future-making today, as well as being an often fetishized saviour of contemporary society, the very essence of twenty-first century progress: innovation. To get at what socio-technical worlds are being made today, however, we explore innovation as a process of power/knowledge, as politics, situated within and co-produced with complex systems of social relations (Chapter 3). Specifically, we here examine innovation along two key dimensions that direct us to instances that promise to be particularly illuminating from a systemic perspective: first, as innovation in response to the Four Challenges; and second, as it is actually taking shape in a political community of sufficient territorial and geopolitical heft to take over and expand capitalism beyond the capacities of the incumbent but troubled hegemony of the United States. There is just one place that has the slightest chance of fulfilling these basic criteria in the early twenty-first century: China.

Together these direct us to the key domain of Chinese urban mobility innovation … and to our proverbial electric two-wheelers. But, as such, our focus here on China also marks something of a departure – in specific subject matter as well as theoretical perspective – from the majority of the growing literature on the Grand Question of whether or not China will 'rule the world' in the twenty-first century. This is not just because of the specific focus on innovation, rather than jumping straight into 'big' (and, no doubt, essential) themes of (geo-)politics, (military and/or 'soft') power and political economy. But also because working 'up' from the concrete remaking of socio-political relations, domestic and international/ geopolitical, that is happening in parallel with ongoing Chinese efforts at 'low carbon transition' affords a more emergent, qualitatively new and surprising set of responses to the Grand Question; answers (and new questions) that do not just extrapolate existing structural tendencies and cemented definitions of politics and international relations (and their theories).

Such openings really matter. For whatever the actual future may hold, for China and the world, it will surely be dramatically conditioned by China's ongoing titanic attempts to grapple with the Four Challenges. Not only are these manifest in China with an exceptional intensity, but also in ways that are new. In other words, 'we' in the West have not been there before, however much both Western and Chinese decision-makers wish to believe the contrary. This qualitative novelty in itself would demand a commensurate novelty in our synthetic and synoptic (perhaps theoretical) attempts to understand and intervene in them. But this imperative is compounded further by the extraordinary conjunction of the ostensibly familiar and utterly, bafflingly unfamiliar (that is, vis-à-vis established Western-centric social scientific understanding) that characterizes contemporary Chinese society and its historically momentous ongoing transformation; a society that, again whatever happens, will undoubtedly be increasingly influential to life globally, including in the West itself. In short, to understand the future, we have no option but to grapple with what I call here (Chapter 4) the 'conceptual challenge of China' – and this means an equal and parallel imperative to study contemporary China empirically, as it is, allowing that process to test, problematize and reform our theoretical common-senses.

Following the first section, in which we explore 'where we are' globally and the theoretical approach adopted here, we thus turn to the substance of the book, in analysis of the contemporary co-production of a China-in-the-world and low-carbon urban mobility innovation (Parts II and III). While this reveals an essentially contested domain of innovation-as-politics, it also uncovers an ongoing process of exceptional socio-political dynamism that affords qualitative extrapolation into a new and resettled twenty-first century capitalism. This new global capitalism, it is argued, will be centred on a Chinese polity that has itself been significantly transformed in the process, especially through the rise of the embryonic historic bloc of the 'middle class'. In short, Chinese urban mobility innovation-as-politics promises to be as significant and influential a domain of twenty-first century global capitalism as was the steel-petroleum automobility of post-War American suburbia to US capitalist hegemony in the twentieth century, and for similar reasons of system government and its domination.

But what will this new hegemonic regime be like, qualitatively and socially? Who will it benefit and empower, and who newly burden, penalize, exclude and disable? What, in short, is currently unfolding out of the terminal crises of neoliberalism? And where does strategic opportunity thus lie for a more equitable and emancipatory politics? Only by addressing these questions can we actually begin to move definitively beyond disheartening and strategically self-defeating critical explanation of the current global predicament. It is for this reason, therefore, that it is from this issue, the final stage of the argument (Part IV/Chapter 11), that the book takes its title. For the emergent power/knowledge momentum of a complexity-adept and innovation-enabled but sharp-elbowed new 'global' middle class (centred on China) most likely conditions the emergence of a new *laissez faire* 'classical liberalism' – thereby marking a significant break with market *fundamentalist* neoliberalism – but fit for the epistemic and governance challenges of an age of complexity rather than marked by the overweening confidence in rational progress of its original incarnation in the early nineteenth century.

This 'Liberalism 2.0' or complexity liberalism thus emerges as the incipient system logic of the most strategically enabled political project of this turbulent (and, *as such*, quintessentially liberal) age, including as political patron and sponsor of a new complexity episteme and breakthroughs in 'complex' systems sciences. It is also, as we shall see, one that is perplexing and problematic: on the one hand, enabling significant action regarding the Four Challenges and in ways that make it clearly preferable to neoliberalism – let alone the chauvinistic nationalist, regressive, even fascistic, *il*liberal populism into which backlash against its terminal crisis is now clearly mutating and that is presenting itself ever-more insistently as the clear-and-present political antagonist of the moment; but also, on the other, most likely itself constructing new and possibly even more egregious but newly 'legitimate' inequalities.

In other words, if correct, the analysis that follows suggests that it is liberalism 2.0 – not any longer neoliberalism – that today most merits our critical attention, as simultaneously the (medium-term) political opportunity (vs zombie neoliberalism and insurgent Trumpism) *and* danger of the day. For in this way we may maximize the efficacy with which this embryonic hegemonic regime can be shaped, directed, ridden, harnessed and resisted towards brighter, more equitable futures, and at this early stage when it is still in-formation and most amenable to political action. Insofar as the Four Challenges put the world quite literally at stake (and in multiple senses), this could scarcely be a more important undertaking and responsibility.

Part I

The problem

The global system crisis of neoliberalism

1 Neoliberalism, knowledge and the global system

A global system in crisis

"The world has gone mad." So think many people as they wake today, 9 November 2016, according to social media. There is plenty of evidence of terrifying global disorientation in the day's newspaper (*Guardian* 2016). Page 21 relates the latest mass 'cyberheist', of Tesco Bank, in which over 40,000 customers have been targeted, half losing money. Increasingly dependent on online platforms, we are also exposed and insecure. Meanwhile, a 'gig economy' delivery company is in dispute with its workers over whether they will be acknowledged as such or treated as independent self-employed contractors (p. 25). At stake is potentially the very future of employment – or mass human redundancy. Directly below, climate change is brought to our attention in a story about wind power (p. 25) – welcome news, but still inadequate for the urgent shift to decarbonized energy needed to stay within 2°C (let alone 1.5°C) of global warming. Finally, philosopher Kwame Anthony Appiah (p. 27) argues against the very idea of a timeless and specifically Western civilization. Rather, in a world beset by turmoil – including ISIS (p. 18) and the exodus of Syrian refugees to a hostile, anti-immigrant Europe – 'civilization' is a title the still-dominant West must work hard to earn anew.

Yet the biggest story of the world's madness, of course, is one that broke too late for today's newspaper: the astonishing election as US president of Donald Trump, "a racist, sexist tax-dodging bully" (Freeman 2016) seemingly at ease with playing fast-and-loose with facts, who was nonetheless still supported by over 50 per cent of white women and 49 per cent of white college graduates. "The most qualified candidate in a generation was defeated by the least qualified of all time" (Freeman 2016). Today, no-one knows what this will mean for America, especially its vulnerable and minority communities, or for the world. It is clear, however, that, to his bitter opponents and differently bitter supporters alike, this changes everything, in a moment of profound disorientation. "The world has gone mad" seems to capture the mood perfectly.

There is a surprising common thread to all these seemingly disparate stories, in that in each case they centrally feature innovation, the ubiquitous buzzword of the day, often as supposed saviour. Cybersecurity, the 'future of work' (Susskind and Susskind 2015), low-carbon energy … all these problems will be solved with – *are* – innovation. With not too much of a push, this is even true of ISIS (who will be *beaten* with innovation), unsettled geopolitical common-senses and moral orders (where new, rising powers are challenging the hegemonic cultural monopoly of the

West *as* rising powers of innovation), and Trump (the most famous reality TV 'entrepreneur' for whom innovation will create the promised return of good American jobs). We will return to this innovation theme. But, more importantly still and our starting point here, these analytically distinct issues all have in common that they are manifestations of a global system in chronic terminal decline. Our task in this opening chapter is thus to clarify what is meant by this grand claim.

In doing so, we must first explain what each of these terms means: 'system', '*global* system' and 'crisis' thereof. And, indeed, *which* specific global system we are talking about (and not), its major salient characteristics and system logic. For the time being, it will suffice to use the following working definition of 'system': the complex multi-factorial and open assemblage of human and non-human (both natural and technological) agencies and the social relations bringing them together, where agencies and dynamic, inchoate structures of social relations ceaselessly condition and perform each other. The *global* system is the growing unification of life on earth, human and non-human, into a single planetary socio-natural system, particularly driven by the endless expansion of capitalism from Europe across the globe over the last 500–800 years and reaching ever new heights of integration (Arrighi 1994).[1]

Every such system involves distributed processes, themselves conditioned and shaped by the pre-existing relations and agencies, through which that system is *governed* and managed to preserve its integrity *as* a system as it evolves and inevitably encounters challenges. These we will call the processes of system government, and they are never a unified, coherent and purely intentional whole. Finally, a system 'crisis' is thus a crisis in the normal workings of that specific *regime* of system government as it encounters a (set of) problem(s) it *cannot* successfully manage, given the limitations and specific ways and sedimented common-senses of that system; hence as a 'crisis of crisis management' (Jessop 2013) of that system.

With that introduction, then, we may now proceed to explore in more detail the concrete regime of global system government that is now in terminal decline. What is this dominant regime, now in crisis? Neoliberalism.

The radical epistemic foundations of neoliberalism as global regime of system government

"Not 'neoliberalism' again!", I hear you cry. A copious and still fast-growing literature has detailed how neoliberalism has, contingently, dominated the evolution of global capitalism since the late 1970s (Harvey 2005; Peck and Tickell 2002; Rose 2007; Crouch 2011; Springer *et al.* 2016; Birch 2015; Birch and Mykhnenko 2010; Mirowski 2014; Davies 2016). This regime has several key characteristics: a political dogma of global economic liberalization and domestic privatization, unleashing and/or introducing markets to replace state provision and protection; financialization in terms of growing economic and political heft of the financial sector, debt and new financial products (Krippner 2011; Tabb 2012); growing corporate power, on a global scale (Soederberg 2010); and take-over and use of state power to drive forward marketization, financialization and concentration (Crouch 2011). Thus neoliberalism is a *dynamic* and distributed project, evolving through different phases and in different ways in different places (Ong 2006; Peck and Tickell 2002) and drawing on multiple sources of intellectual inspiration (Mirowski and Plehwe 2009; Peck 2010; Stedman Jones 2014).

In recent years academic literature has sought to analyse and categorize neo-liberalism – and neoliberalism*s* – with ever-increasing detail and nuance, and hence also disagreement. Meanwhile, in the public sphere, 'neoliberalism' has, at worst, been reduced to a politically progressive swearword, often without substance. This is not a propitious climate on which to draw on a concept. The latter can be repudi-ated by deploying a sufficiently rigorous definition, but the former can stand in the way of that too. As such, we are not here seeking to summarize or displace that ongoing discussion and its many important insights – though it does have a distinct flavour of the Owl of Minerva about it, taking flight at dusk. Rather, we call on the terminology of 'neoliberalism' as an *essential* term, but when conceptualized in two specific ways.

First, we are specifically interested in neoliberalism as a *regime of (global) system government* – in order to get *beyond* it. This regime can be analysed both in terms of the abstract system logic of neoliberal government and a more concrete history. In both cases, what stands out is the essential interconnection of neoliberalism and the specific model of knowledge production – research and innovation (R&I) – that it has conditioned. In other words, our focus is on the specific *relations to knowledge* (production) built into neoliberalism as regime of system government. Fleshing out this conceptualization allows us to see how – and how profoundly – the current global system is in crisis. These insights, however, require some inescapable concep-tual heavy-lifting – for which, apologies from the outset.

Second, the regime called 'neoliberalism' is here defined as a dynamic and vora-cious power/knowledge regime built upon a political project and ideology of *epi-stemic market fundamentalism* (Mirowski 2011; Mirowski and Plehwe 2009). This means that it elevates the 'market' from optimal mechanism of *allocation*, as in neo-classical economics or 'classical liberal' thought à la Adam Smith, to optimal and supra-human decision-maker or producer of *knowledge*, hence 'epistemic' (Mirowski 2009; Peck 2010; Brown 2015). This point is crucial and merits emphasis. It is common-sense today – obvious to the point of banality, and not just in the dominant West – that optimal government of society hinges on (social mechanisms for) making the best, and best-informed, decisions about the trajectory of societal development. Hence, for a sceptical, scientific Western epistemology, say, democracy and an open public sphere supplemented with public (scientific) expertise are deemed the best (or least worst) organization of society because the open contesta-tion of argument is presumed to lead to reasoned and evidence-based decisions. But this epistemic foundation is equally true of other, say authoritarian or Confucian, forms of modern government, just using different theories of knowledge: e.g. where the strong man dictator or Party *knows* better than the people, or where the enlight-ened Emperor and bureaucracy know better given their Confucian virtue, respectively.

Neoliberalism has ascended to deepening system dominance by working with such common-sense epistemic understanding of legitimate political order. It is thus *founded* on an epistemic cornerstone. But it also represents a radical break, albeit systematically concealed and exploited as such, with that common-sense since, for neoliberals, it is the market that can best achieve such knowledgeable decision-making. The market, on this conception, aggregates the individually-limited choices and information of market players in ways that then automatically reach the best possible outcome, in terms of both maximal realization of human negative freedom and aggregated knowledge, with the latter the priority.[2] The market thus *knows* in ways that systematically exceed limited human intelligence, individual and/or collective/institutional.

To see the political radicalism of this position, note first that it follows that the market and its outcomes *cannot in principle* be bettered. This thus makes impossible by definition all attempts to generate what may seem more 'rational' or 'acceptable' outcomes than those generated by markets. Indeed, it robs of all epistemic legitimacy even attempted criticisms of markets and market-organized processes. In short, therefore, neoliberalism as *market fundamentalist epistemology* becomes a pro-market political project that is incalculably more profound than even the more familiar (if still politically controversial) doctrines of *laissez faire* classical liberalism.

First, if the market is the best of all possible decision-makers, *all* processes involving or susceptible to decision processes would be optimized if arranged as markets. Even today, this is a radical political programme, but it was undoubtedly and obviously so at the moment of neoliberalism's ascendancy from the late 1970s. In that period, the 'core' capitalist societies from which neoliberalism most strikingly emerged were arranged as Keynesian welfare states, with significant and explicitly carved-out spheres of social life publicly provided; a development itself emergent from the turbulent attempts to preserve capitalism against fascism and communism in the early/mid twentieth century. Neoliberalism as a project of unbounded marketization, therefore, entails the fundamental reconstitution of socio-political order.

Second, the radicalism of its project of marketizing society inevitably raises significant and heated objections, together with arguments regarding the allegedly catastrophic nature of such reforms. The result is that markets must be *forcibly* constructed, in the face of what, by definition from a neoliberal perspective, is 'irrational', 'partial' and/or 'short-sighted' opposition. Yet the primary agency for such forcible construction is the apparatus of state power. State coercion of marketization, however, remains fundamentally legitimated in epistemic terms: insofar as it serves the market, state power is rational and is so *without* limit. Again, the political radicalism of this position is in marked contrast even with classical liberalism. For the latter, markets, conceptualized as spontaneously and 'naturally' emergent forms of social organization, provide the rational argument for the *limitation* of state power. Moreover, to the extent that a specific market appears to 'fail', for neoliberalism this is evidence for the need for more state intervention to ensure the market 'works' (Clarke 1982). Conversely, for classical liberalism, market failure is the crucial limit case of the rationality of markets; the exception in which a specific good or service is constituted such that its optimal allocation depends on state provision.

Third, since the market can neither be rationally gainsaid, nor reach a sub-optimal outcome, there are necessarily no limits to its rational application. Thus, not only the provision of all things but, further, all processes involving socially-significant judgements should be (re-)arranged as markets. With no limits to markets in principle (and in practice always receding behind new market 'fixes'), nor are there 'real' limits imposed by the (more or less scientifically knowable) 'natures' of particular goods, services or valued phenomena that would render their provision and management incompatible with marketization (cf. the Polanyian (1944/1957) argument that markets tend to the destruction of social bonds and stewardship of 'nature').

Neoliberalism is fundamentally dismissive of all such apparent ontological limits, seeing them as the contingent *epistemic* limits of, at best, current (scientific) 'knowledge' that credulous, limited humans dress up in ontological costume. This includes a productive disregard also for any neat ontological division of the world according to given (scientific paradigm-informed) common-senses; as in assigning a supposed reality to distinctions of 'nature' vs 'society', 'science' vs 'technology', 'world' vs 'nation' or, crucially, 'reality' vs 'knowledge' and 'truth' vs 'power'. To the contrary, only through active experiments of subjecting a phenomenon to market-based entrepreneurship will we find out what that thing is, or rather, what it may become. Inured to *ex ante* warnings about dangers inherent in a specific innovation (or trajectory thereof) as the weak-willed chattering of Cassandras and busybody regulators, neoliberalism thus empowers innovation that is particularly and proudly productive of social turbulence, inverting Schumpeter with its destructive creation.

Fourth, neoliberalism always has a compelling, easily memorized and world-making – not merely abstract, nuanced and critical – response to the obvious question of what should be done when confronted with unwanted outcomes from projects of marketization: more market! In its active commitment to such projects and associated (particularly technological) entrepreneurship, neoliberalism's epistemology counsels and legitimates a specific form of activity that often proves to be self-confirming: unshackled, the successful entrepreneur has undoubtedly introduced something new into the world (whether trivial or profound) that thereby both legitimates the neoliberal credo that the naysayers were wrong *and* generates the only relevant criterion of 'winning the (political) argument' – so "*Stick it to* the naysayers!" whether they are persuaded or not – namely actual business success on the market.

From this perspective, though, the emergence of problems that projects of marketization and proudly *irresponsible* innovation *themselves* produce is, for neoliberalism, simply the frontier of opportunity for further Promethean entrepreneurship (cf. Klein 2007; Pellizzoni 2011). 'Innovation' is thus crucial in neoliberal system maintenance because it is always and only the *next* round of neoliberal innovations that prevents the novel system challenges that neoliberal innovation *itself* produces from engendering broader system disintegration: an accelerating treadmill of innovation and novel risks (of growing scale and depth) that propels the exceptionally dynamic and ever-expanding construction of a society and a dominant model of R&I of a *specific*, i.e. neoliberal, type.

In this way, neoliberalism's epistemic fundamentalism – i.e. its apparently flagrant *a*rationalism – proves (in many, but not all, circumstances and in the short-to-medium term) to be strategically self-empowering and not, as many critics of neoliberalism have hoped, transparently contradictory and self-defeating. In short, the peculiar *epistemic* character of neoliberalism's political radicalism helps us understand how neoliberalism emerges and is sustained, even as its explicit disregard for 'limits' necessarily generates proliferating crises.

Finally, the epistemic nature of neoliberalism's political radicalism entails that projects of knowledge production, and the reframing of their institutions as markets, assume a centrality to the broader political project. This follows directly from the redefinition of the market as a primarily epistemic device, a 'marketplace of ideas' (MoI) (Mirowski 2011; Nik-Khah 2017). For 'ideas' or knowledges become the privileged medium of politically reconstructing societies, particularly in two key forms of the novel mediation of social relations by profit-seeking technological innovations and/or market-supporting government regulation and coercion. Moreover, 'ideas' themselves become a key sphere of social life to be subjected to marketization. The result is, respectively, the construction of ever-greater systemic demands for, and fetishization of, 'innovation' together with the tendential conflation of science with ('hi-tech') commercialized innovation (Tyfield 2012a).

This leads directly to the other face of neoliberalism: its explicitly political (and thus *anti*-epistemic) *epistemic radicalism*. This refers to the way in which, as a market fundamentalist epistemology, neoliberalism necessarily both wears the garb of rational legitimacy but also hands questions of rational judgement over to the market. Neoliberalism, therefore, is foundationally inimical to institutions of knowledge production that claim for themselves intrinsic criteria of rational argument; arguments that may well (often, if contingently) contradict the judgements of marketized outcomes and so represent intransigent critical outposts against the market rule.

Committed to the truth of the market as optimal knowledge-producing process and *thereby* opposed to truth as a matter of rational knowledge, neoliberalism (and especially neoliberal think-tanks) deploy a 'double truth' regime (Mirowski 2012). Here one truth is used for public, political consumption (e.g. 'market *vs* state' or 'the more knowledge the better') while another is presupposed or acted upon (e.g. 'subsume the state by the market' or 'more knowledge = more market'). More specifically, by vesting the construction of legitimate knowledge *not* in critical, rational and (quasi-)public debate, or an idealized 'republic of science' (Polanyi 1962), but in the outcomes of market-based entrepreneurialism, neoliberalism is also foundationally opposed to the structures and professionalized institutions of modern scientific research and education in the 'public good' form they had taken in the post-war period. While often prosecuted in the language of economics, efficiency, optimized 'output' etc.... the neoliberal commercialization of science is thus primarily a political project, and one founded upon the destruction of existing scientific institutions and their replacement with the 'marketplace of ideas' (Nik Khah 2017).

R&I and the history of neoliberalism

Science and innovation featured significantly in the historical emergence of neoliberalism to political dominance. This centrality takes several overlapping forms. First, consider the neoliberal reform of the political economic structures of knowledge-production, in the 1994 TRIPs (Trade Related Intellectual Property) agreement of the World Trade Organization. This global legislation is a pillar of the neoliberal Washington Consensus and was a highly controversial treaty. Yet it was implemented despite having been drafted by, and overwhelming beneficial to, a handful of (primarily US-based) corporations who, not being sovereign governments, were not even signatories. As Sell (1999, 171; see also 2003) described this unprecedented global coup, 'twelve corporations made public law for the world'.

To understand this unprecedented global coup, one must attend to its political economic context, and in particular to the rise of neoliberalism. The crises of the post-war Keynesian welfare state demanded new spheres of socio-economic life to be opened up to profitable capitalist enterprise. This involved projects of privatization and marketization, a distinctively neoliberal (and financialized) 'globalization' and a seemingly unanimous political consensus supporting new technological innovations. Bringing these three developments together, and thus at the core of neoliberal restructuring of the regulatory architecture of global capitalism, lay the 'globalized construction of knowledge scarcity' (May 2006, 53) through accumulation by dispossession (Harvey 2005), subjecting 'knowledge' production to private appropriation in search of profit in the form of strong and global intellectual property rights (IPRs).

In particular, allegedly wondrous, and massively hyped, new prospects of innovation that were supposedly dependent on the private appropriation of knowledge through IPRs and on a global scale were deployed to argue for a global knowledge 'enclosure' movement (Boyle 2003; Lessig 2001; Zeller 2008). The key example, and political agent, in this regard was an emerging coalition of massive pharmaceutical firms, start-ups and elite universities all interested in 'biotech' and all dependent, if in different ways, on private ownership of research results in the form of patents (Tyfield 2008). Justified, thus, in terms of needed and 'life-saving' innovation, TRIPs was primarily the means for this key neoliberal project, constructing a global regulatory architecture suitable for the marketization of knowledge.

Alongside these knowledge-focused transformations to the political economy, the *concept* of 'science' was also crucial to formulation of discourses that served to legitimate neoliberal transformations. Particularly in economics (Nik-Khah and Van Horn 2016), work through the post-war period, centred on the Chicago school, investigating the 'economics of science' laid the groundwork for the reconceptualization of the 'market' as the ideal information processor; concepts that ascended from *academic* respectability, if unorthodoxy, to *political* dominance in the late 1970s. Claiming to show how the 'production of knowledge', like that of goods and services, could be best arranged as a free market supposedly established the 'rational' dominance of projects of marketization over the (often intransigent) expertise of post-war 'public' (and publicly-funded) intellectuals, rather than presumed relations in the opposite direction.

Nor has this work in economics just been crucial to constructing power/knowledge technologies of *legitimation* of the neoliberal project. It has also provided key arguments that have shaped the regulatory framework for a specific model of innovation that fetishizes 'innovation' per se – itself identified with (high) 'technology' – especially in 'cutting-edge' industries. Alongside the state-sponsored knowledge enclosure, this model of innovation privileges innovation that:

- promises high, short-term returns, especially as financial(izable) assets;
- focuses on products servicing market demand of corporate/individual consumers, as opposed to publics or states;
- supports projects of corporate enclosure of bodies of knowledge that promise to maximize global corporate control of particular (technoscience-intensive) markets;
- revolutionizes (or 'disrupts') business processes (and the means of production) by substituting ever-more-skilled labour with capital-owned technologies and machine intelligence;
- is constitutively dismissive of ontological limits and risks.

The quintessential example here of this broader neoliberal innovation model is genetically-modified (GM) agriculture (Kinchy 2012; Bronson 2009; Levidow and Carr 2009; Bonneuil *et al.* 2014). For GM food staples that are genetically owned (both via global IPRs and, more effectively, through technologies only feasibly developed at great expense in well-funded corporate laboratories) by major transnational corporations present a paradigm case of a commodity that can achieve almost universal, unquestionable consumer demand together with all-but-unbreakable monopoly control of supply. This simultaneously drives a further corporate concentration of global agriculture (Weis 2007; McMichael 2009; Kloppenburg

1988). Moreover, this was an industry associated with the most 'promising' break-throughs in science of the day, namely in genomic biotechnology (Jasanoff 2005); while, conversely, the industry also systematically belittled any and every objection, regarding not just the potential risks but rather undeniable and irreducible uncertain-ties and unknowns involved in the introduction of these technologies: to complex ecologies, food chains, food quality and human health, social relations of farming, (traditional) farming knowledge, control of seeds and food sovereignty – and all these over the long-term. Together with medical or 'red' biotech, this has generated official discourses of 'knowledge-based bio-economies' as instances of the neoliberal fetishization of innovation (Birch *et al.* 2010; Boyd *et al.* 2001).

A key third example concerns the commercialization of science and the academy, particularly in the US and UK, but increasingly also across the rest of the world (McGettigan 2013; Collini 2012; Losh 2014; Best and Rich 2017; Xu and Ye 2017). This has taken multiple forms (Tyfield 2012a: 11; Slaughter and Rhoades 2004; Radder 2010) including:

- increased privatization of research funding;
- commercial 'accountability' and 'relevance'/'impact' criteria in competitive public funding;
- growth in university-industry relations and direct incorporation of science into commerce;
- growth in patenting, especially at universities and especially in life sciences;
- commodification of higher education.

While the general direction of change is not disputed, debate focuses on the *extent* of these changes and their positive or negative effects. Shapin (2008), for instance, argues convincingly that there is no clear *ex ante* reason why increasing scientific research done within or funded by private industry should be seen as problematic (see also Edgerton 2017). Publicly-funded university labs can do 'applied' research, just as, conversely, corporate labs have attained Nobel prizes for fundamental scient-ific insights. Such objections, nonetheless, miss much of what is seen to be troubling about the commercialization of science. These include concerns regarding the effects of a deeper penetration of commercial logics into scientific decisions (e.g. pressures to interpret or even massage data), research agendas, sharing findings (or not, against demands of commercial confidentiality) and other conflicts of interest, and the resulting dangers of loss of social trust in, and epistemic authority of, science more broadly (Radder 2010: 14; Kleinman 2010).

High-profile examples of all the objections have undoubtedly emerged (Mirowski and Sent 2002), though it is also clearly the case that the institutions of 'science' have not (yet) collapsed. Focusing exclusively on supposed evidence of an abstract incompatibility between 'business' and 'science', however, paradoxically serves to miss what is arguably most troubling. For in conceptualizing 'science' as a pristine sphere of knowledge production entirely for its own sake, one is unable to explore empirically two key questions. First, how what counts as 'science' is itself a political battle-ground in which diverse, if often tacit, political commitments are always in play; hence affording the potential for distinctively *neoliberal* science to emerge (Lave 2012; Lave *et al.* 2010; Busch 2011). And, second, how the foundationally inimical (as instrumental) relation of neoliberalism to the 'Republic of Science' can also be seen in various recent developments that are both deeply problematic for science and rational public debate, *and* both *un*problematic and, indeed, positively productive for neoliberalism and the construction of a social acceptance of its epistemology.

For instance, neoliberalism has deliberately cultivated public suspicion regarding the undisclosed 'political' motives underlying all discourses of rational objections, and especially to projects of marketization and ontological 'limits'. This has produced a cultural discourse intolerant of scientific conclusions identifying the emergence of potentially existential systemic threats, especially on ecological issues such as climate change. Moreover, such popular mistrust is often *justified* given the emergence of a 'marketplace of ideas'. In particular, sponsored by the 'double truth' regime and its active cultivation of 'truthiness', a new regime of knowledge production has emerged: agnotology (Oreskes and Conway 2011; Fernández Pinto 2017). Here 'knowledge' is deliberately treated as, first and foremost, a tool in political or commercial strategic projects within the marketplace of ideas; a device, moreover, whose effectiveness is parasitic upon the 'scientific' status and epistemic (and hence *political*) authority of such claims in winning high-stakes contests in the public sphere.

A key element of this process is the production of *ignorance* (Davies and McGoey 2012), and of three kinds:

- as obstacles to scientific findings that are politically disadvantageous to specific and (R&I-) empowered interests (e.g. quintessentially regarding tobacco, nutrition or climate change);
- its converse of 'science-as-PR', not science-as-truth, where the primary goal of the knowledge work is to secure some credibility for a particular strategic project, not to establish actual knowledge (e.g. regarding ghost-written and carefully curated literature on new pharmaceuticals (Sismondo 2017));
- ignorance regarding a systematically unaccountable scientific process (e.g. again pharmaceuticals, with negative results withheld due to 'commercial confidentiality').

The last of these is crucial and arguably the most self-destructive of the three. For the epistemic authority of science actually reposes upon a foundation of broad-based and generally unquestioned – but laboriously produced – public trust in the (supposedly) open, sceptical and unaligned *process* of its production. We suppose we *can* hold 'science' to account, even if we personally do not do so in every (or perhaps, any) particular instance. A dawning cynicism, if not rejection, regarding that trust and a deepening rejection of all forms of epistemic expertise (cf. Collins and Evans 2008) threatens this key pillar of the elevated political status of knowledge.

The popular distrust of science is paradoxically heightened further by the neoliberal instrumental deployment of science in the attempted *depoliticization* of political debate, instead leaving the field of political decision-making free for a seemingly 'objective' government by the market. Again, GM agriculture is a classic case. For as described by Levidow *et al.* (2007), the trans-Atlantic controversy thrown up by how, or if, to regulate GM crops took the form of pro-GM (American) denunciations of (European) objections as based on 'junk', not 'sound', science and the argument, before the WTO, that this was thus an illegitimate basis for an obstacle to free global trade. This thus constitutes an attempted 'scientification' of politics, transforming broad questions of technological uncertainty purely into specific and answerable scientific questions of risk assessment (cf. Stirling 2010). But, going further, it is also an attempted 'scientization' of politics and the political process per se, attempting to bypass and neutralize with 'sound' (i.e. neoliberal-supportive) science all *public and political* objection to the commercial introduction of new innovations (e.g. concerns about food sovereignty).

The actual effect of this process, however, has been precisely the opposite, as political controversy has leached in the other direction, ever-deeper into the science itself, as regarding GM agriculture or climate change (Grundmann 2012). To the extent this penetrates to issues that remain essentially undecided and uncertain, as is necessarily the case towards the forefront of scientific advance, this also can then pollute and frustrate the whole enterprise. For, caught up in political suspicion and recrimination, reasoned argument becomes practically impossible. In other words, the attempt to foreclose political debate with science simply leads to the *politicization* of science. Yet this hardly entails a defeat for the neoliberal project. To the contrary, as discussed above, by fragmenting further the 'Republic of Science' and the (largely overlooked) social preconditions for its epistemic authority, this is precisely to inoculate the project of ever-deeper marketization from even the *possibility* of concerted and 'reason-based' objection.

Complex power/knowledge systems and innovation-as-politics

In all these ways, the neoliberal regime of system government conditions cycles of a specific model of socio-technical and political-cultural change that tends to the incremental propagation of particularly controversial and socially divisive innovation, driving ever-greater marketization of the world for private gain and ever-more encompassing turbulence, in positive feedback loops, alongside the parallel gutting of the epistemic resources with which to hold it politically to account.

This key dynamic of neoliberalism tells us several crucial things. First, it shows just how important R&I is to the system government of neoliberalism; not just a crucial instrument of managing system processes to given neoliberal ends, but rather the means *and end* of this process, as the specifically neoliberal model of innovation mediates into existence the broader reshaping of the *world* – the *global system* – as neoliberal too. Second, this means that to understand neoliberalism *as* a regime of system government and political project, we must fundamentally rethink our default conception of knowledge (production), including (socio-technical) innovation (on a broad definition of knowledge). The central governmental and political role played by neoliberal-fashioned R&I is crucial to how this regime *works* and grows, not just *despite* but *through* its multiple contradictions and socio-political transgressions. To think *beyond* neoliberalism then, we must first think *with* it, recognizing that each and every deployment of knowledge – discursive, as ideas or texts, or material, as innovations – is always also a political, strategic intervention, turning them into specific 'technologies' of *power*/knowledge.

Finally, it shows how deeply immersed we are in this neoliberal predicament today. This is not a problem for neoliberalism, but for us. The very dynamic of neoliberal system government *through* neoliberal R&I tends not just to more, and more-system-enabled, 'Promethean' innovation, but also, thereby, to the progressive incubation of the ensuing turbulence beyond isolatable, exploitable 'risks' to uncontrollable and existential, system-threatening dangers. It is the overlapping proliferation of these that together make up the epochal challenge of the global system crisis of crisis management.

But what is the solution to all these problems today? The mainstream answer, of course, is precisely more innovation. Innovation to the rescue! The challenge of comprehending our current global predicament so as to intervene effectively in charting courses to better futures thus hinges to a great extent on how we are conceptualizing *this key term*; just as, conversely, understanding the *problem* as one of global system crisis of neoliberalism demands attention to its key role. In short, concerned about global system crisis, it emerges that our primary interest is regarding how to think about R&I, and conversely how R&I is central to all of these prior issues and lenses. This leads to the specific theoretical perspective of this book: a cultural political economy of research and innovation that explores the co-production of R&I and socio-political regimes in terms of dynamic emergent systems of relations and technologies of power/knowledge; or complex power/knowledge systems (CP/KS) for short.

This CP/KS perspective draws together cultural political economy (CPE) (Jessop and Sum 2006), political ecology (Lawhon and Murphy 2013), theories of socio-technical systems transition (Smith *et al.* 2010) and Foucauldian analysis of government, regarding the 'conduct of conduct' of polities and selves (Dean 2010; Lemke 2011), with a specific focus on issues of research and innovation (Tyfield 2012b). Asking primarily 'how?' and 'which?', not 'what?' or 'why?', X is the case regarding contemporary innovation and social order, this builds on Foucault's discussion of power to explore ' "the total structure of actions brought to bear" by some on the actions of others' (Hindess, 2006: 116, quoting Foucault, 2001: 336).

This leads to a conception of power/knowledge akin to that from discussion of neoliberalism above, albeit approaching from the concept of 'power' not 'knowledge'. Against contemporary common-sense understandings, power here is not conceived as a zero-sum and brute capacity held by the powerful over the powerless. Nor is power presumed to be normatively bad unless and until it is tamed by reasoned acceptance and legitimation. Instead, power is dispersed, ubiquitous, strategic, relational and productive or *constitutive* – of both larger 'systems' of government and of subjectivities within them. Importantly it is also normatively ambivalent – dangerous not necessarily bad (Foucault 2001), but also to be celebrated since living subjects and systems are themselves constituted and substantively shaped by power relations with others.

But these power relations are also relations of power/*knowledge*, for they constitute the very sense and meaning of the world – for thinking, strategic beings like ourselves – and specific roles and practices within it. Both the knowledges that understand, are interpreted and are applied in human action in the world *and* the very *world* thereby produced, therefore, are situated within and constitute such systems of relations and technologies of power/knowledge. While neoliberalism forces us to see that 'knowledge' is always an ontologically-productive, strategic intervention, then, Foucault's concept of power shows us that this is an insight about the human-shaped world per se not just the neoliberal-conditioned world at the start of the twenty-first century.

Here, in other words, is a way to think *with* neoliberalism without thereby being trapped within its destructive worldview. This is evident, for instance, in application of this perspective to the key contemporary challenge of understanding sociotechnical systems transition to more sustainable ways of life (Tyfield 2014). Shifting from a structural account of power, as something possessed by some over others in zero-sum relations of pure domination, to a relational, capillary and constitutive conception immediately loosens up and dynamizes concepts of system 'lock-in' and transition (Garvey *et al.* 2015; cf. Unruh 2000; Geels 2014). Instead of an analytical and practical paralysis, in which those structurally enabled today seem immoveable, openings are presented to explore (and perhaps assist) emergent alternative regimes. Reframed in CP/KS terms, transition becomes a process in which power/knowledge relations mediate strategic agency that is, in turn, qualitatively shaping *new* power/knowledge relations and technologies (and hence power systems) (see Figure 1.1). Furthermore, with innovation itself conceived as a process of socio-technical power – i.e. as *politics* – it also becomes a privileged window into this process, precisely as the key reflexive moment of *power/knowledge acting on itself.*

This concerted attention to reconceptualizing innovation is particularly important at this moment of system crisis. On the one hand, because – from a CP/KS perspective – the challenge of system crisis is to establish new system-adequate processes of system government that will condition the self-sustaining cycles of action and system (re)production that just *are* stable system government – and where such processes are de facto unlikely to be purely intentional or totally planned. This is precisely the challenge of formulating and constructing new power/knowledge technologies and relations. But, on the other hand, because today, in the incumbent CP/KS, 'innovation' is widely deemed to be the panacea, as just discussed. Discourses of how research and innovation promise to tackle and eliminate the multiple problems of the present – e.g. squaring 'green' and 'growth', or Big Pharma profits and global public health, or … – have reached new heights, manifesting almost a fetishism of innovation (Godin 2006; Tyfield 2012b).

When praising innovation as saviour in this way, the vast majority of conventional political and policy debate implicitly deploys a default contemporary definition of 'innovation' as the high-technology, privately-owned and research-intensive kind, i.e.

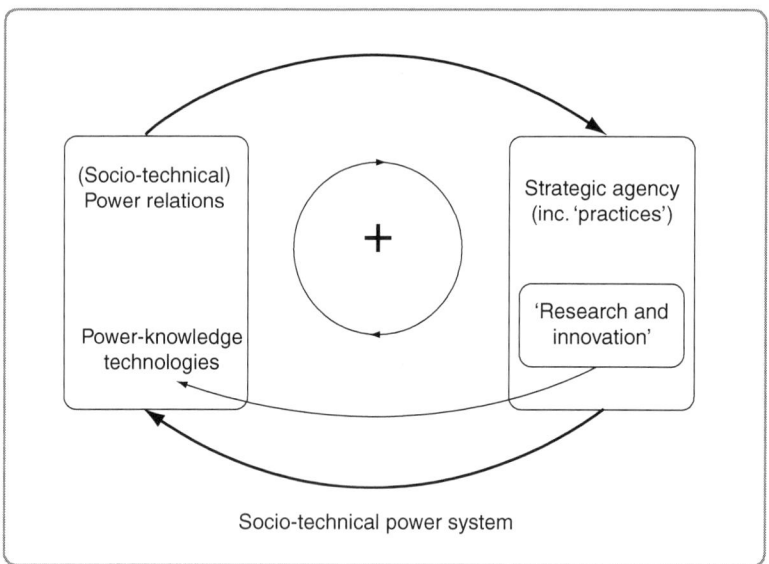

Figure 1.1 Complex socio-technical power/knowledge systems.

precisely the *neoliberal* model of innovation. In doing so, the questions they pose themselves and the challenges they pose for innovation also take the qualitative and apparently 'common sense' social characteristics of contemporary life for granted, searching for discrete interventions that will 'solve' the problems, while leaving social life as it is today substantially intact. At best, attention is paid to how innovation may change the macro-economic profile of society, as in economic debate between 'Austerians' and 'Keynesians' (e.g. Krugman 2013; Janeway 2012).

Other more broad-based and systemic approaches discuss 'long waves' of techno-economic change (Perez 2002, 2016; Mason 2015) or (still largely stalling) efforts at socio-technical system transition (Elzen *et al.* 2004). The latter in particular offer much insight *post hoc* into the complexity, multidimensionality and irreducible *social* aspects of historical transitions. But they struggle to illuminate urgently needed ways forward in the here and now (Smith *et al.* 2010). What all these approaches have in common, however, is the presumption that 'innovation' is both crucial to these solutions and a key factor in the flourishing of twenty-first century societies more generally. In themselves, these are unobjectionable conclusions. But even for those that do not simply add 'innovation' on at the end of their narrative, scant attention is paid to how innovation is already making the world and its problems; let alone to how specific dominant *models* of innovation are intricately associated with, and themselves situated within and constitutive of the dynamic *socio-political* structures – of power/knowledge relations – that are the primary locus of these problems at their most intractable, intense and 'wicked' (Verweij and Thompson 2006).

Conversely, this book *starts* with rethinking the crucial twenty-first century concept of 'innovation' and argues it is best conceived as an irreducibly social, political and cultural process, or innovation-as-politics. In particular, as just outlined, innovation is a process that both constructs and is itself mediated by the technologies (in a broad sense of power/knowledge) that enable and constrain the particular strategic power relations that constitute society as socio-technical and political systems. Understood in this way, we cannot miss how a dominant, incumbent model of innovation is not only *not* the solution but is itself a profound aspect of the problem – *as* a problem of the processes of contemporary system government.

Indeed, this specific regime of innovation is seeding, while simultaneously obstructing effective responses to, four key system challenges – what we will call the Four Great Challenges – that are compounding and coproducing neoliberal system crisis, and that are its most significant manifestations precisely *as* global complex system challenges. To stress, these Four Great Challenges are produced by the very *success* of neoliberalism and its regime of global system government through neoliberal innovation, actively *integrating* the world in its many analytically distinguishable aspects in ever-growing storms of global turbulence from which neoliberalism may feed itself yet further. But as such they are only ever exacerbated by it, and this includes through its crises and in its responses to those crises. Moreover, amongst these Challenges is the crisis of *R&I itself*, further problematizing any simplistic or pat allusion to innovation as the 'answer'.

Conclusion

What are these Four Great Challenges, for global system government and innovation at the beginning of the new millennium? We turn to these now. Throughout, our focus on each of these massive topics is to highlight how the regime of system government that is neoliberalism as CP/KS-*through*-innovation is key in all cases. As such, not only do they overlap substantively in many complex and 'wicked' ways, but they are also united by the single meta-problem of the system crisis of neoliberal government. Meanwhile, they are still being addressed, if at all, as separate and discrete challenges in ways that are still proving starkly – and from this perspective, unsurprisingly – inadequate. But just as it is inextricably situated within the challenges of system government that we may hope it can resolve, so too it follows that innovation, as a 'world-making' process, is also a key analytical window into both the reproduction of current societies and their problems, and the emergence of processes and new powerful groups that may disrupt, upend or otherwise transform existing systems. In other words, as we will discuss in more detail below (Chapter 3), through innovation-as-politics in complex power/knowledge systems we have a way not just to *think* beyond neoliberalism, and its appalling version of a global knowledge society – but also perhaps to begin to *see* beyond it too.

Notes

1 We use the language of 'global' system to differentiate it from the established 'world systems' literature (Wallerstein 2004), with which the approach has both considerable overlaps and significant differences.
2 'Negative freedoms' are the essentially liberal freedoms to be left alone (by the state and others) so long as one is not breaking explicitly legislated rules, as against the 'positive' freedom of rights to collective or state support actively to enable one's capacity for enjoyment of civic and socio-economic goods (Berlin 1959).

References

Arrighi, G. (1994) *The Long Twentieth Century*, London: Verso.
Berlin, I. (1959) *Two Concepts of Liberty: An Inaugural Lecture Delivered Before the University of Oxford on 31 October 1958*, Oxford: Clarendon.
Best, E. and D. Rich (2017) 'The political economy of higher education and student debt', in D. Tyfield, R. Lave, S. Randalls and C. Thorpe (eds), *The Routledge Handbook of the Political Economy of Science*, London: Routledge: 144–155.
Birch, K. (2015) *We Have Never Been Neoliberal*, London: Zero Books.
Birch, K. and V. Mykhnenko (eds) (2010) *The Rise and Fall of Neoliberalism: The Collapse of an Economic Order?*, London: Zed Books.
Birch, K., L. Levidow and T. Papaioannou (2010) 'Sustainable capital? The neoliberalization of nature and knowledge in the European "Knowledge-Based Bio-Economy"', *Sustainability* 2: 2898–2918.
Bonneuil, C., J. Foyer and B. Wynne (2014) 'Genetic fallout in bio-cultural landscapes: Molecular imperialism and the cultural politics of (not) seeing transgenes in Mexico', *Social Studies of Science* 44(6): 901–929.
Boyd, W., S, Prudham and R. Schurman (2001) 'Industrial dynamics and the problem of nature', *Society and Natural Resources* 14: 555–570.
Boyle, J. (2003) 'The second enclosure movement and the construction of the public domain', *Law and Contemporary Problems* 66: 33–73.
Bronson, K. (2009) 'What we talk about when we talk about biotechnology', *Politics and Culture* 2. Retrieved from: www.politicsandculture.org/issue/2009-issue-2/.
Brown, W. (2015) *Undoing the Demos: Neoliberalism's Stealth Revolution*, Cambridge, MA: MIT Press.
Busch, L. (2011) *Standards, Recipes for Reality*, Cambridge: MIT Press.
Clarke, S. (1982) *Marx, Marginalism and Modern Sociology*, Basingstoke: Macmillan.

Collini, S. (2012) *What are Universities For?*, London: Penguin.

Collins, H.M. and R. Evans (2008) *Rethinking Expertise*, Chicago: University of Chicago Press.

Crouch, C. (2011) *The Strange Non-Death of Neoliberalism*, Cambridge: Polity.

Davies, W. (2016) *The Limits of Neoliberalism*, London and Thousand Oaks, CA: Sage.

Davies, W. and L. McGoey (2012) 'Rationalities of ignorance: On financial crisis and the ambivalence of neoliberal epistemology', *Economy & Society* 41(1): 64–83.

Dean, M. (2010) *Governmentality: Power and Rule in Modern Society* (2nd edition), London: Sage.

Edgerton, D. (2017) 'The political economy of science: Prospects and retrospects', in D. Tyfield, R. Lave, S. Randalls and C. Thorpe (eds), *The Routledge Handbook of the Political Economy of Science*, London: Routledge: 21–31.

Elzen, B., F. Geels and K. Green (eds) (2004) *System Innovation and the Transition to Sustainability: Theory, Evidence and Policy*, Cheltenham: Edward Elgar.

Fernández Pinto, M. (2017) 'Agnotology and the new politicization of science and scientization of politics', in D. Tyfield, R. Lave, S. Randalls and C. Thorpe (eds), *The Routledge Handbook of the Political Economy of Science*, London: Routledge: 341–350.

Foucault, M. (2001) 'The subject and power', in M. Foucault and J.D. Faubion (eds), *Power: the Essential Works*, Vol. 3., London: Allen Lane: 326–348,

Freeman, H. (2016) 'I've heard enough of the white male rage narrative', *Guardian*, 10 November.

Garvey, B., D. Tyfield and L.F. de Mello (2015) 'Meet the new boss ... same as the old boss?: technology, toil and tension in the agrofuel frontier', *New Technology, Work & Employment* 30(2): 79–94.

Geels, F. (2014) 'Regime resistance against low-carbon transitions: Introducing politics and power into the multi-level perspective', *Theory, Culture & Society* 31(2): 21–40.

Godin, B. (2006) 'The Knowledge-Based Economy: Conceptual framework or buzzword?', *Journal of Technology Transfer* 31: 17–30.

Grundmann, R. (2012) '"Climategate" and the scientific ethos', *Science, Technology and Human Values* 38(1): 67–93.

Guardian (2016) 10 November.

Harvey, D. (2005) *A Brief History of Neoliberalism*, Oxford: Oxford University Press.

Hindess, B. (2006) 'Bringing states back in', *Political Studies Review* 4, 115–123.

Janeway, W.H. (2012) *Doing Capitalism in the Innovation Economy*, Cambridge: Cambridge University Press.

Jasanoff, S. (2005) *Designs on Nature*, Princeton: Princeton University Press.

Jessop, B. (2013) 'Revisiting the Regulation Approach: Critical reflections on the contradictions, dilemmas, fixes and crisis dynamics of growth regimes', *Capital & Class* 37(1): 5–24.

Jessop, B. and N.L. Sum (2006) *Beyond the Regulation Approach: Putting Capitalist Economies in their Place*, Cheltenham: Edward Elgar.

Kinchy, A. (2012) *Seeds, Science, and Struggle*, Cambridge: MIT Press.

Klein, N. (2007) *The Shock Doctrine*, London: Penguin.

Kleinman, D.L. (2010) 'The commercialization of academic culture and the future of the university', in H. Radder (ed.), *The Commodification of Academic Research: Science and the Modern University*, Pittsburgh: University of Pittsburgh Press: 24–33.

Kloppenburg, J.R. (1988) *First the Seed: The Political Economy of Plant Biotechnology* (2nd edition), Cambridge: Cambridge University Press.

Krippner, G.R. (2011) *Capitalizing on Crisis*, Harvard University Press.

Krugman, P. (2013) *End this Depression Now!*, New York: W.W. Norton.

Lave, R. (2012) *Fields and Streams: Stream Restoration, Neoliberalism and the Future of Environmental Science*, Athens and London: University of Georgia Press.

Lave, R., P. Mirowski and S. Randalls (2010) 'Introduction: STS and neoliberal science', *Social Studies of Science* 40(5): 659–675.

Lawhon, M. and J.T. Murphy (2013) 'Socio-technical regimes and sustainability transitions: Insights from political ecology', *Progress in Human Geography* 36(3): 354–378.

Lemke, T. (2011) *Foucault, Governmentality, and Critique*, Boulder, CO: Paradigm Publishers.

Lessig, L. (2001) *The Future of Ideas*, New York: Random House.

Levidow, L. and S. Carr (2009) *GM Food on Trial: Testing European Democracy*, London: Routledge.

Levidow, L., J. Murphy and S. Carr (2007) 'Recasting "substantial equivalence": transatlantic governance of GM food', *Science, Technology & Human Values* 32(1): 26–64.

Losh, E. (2014) *The War on Learning*, Cambridge, MA: MIT Press.

Mason, P. (2015) *Post-capitalism* London: Allen Lane.

May, C. (2006) 'The denial of history: Reification, intellectual property rights and the lessons of the past', *Capital & Class* 88: 33–56.

McGettigan, A. (2013) *The Great University Gamble: Money, Markets and the Future of Higher Education*, London: Pluto Press.

McMichael, P. (2009) 'A food regime analysis of the world food crisis', *Agriculture and Human Values* 4: 281–295.

Mirowski, P. (2009) 'Postface: Defining neoliberalism', in P. Mirowski and D. Plehwe (eds), *The Road From Mont Pelerin: The Making of the Neo-Liberal Thought Collective*, Cambridge, MA: Harvard University Press: 417–456.

Mirowski, P. (2011) *Science-Mart: Privatizing American Science*, Cambridge, MA: Harvard University Press.

Mirowski, P. (2012) 'The modern commercialization of science is a passel of Ponzi schemes', *Social Epistemology* 26(3/4): 285–310.

Mirowski, P. (2014) *Never Let a Serious Crisis Go to Waste: How Neoliberalism Survived the Financial Meltdown*, London: Verso.

Mirowski, P. and D. Plehwe (eds) (2009) *The Road From Mont Pelerin: The Making of the Neo-Liberal Thought Collective*, Cambridge, MA: Harvard University Press.

Mirowski, P. and E.-M. Sent (2002) 'Introduction' in P. Mirowski and E.-M. Sent (eds), *Science Bought and Sold*, Chicago: University of Chicago Press: 1–67.

Nik-Khah, E. (2017) 'The "Marketplace of Ideas" and the centrality of science to neoliberalism', in D. Tyfield, R. Lave, S. Randalls and C. Thorpe (eds), *The Routledge Handbook of the Political Economy of Science*, London: Routledge: 32–42.

Nik-Khah, E. and R. Van Horn (2016) 'The ascendency of Chicago neoliberalism', in S. Springer, K. Birch and J. Macleavy (eds), *The Routledge Handbook of Neoliberalism*, London: Routledge: 27–38.

Ong, A. (2006) *Neoliberalism as Exception: Mutations in Citizenship and Sovereignty*, Durham, NC: Duke University Press.

Oreskes, N. and E.M. Conway (2011) *Merchants of Doubt*, New York: Bloomsbury Publishing USA.

Peck, J. (2010) *Constructions of Neoliberal Reason*, Oxford: Oxford University Press.

Peck, J. and A. Tickell (2002) 'Neoliberalizing space', *Antipode* 34(3): 380–404.

Pellizzoni, L. (2011) 'Governing through disorder: Neoliberal environmental governance and social theory', *Global Environmental Change* 21: 795–803.

Perez, C. (2002) *Technological Revolutions and Financial Capital – The Dynamics of Bubbles and Golden Ages*, Cheltenham and Northampton (MA): Edward Elgar.

Perez, C. (2016) 'Capitalism, technology and a green global Golden Age: The role of history in helping to shape the future', in M. Jacobs and M. Mazzucato (eds), *Rethinking Capitalism*, Malden, MA and Chichester: Wiley Blackwell: 191–217.

Polanyi, K. (1944/1957) *The Great Transformation*, Boston: Beacon Press.

Polanyi, M. (1962) 'The republic of science', *Minerva* 1: 54–73.

Radder, H. (ed.) (2010) *The Commodification of Academic Research: Science and the Modern University*, Pittsburgh: University of Pittsburgh Press.

Rose, N. (2007) *The Politics of Life Itself*, Princeton: Princeton University Press.

Sell, S. (1999) 'Multinational corporations as agents of change: The globalization of intellectual property rights', in A. Cutler, V. Haufler and T. Porter (eds), *Private Authority and International Affairs*, Albany: SUNY Press: 169–198.

Sell, S. (2003) *Private Power, Public Law: The Globalization of Intellectual Property Rights*, Cambridge: Cambridge University Press.

Shapin, S. (2008) *The Scientific Life: A Moral History of a Late Modern Vocation*, Chicago: University of Chicago Press.

Sismondo, S. (2017) 'Non-randomized controlled flows of pharmaceutical knowledge', in D. Tyfield, R. Lave, S. Randalls and C. Thorpe (eds), *The Routledge Handbook of the Political Economy of Science*, London: Routledge: 119–131.

Slaughter, S. and G. Rhoades (2004) *Academic Capitalism and the New Economy*, Baltimore and London: Johns Hopkins University Press.

Smith, A., J.P. Voß and J. Grin (2010) 'Innovation studies and sustainability transitions: the allure of the Multi-Level Perspective and its challenges', *Research Policy* 39(4): 435–448.

Soederberg, S. (2010) *Corporate Power and Ownership in Contemporary Capitalism*, Abingdon and New York: Routledge.

Springer, S., K. Birch and J. MacLeavy (eds) (2016) *The Routledge Handbook of Neoliberalism*, London and New York: Routledge.

Stedman Jones, D. (2014) *Masters of the Universe: Hayek, Friedman, and the Birth of Neoliberal Politics*, Princeton: Princeton University Press.

Stirling, A. (2010) 'Keep it complex', *Nature* 468(7327): 1029–1031.

Susskind, R. and D. Susskind (2015) *The Future of the Professions*, Oxford: Oxford University Press.

Tabb, W. (2012) *The Restructuring of Capitalism in our Time*, New York: Columbia UP.

Tyfield, D. (2008) 'Enabling TRIPs: The pharma-biotech-university patent coalition', *Review of International Political Economy* 15(4): 535–566.

Tyfield, D. (2012a) *The Economics of Science: A Critical Realist Overview. Volume 1: Illustrations and Philosophical Preliminaries*, Abingdon and New York: Routledge.

Tyfield, D. (2012b) 'A cultural political economy of research and innovation in an age of crisis', *Minerva* 50(2): 149–167.

Tyfield, D. (2014) 'Putting the power in 'socio-technical regimes'– E-mobility transition in China as political process', *Mobilities* 9(4): 585–603.

Unruh, G. (2000) 'Understanding carbon lock-in', *Energy Policy* 28: 817–830.

Verweij, M. and M. Thompson (2006) *Clumsy Solutions for a Complex World: Governance, Politics and Plural Perceptions*, Basingstoke: Palgrave Macmillan.

Wallerstein, I. (2004) *World Systems Analysis: An Introduction*, Durham, NC: Duke University Press.

Weis, T. (2007) *The Global Food Economy: The Battle for the Future of Food and Farming*, London: Zed Books.

Xu, H.G. and T. Ye (2017) 'Changes in Chinese higher education in the era of globalization', in D. Tyfield, R. Lave, S. Randalls and C. Thorpe (eds), *The Routledge Handbook of the Political Economy of Science*, London: Routledge:156–168.

Zeller, C. (2008) 'From the gene to the globe: Extracting rents based on intellectual property monopolies', *Review of International Political Economy* 15(1): 86–115.

2 Four Great Challenges

1 Environment and anthropocene

The first and most existentially threatening of the Four Challenges concerns the host of environmental problems that are increasingly summarized by the neologism of the 'Anthropocene' (Bonneuil and Fressoz 2016); a new geological age in which humanity's – profoundly destructive – mark upon our planet has reached planetary and potentially catastrophic proportions. Most obviously regarding anthropogenic global warming and climate change (IPCC 2013; Monbiot 2006; Lynas 2007), but also in a series of ecological processes that humanity – and especially *rich* humanity (Roberts and Parks 2007) – is pushing towards, or over, boundaries of planetary health (Rockström *et al.* 2009) and in the ongoing destruction of ecosystems and species all around the world (Butchart *et al.* 2010), today our relations with our natural, planetary home are in a terrible and terrifying state.

It is clear that the primary cause of this ecological despoliation has been an untrammelled growth in the consumption of natural resources and the waste and pollution produced in these processes. In fact, the route to global planetary impact was sown earlier than the neoliberal period, particularly with the post-war oil- (and hence high-carbon-) based energy system and the much greater energic and resource demands of an industrial and mass-consumerist system and geopolitical order centred on the global North of US–Western Europe and East Asia (Huber 2013; Mitchell 2011).

But this was compounded, from the late 1970s with crisis of that Keynesian regime, by the ascendancy of neoliberal system government and its innovation model, including the specific model of globalization this involved. As alluded to above, the neoliberal "counter-revolution" (Arrighi 1994) primarily involved the reorganization and thence profound integration, through liberalization, of the global economy by and for corporations whose primary goal was increasingly to deliver maximized shareholder value for their new finance capital taskmasters. This conditioned a model of offshored globalization, shifting industry and manufacturing to the cheapest locations anywhere in the world and driving a boom in fossil-fuel-powered global mobility of both things and people by slow-moving container ships (perhaps the single most important innovation of the age (Cudahy 2006)) and supersonic planes.

The resulting thirst for mobility was not limited to episodic, long-distance travel. As market societies, dependent on accelerating circulation (Foucault 2009; Virilio 1986; Rajan 2006), everyday mobility also massively increased (Urry 2007), and particularly in the wasteful form of mass privatized automobility (Paterson 2007), again building on and greatly expanding a legacy from earlier in the century. With cheap, debt-fuelled consumerism and a culture of celebrating materialist individualism and extravagant conspicuous self-assertion, cars themselves grew not just in number (encouraged further by deliberate neglect of public transport) but in size to SUVs, pick-ups and even mini-tanks. Even relative improvements in energy consumption efficiency were thus lost in absolute increases of energy use. And it was not just cars that grew, as house sizes and household energy use also took off in the global North.[1] Finally, all of this new profligacy was based on a new age of cheap oil after the turmoil of the 1970s, preserved by a US military–industrial complex again supreme by the 1980s as the USSR imploded (Mitchell 2011).

Neoliberal innovation was crucial in this process. The focus of neoliberal innovation on increasing concentration and capitalization of industries (including agriculture) directly increased environmentally unsustainable resource extraction, as in cheap food and goods from high-input monocultures (Tansey and Worsley 2014). Conversely, a focus at the high-profile summit of 'innovation' on branded consumer technology and other goods, affording short-term proprietary global profits, drove demand to further horizons of consumerism and energy use (e.g. in using digital gadgetry (Fettweis and Zimmerman 2008; Walsh 2013)) and enabled and fed insatiable appetites for mobility, tourism and consumption of places and places of consumption (Gottdiener 2000; Urry 1995). Today these trends are exemplified by Dubai, Las Vegas or Macao, and pretty much every international airport (Simpson 2016; Urry *et al.* 2016).

On the flipside, neoliberalism sponsored a regime of innovation that was dismissive of limits and knowledges thereof, including the effects of new technologies, such as GM crops (discussed above), while positively obstructing any political action of regulation (apart from those presupposed by marketization), in defence of unfettered opportunities for market-based entrepreneurship (e.g. Kosek 2010). Rather, any localized environmental problems were embraced as opportunities for further (proprietary, hi-tech) innovation, as in the perennially favourite case of GM crops supposedly curing global hunger through drought resistance.

Today, however, there is increasingly secure and compelling evidence for many of the global and local environmental crises that this process has generated (e.g. IPCC 2013; Steffen *et al.* 2015), *and crucially*, increasing political concern about them around the world, especially from strong and growing environmental movements but also, if far behind, national governments (e.g. Paris COP 2015 and now Marrakech 2016 (McGrath 2016)). It is thus both increasingly clear, if we choose to see the evidence, that these problems are indeed crises of the human-planetary system; and, no less importantly, that there is increasing political intransigence (bottom-up, if not yet top-down) to defend our ecological preconditions against the onslaught of a heedless, gorging neoliberal system.

Signalling the growing weakness of neoliberalism itself in regards to these system challenges is the growing body of financial evidence concluding that increasingly unsettled climates of 'BAU' (business as usual) emissions growth would be so expensive from a reinsurance perspective, creating a growing 'protection gap', that it threatens 'the insurance industry's role as society's risk manager' (Carrington 2016); or that current valuations of fossil fuel reserves (especially coal and oil) are incompatible with global policy commitments to stay within 2°C of global warming and the consequent need to leave these in the ground (*The Economist* 2013; Carrington

2015). These financial warnings signal how profoundly these system crises are now exceeding the capacity of the financialized neoliberal model of innovation to exploit them even for some short-term and private advantage.

We can also see how attempts to harness climate change to a new project of neoliberalism (Mirowski 2014) have to date spectacularly failed. Whether regarding the marketization of the atmosphere in carbon and other weather markets (Lohmann 2017; Randalls 2017), or the attempted technical fixes of carbon capture and storage (Tyfield 2014) and geoengineering (Markusson *et al.* 2017; Szerszynski *et al.* 2013) – both *specifically* pursued as neoliberal programmes of innovation – we find not just utter failure but also significant and growing empowerment of counteracting forces. The neoliberal solution to ecological crisis of 'more market' (via/as more hi-tech innovation) is thus increasingly met with forceful backlash.

But this increasing system rejection of neoliberal government of the planet and the energy and resources flows of global society do not in themselves spell a turn to a better system. New innovations are needed that will utterly transform the ecological footprint of social life, regarding systems of production, circulation, provision and consumption. These new low-carbon, sustainability-oriented (Altenburg and Pegels 2012) systems, however, must be constructed from within the existing neoliberal-governed system in crisis, including needing to conduct *themselves*, up to at least the point of their emergence to system dominance, within the energic cycles of high-carbon neoliberalism and associated *socio-political* relations of system government. Failures of neoliberal-shaped systems of low-carbon innovation thus do not serve to mitigate the Challenge of the Anthropocene – at least in the short-term – but rather to *exacerbate* it by deepening the dysfunction at system level of the dominant neoliberal regime itself.

This includes how incumbent neoliberal conditions – dogmatically backing the market – forestall any meaningful response to climate change by (nation-)state Governments,[2] especially in the pivotal instance of the US and in many of its satellites (including the UK, Australia, Canada and the Gulf states), and especially in post-crash austerity. Yet state support for these big, public infrastructural projects is all but essential (if certainly not in itself sufficient (Goldstein and Tyfield 2017)). Moreover, notwithstanding the 'success' of the 2015 Paris COP in agreeing global and legally binding GHG emission targets, a global deal of adequate ambition, let alone its implementation, is still far off; which takes us to the next Great Challenge.

2 Cosmopolitized globalism

The second Challenge is that of *global* system government in a world still primarily governed as sovereign – and mutually competitive – nation states (Corry 2013). This is epitomized by the difficulty of reaching agreement to tackle global environment risks (Duara 2014), but is no less clear across a range of other issues: the global financial system; pandemics; food and water security; regulation of the Internet; or global terrorism. The particular imperative of a new *global* level of system government, however, does not come solely from the emergence of new global risks (Beck 2009), i.e. challenges confronting the planet as a whole created by the very success of incumbent socio-technical systems and socio-political orders. The broader process of globalization is also forging a new socio-political condition and new subjectivities that are intricately and intimately interdependent on multiple distant others and elsewheres.

This 'cosmopolitization' of society (Beck 2006) affects everyone, from the highest to the lowest social strata, in multiple different ways: the jet-setting executive or celebrity artist with homes in three continents; the Australian service worker taking regular holidays in south east Asia; the Chinese teenager avidly consuming Korean pop music, American TV and Parisian fashions; and Syrian refugees, seeking a new home in Germany as they flee the multi-national forces of ISIS. All are profoundly cosmopolitized, signalling the ongoing emergence of a new globalizing world that crosses territorial borders so often and in such complex patterns as to render entirely problematic the concept of 'society' itself, in the twentieth century common-sense meaning of an integral, unified, largely ethnically homogeneous social group 'contained' within the boundaries of a nation-state.

But, as the CP/KS perspective makes abundantly clear, if 'societies' are now evolving in profound inter-connection, with thoroughly cosmopolitized subjects and agencies, it is not only a grievous violence to social analysis to exclude and sever these ties. These interlocking geographically dispersed systems are also working or not – being governed and conducted functionally or not – through equally cosmopolitized power/knowledge relations and technologies. That the very existence of a cosmopolitizing global system in itself entails the existence in turn of emergent processes of *global* system government (but not Government), however, does not mean the latter is being done well. Indeed, across the list of global risks above we see clear evidence that it most certainly is not. Rather, the present confronts us with the challenge of an emergent global level of social organization and its attendant challenges that must be addressed at that level, but currently without anything approaching adequate institutions of government, whether formal or informal.

Again, this is an issue with a massive literature that one cannot hope to summarize adequately here (Held and McGrew 2003; Scholte 2000). But our focus here is once more on how neoliberal system government and innovation condition this profound epochal Challenge. Our starting point is the neoliberal-sponsored global expansion of capital relations from the 1980s, reviving this crucial system process – crucial, that is, to the incumbent global system of capitalism (see Chapter 3) – after the periodic crisis of the 1970s (Harvey 1989).

Global finance capital, centred on Wall St, and much more focused on short-term financial returns than on business operations, drove the constant reorganization of productive capital to spread across the world in search of cheaper manufacturing costs (especially of skilled, disciplined labour) and new resources and markets for profitable exploitation. Crucial to this process was the incremental fusion of a new global business model (Nolan 2004) consisting of modular production in global commodity chains and global production networks (Gereffi *et al.* 2005; Sturgeon 2002). These new networks were coordinated by, and to the primary strategic and financial advantage of, the transnational corporation (TNC) owning the proprietary branded intellectual property (IP) of the final product and controlling the most specialized processes of their production and continual innovation upgrade (Dicken 2014).

This created, in the short-term, positive feedback loops generating a distinctive international division of labour of innovation (IDLI), and its super-rent profits; a 'spiky' world (Florida 2005) of localized socio-economic development for the emerging global hubs and nodes of these global networks (Bathelt *et al.* 2004), and stagnation and decay elsewhere. High value, IP-intensive businesses in the global North could offshore their manufacturing to low-cost companies overseas in a growing tidal wave of foreign direct investment, perhaps gratefully received by their developing country hosts in free trade zones. This further weakened Northern industrial (and thence other kinds of) wages, further empowering the TNCs for yet more offshoring, and diminishing the pressures for investment in their own manufacturing capacities, especially on home soil. And, of course, the emerging ICT innovations of this process were themselves crucial for the space-time compression and distanciation (Harvey 1989; Dicken 2014) – simultaneously shrinking the world and extending the reach of ever-more-instantaneous communication – that specifically empowered the beneficiaries of this globalization (Massey 2005) and enabled the operation of organizations across the world.

Shaped by, and in the interests of, TNCs from the global North (and the US especially) and their financial(ized) investors, therefore, this model of globalization also took on a specific one-size-fits-all form, where the size in question was the one fitting these TNCs, as in the so-called Washington Consensus. The result was their progressive empowerment to the point that these private enterprises could even themselves set new public law for the world, and in the teeth of bitter opposition, as in the case of the WTO's 1994 TRIPs agreement. Moreover, as this stark instance exemplifies, these power/knowledge dynamics were redoubled with the US unipolar moment, as the resurgent growth and innovation of a Reaganite USA proved a step too far for the USSR.

In subsequent decades, however, the very success of this model of footloose globalization and hi-tech proprietary innovation has led to further evolution of, and integration of, the global system, *and* in ways that are increasingly at odds with neoliberal system government. This is perhaps clearest precisely in the domain of (hi-tech) innovation itself. First, identifying hi-tech innovation as a key aspect of national political economic heft, countries around the world have introduced techno-nationalist policies attempting to catch-up with the US in *this* proprietary, hi-tech game. Of course, many countries in the global South are failing, confirming (if also transforming) the neoliberal IDLI (e.g. Delvenne and Kreimer 2017); but not all and not in every industry. This thus unsettles neoliberal globalization in its very success of deeper global integration, but on terms that are explicitly competitive between nations, making even more complicated and intense the conflict between a global system and nation state-based Government of hi-tech innovation. In particular, with the surge of economic growth since 2000 in the Global South (Mason 2015: 94–104) (and especially its most massive and populous countries, such as China, India, Brazil, Indonesia etc....) and commensurate growth in R&I investment (public and private), the geographical global centre of R&I and the sites of greatest global influence are shifting demonstrably away from the trans-Atlantic axis of the twentieth century (Leydesdorff *et al.* 2013).

Second, there is the ongoing emergence of a qualitatively unprecedented novelty in the continuing construction of a 'global' geography of knowledge, via globalized and globalizing innovation networks (GINs) (Ernst and Kim 2002). These GINs have taken shape in large part under the imperative of maintaining a competitive edge in hi-tech (or otherwise innovation-intensive) businesses, which are also tackling ever more challenging vistas of innovation with increasingly sophisticated products, services and processes. Combined, these twin pressures place a growing business imperative on collaborations of the best teams, wherever they may be from and/or based. This plays out precisely through forms and process of globalization and cosmopolitization (Beck *et al.* 2013) that problematize the crude conception of a 'shift' from 'West' to 'East', 'North' to 'South'. Instead we see qualitative changes in which, for instance, leading global mega-cities and their R&I clusters and campuses are more closely connected to each other than with the rural or peri-urban and co-national cities in their hinterland.

Combined with the fast-changing geography of knowledge just described, this presents a fascinating and unstable conjunction. Notwithstanding the profound influence of a neoliberal globalization on *domestic* politics across the global South, there remain significant and enduring differences in cultural, political and socio-economic processes, practices and tacit knowledges that underpin and enable political regimes and their co-production with R&I across the world. And, to repeat, these dynamics are compounded by specifically *techno-nationalist* states and corporations. This is thus to raise new and globally significant challenges as a changing and emerging *global* geography of knowledge clashes with an incumbent global regime heavily dependent on a *geographically-specific* (if globally ambitious) Euro-American (and neoliberal) understanding of 'knowledge'.

Regarding an unfolding crisis of crisis management, in the specific issue of innovation and its global regulatory architecture and the political economy more broadly, we have already witnessed since 2000 how a neoliberal (-cum-neoconservative), unipolar US was incapable of hearing and accommodating *either* new predicament – of new rising powers, or emerging global system. Rather, it was committed to the essentially neoliberal, Promethean logic of opportunistically (and with just as much state violence and system turbulence as necessary) *making* reality (Fisher 2009) so as to maximize global market opportunities for US TNCs and finance that would then further strengthen the US-*as*-neoliberal/neocon power. Of course, the actual outcome of that debacle has been the construction of a new and more virulent global system threat in the protean form of ISIS – a development also profoundly conditioned by innovation, both military (e.g. drones) and civilian (social media).

Similarly, regarding economic globalization, the less outrageously aggressive Obama administration (Ali 2011) has been no more successful in driving further liberalization. The Doha round of multilateral WTO trade negotiations is stalled to paralysis, and even the more regionally-focused, but still sectorally-ambitious, TTIP and TPP agreements are struggling against significant political opposition – and surely now, with President Trump, all but dead. It is hard to see how an anti-globalization and/or self-destructive Trump-led US, aggressively self-conscious of its own decline (namely, "Make America Great *Again*"), will prove any more adept or successful in shaping globalization once more to America's overwhelming advantage.

In short, the Challenge of cosmopolitized globalism reveals a social and economic world system of increasing interdependence alongside the deepening competition of a waning superpower thoroughly shaped by neoliberalism and resurgent nation-state rivals that are increasingly assertive in response. The very attempt to reassert the neoliberal stamp on the former has thus only served to inflame the latter dynamic, now to the point of exhaustion in American nationalist isolationism; the icing on the cake of a world now dominated by a new chauvinistic nationalist populism (*The Economist* 2016a).

Moreover, we can also see cognate and intra-active cycles of dysfunction regarding global (movements of) labour. For some time, global integration and concentration of opportunities in the global North together with new global mobility afforded mass immigration of skilled and unskilled alike. Meanwhile, globalization offered new opportunities, especially for the high-skilled from the global North, generating new cosmopolitan innovation networks (Saxenian 2005; Tyfield and Urry 2009). Especially after the great financial crash (GFC) of 2007/2008 and the ensuing austerity, though, the combination of unskilled migration with an innovation and business model ever more committed to destroying working class jobs in the global North has unsurprisingly generated deepening tensions, now boiling over in flagrant racism and/or anti-immigrant sentiment across the US and Europe. Yet, on the flipside, cosmopolitization of the 'skilled' workforce continues – in part responding to exactly the same domestic economic pressures – feeding a growing divergence between such 'high' and 'low' skilled (knowledge) workers, not just in terms of incomes and assets, but also as rich or poor in mobility (Urry 2007) and motility (Kaufmann 2010), network-capital (Urry and Elliott 2010) and in positive or negative attitudes to the global system forming around them.

These dynamics thus feed a deepening crisis of globalization itself (*The Economist* 2016b; Elliot 2016; Mason 2012), with powerful popular anti-globalization forces furious with and excluded from an emergent global system that not only now inexorably exists – is *constitutively* cosmopolitized – but which urgently needs *better* government for the sake of planetary survival. Anger with this system, much of it thoroughly justified and long-in-the-coming, however, cannot achieve its goal of return to a 'simpler' less globalizing age, and so profoundly complicates the process of forging that better global government. But this, too, is the direct legacy of neoliberalism.

There surely can be no greater evidence of this implosion of neoliberalism than the figure of Donald Trump himself: amongst the most privileged and archetypal progeny of a free-wheeling asset-based neoliberalism yet now the figurehead of the American movement of anti-globalization fury, and through the most shameless display of narcissistic, vitriolic, post-truth, reality-TV sensationalist politics to date. In this, though, he epitomizes the increasing internal divisions within neoliberalism itself. These fractures have been clear since the GFC, if not before in the post-9/11 insurgency of neoconservatism. Regarding the former, just consider the unanswerable question of whether austerity or the *financial* stimulus of quantitative easing (QE), keeping Wall St afloat, is the 'neoliberal' policy; or attempts to distinguish anew the 'market' from corporations and/or the revolving door to state power (Zingales 2014).

3 Post-human innovation

These radical mutations of neoliberalism itself, including now the bastard prodigal son of Trump – the man *as* the movement – illustrate the crisis of neoliberalism in the Great Challenge of cosmopolitizating globalism, Another (specifically political economic) variation of neoliberalism – most clearly emergent through the Obama years – captures the third Great Challenge of post-human informationalization and its concomitant stacceleration. By this mouthful (abbreviated henceforth as 'post-human innovation') we refer to the ways in which the neoliberal 'knowledge economy' has progressively taken the specific form of labour-substituting innovation through digital, networked informationalization. Not only has this increasingly shaped the broader economy, but with the combination of the post-GFC economic malaise and continuing advancements in artificial intelligence (AI), virtual reality and networked learning robotization, it is now threatening much more than industrial work in specific industries. Rather, we now seem to face the prospect of mass human redundancy, not just in the sense of a huge reserve army of 'unskilled' (but also increasingly 'skilled') labour, but also as human beings per se: a hi-tech dystopia of a few massive corporations controlling production chains staffed almost entirely by robots, while most of humanity scrabbles for the few remaining sources of income and livelihood.

What has conditioned this prospect is again precisely the neoliberal innovation model co-produced with the regime of system government of financialized maximization of short-term shareholder value return, together with progressive empowering of state forces for, and disempowering of state forces against, the destruction of organized labour. Through the 1980s and 1990s, this led to self-fulfilling advantage and system-productiveness in positive feedback loops: weakening Northern labour (compounding the offshoring threat, above); revolutionizing existing industries for short-term profit maximization (to serve the corporations' speculative financial masters); as well as new opportunities for businesses and hi-tech innovations offering just these labour-substituting productivity-boosting and/or consumer technologies. This in turn forged unassailable competitive advantages in hi-tech innovations, producing a reboot in the unquestionable corporate strength of the TNCs, feeding a self-propagating boom in corporate restructuring (mergers and acquisitions) and other investment banking towards an ever more concentrated corporate control of key industrial sectors (Carroll *et al.* 2010).

The cycling through financialization was also key here, since such hi-tech innovation dominance and concentration of market control afforded the growing corporate valuation in financialized terms. This thereby reflected and propelled the continued and apparently endless growth of the much-hyped post-industrial 'intangible' 'knowledge economy' (Leadbeater 1999; Kelly 1997), founded on supposedly new foundations of economic value itself in corporation-owned IP, branding and goodwill. Such financial backing came with the *quid pro quo* of the expectation of more of the same business strategy. Via debt and speculative financial products, then, industrial transformation and neoliberal innovation went hand-in-hand with growth of the massive (and by the 1990s, massively *greater* (Blackburn 2006)) hinterland of the financialized economy.

There is an immediately obvious flaw to this process – a classic contradiction of capital, no less – that reducing the wage base is not just reducing and 'streamlining' business costs, but also the incremental undercutting of aggregate demand. But the parallel growth of the financialized economy squared this circle in the medium-term through a new boom in consumer debt. The problem, in other words, was indeed harnessed as a further opportunity for neoliberal system growth. Free-flowing credit not only propped up falling or stagnating real wages (and thus an unprecedented new era of steady inflation – the 'Great Moderation', understood by neoliberal economists and regulators as further evidence of their supreme world-making wisdom (Bernanke 2004)) but even led to some popular sense of growing wealth so long as (debt-fuelled) *asset* price booms continued. This both secured a broad-based acceptance in the global North of this new political economic model while feeding back into growth of the financialized economy; a privatized Keynesianism (Crouch 2011) and financialization of everyday life (Martin 2002; Langley 2008).

The growing normalization, and lionization, of financial 'risk-takers' also then served to empower them further politically, forcing through the greatest of political economic coups, in which the Keynesian settlement of *taxing* the rich was replaced by taking *debt* from them (Birch 2015). The unsurprising result of this whole process was the construction of increasingly vertiginous and systemic Ponzi schemes, siphoning wealth from the bottom to the financial elite and their financial institutions – now too big and too systemically central to fail – and the consequent division of society with a yawning inequality not seen for around a century (Piketty 2014; Stiglitz 2013; Sayer 2015). Meanwhile, increasingly becoming veritable Masters of the Universe, global finance was encouraged to take ever-greater risks, straying into outright corruption, fraud and gaming of the system.

Again, the instability and inequity of this system in itself is not necessarily a threat for neoliberalism, to the extent it provides opportunities for innovation (and Government policy) exploiting the system turbulence. Indeed, even as the egregious inequality and financial systemic dysfunction overflowed in the GFC, the response in subsequent years illustrates this perfectly, with both austerity and financial stimulus taking forms that have simply exacerbated the division of society as financial elites have profited even more handsomely and to new heights (Stephens 2010; Duncan 2012), while offloading the costs of system failure onto those already most excluded and penalized (Curran 2013).

Even more importantly for our purposes, however, is how for neoliberalism to prosper from turbulence it must be able to drive specifically neoliberal innovation that boosts productivity in the industries thereby reorganized. Yet, the very success of the essentially parasitic and profoundly unproductive model of neoliberal innovation progressively strangles even this possibility. It is thus in a crisis of *innovation* and *productivity* itself that we find the most stark self-destruction of neoliberalism as political economic regime. For, however free-floating from the 'real' economy the financialized knowledge economy can fly, it must also, ultimately, be tethered to ongoing profitable transformation of productive capital.

Crucially, though, this crisis of neoliberalism takes the form not just of progressive stagnation of the productive, or 'real', economy but also, and in inseparable parallel, a runaway acceleration of a new mutation of neoliberal innovation. These *two* dynamics then feed each other; a paradoxical (and economically impossible, for mainstream economics) combination to rival that of the stagflation (i.e. simultaneous stagnation and inflation) of the 1970s, as a 'stacceleration'. This has emerged clearly since the GFC, but was already crucial in that event, both directly, through macroeconomic imbalances, and indirectly, via accelerating the dynamics of inequality (Rajan 2011).

That there is a contemporary crisis of innovation is evident from widespread discussion, even percolating into the economics mainstream. There is, however, heated disagreement about the nature of this crisis of innovation. On the one hand there are those argue that we are now amidst an era of secular stagnation (Teulings and Baldwin 2014; Gordon 2016), in large part because most of the 'low hanging fruit' of (socio-) technical changes effecting significant opportunities for investment and growth were one-off opportunities in the development of human circumstances and these are now largely exhausted (Cowan 2011). Flushing toilets, household labour-saving appliances, cars and entertainment gadgets etc. … all effected a significant improvement in standards of living which, for a time, led to positive feedback loops of greater prosperity, employment and investment … not least in further innovation.

With these gone now, the whole cycle is drying up, including the supposed 'engine' of this whole process, innovation, as the positive feedback loops go into reverse. Fewer opportunities for profitable innovation and mass roll-out lead to diminishing investment – especially when compounded by an uncertain business climate after 2008 – which dries up the capacities and efforts dedicated to innovation, and so on. Hence while (American) lives in 1980 were dramatically different to those in 1940, and even more so over the 40 years before that, lives today are largely indistinguishable from those of 1980 (Gordon 2012). Such an argument thus more-or-less explicitly disregards the entire "internet revolution" – on the basis that we can see it everywhere, "except in the statistics" (Solow 1987) – which, of course, is also the period of neoliberal dominance. This delivers a damning indictment on the *lack* of innovation in the neoliberal era that dominant rhetoric would trumpet as its greatest achievement. But it also parochially assumes a US history of improving standards of living, in a world where those still lacking even ready access to potable water, sanitation and electricity number 663 million, 2.4 billion and 1.2 billion people (or 10 per cent, 36 per cent and 18 per cent) respectively (Water.org 2015; World Bank 2016).

As Pagano and Rossi (2009, 2017) show, however, the specific neoliberal regime of R&I is conditioning a generalized stagnation of innovation investment (see also Erixon and Weigel 2016). This includes not just ideologically conditioned shrinking of state funding, but also the private sector, which has historically been the dominant site of R&I (Edgerton 2017). Crucial to this process has been the propagation of an overly-proprietary model of innovation that locks up knowledge-intensive products into an increasingly impenetrable thicket of mutually exclusive claims of ownership. This has produced not only an 'anti-commons' (Heller and Eisenberg 1998), in which the shared commons of 'ideas', upon which the generation of further ideas are premised, are increasingly inaccessible. It also has generated a deepening 'investment strike' (Pagano and Rossi 2009), in which private ownership of knowledge ironically disincentivizes investment in its generation, even as strong intellectual property rights are advocated for precisely the opposite reason. This, in turn, substantially underlies the crisis of productivity and stagnation of the broader political economy that culminated in the GFC (Pagano and Rossi 2017; Lazonick *et al.* 2017) as this dynamic emerged in the United States.

In other words, far from yielding unprecedented productivity and advances in R&I, the multiple negative effects from organizing innovation as a financialized market, based on a model of maximizing shareholder value, are increasingly apparent in neoliberal innovation itself. As Lazonick *et al.* (2017) illustrate, this is particularly true in industries that 15–20 years ago were the very acme of neoliberal innovation success stories, such as the pivotal case of the pharmaceutical biotech industry and its drying pipeline of innovations (see also Mittra *et al.* 2011). This political economic model of R&I thus generates 'profits without prosperity' (Lazonick 2014), and in increasingly self-destructive ways.

On the other hand, though, are those who point to the entirely opposite problem, namely a dangerously run-away acceleration of innovation, as the internet – here the *sole* focus of the analysis – is unleashing an entrepreneurial revolution (Brynjolfsson and McAfee 2011, 2016; Straw and Baxter 2014; Mason 2015; see next chapter). With low-cost platforms now available online, barriers to entry for start-ups are exceptionally low and almost anyone (with a bit of coding knowledge, or simply sheer overweening self-belief) can set up a company that offers an online-based product that could 'disrupt' existing industries through forms of labour-substituting and personal consumer hi-tech automated interconnectivity. This involves growing concentration and monopoly capture of the positive externalities and amortized control of the resulting 'big' data assets by venture-capital backed 'Tech' (Morozov 2014; Keen 2015; Lanier 2013). This model of innovation thus deploys business models seeking quick and big financial returns and financialized investment, betting on swift moves to platform dominance.

Amidst the broader political economic malaise of the new century, and as the forefront of a domain of innovation that stands out for its vitality (periodic boom and bust notwithstanding), this model of innovation has crafted a 'late' neoliberalism that often poses as its opposite – a fake anti-neoliberalism or 'Googliberalism' (Tyfield 2013). Here, Silicon Valley sets itself up as the opposite of the neoliberal, highly-proprietary corporate model (think Microsoft) and the parallel 'unproductive' world of financial speculation. Instead it focuses on its free-to-access, open source and 'sharing' innovation that supposedly produces 'real' and 'disruptive' sociotechnical change and knowledge. Yet the reality of this model is even more neoliberal: aiming at the global concentration of all (knowledge-based) industries (which itself increasingly means *all* industries) in the hands of a tiny number of corporations with even more aggressively proprietary, secretive and outright larcenous (Lanier 2011; Darnton 2009) IPR strategies than those of the patent-dependent giants of the 1990s and that exploit the free labour (and leisure and *biographies*) of their billions of users for their private gain, turning the users *themselves* into the commodity. Through venture capital and private equity, this is also a model no less founded upon speculative and self-styled 'risk-taking' (but in fact, remarkably risk-averse (Janeway 2012)) finance capital.

The combination of dominant neoliberal models of innovation and financialization even subverts the key figure of the entrepreneur; just as this political project explicitly celebrates the Randian vision of its champions: Brin and Page (Google), Zuckerberg (Facebook), Bezos (Amazon) and, increasingly, Kalanick (Uber). Instead of the creator of new markets and commodities, he (rarely she) becomes its antithesis, the rentier, developing new technoscientific interventions that aim to exploit the (positive externalities associated with) monopolization of *existing* stocks of (perhaps publicly furnished (Mazzucato 2011)) assets (Birch 2016; Zeller 2008). It is thus *this* avenue of innovation that has bred the growing concern about the future of employment, and humanity to boot, against an unstoppable ascendancy of inter-connected and learning machines (Ford 2015).

Both of these arguments individually would indeed point to profound systemic problems for neoliberalism. On the one hand, the deceleration argument points to an inescapable winding down of the motor of a system of government – namely global capitalism, and particularly *financialized* global capitalism, demanding steadily increasing (or preferably spectacular) annual returns – that depends, to the exact contrary, on continual acceleration. On the other, the acceleration argument also offers little succour to neoliberalism, since it points to a world in which a temporary post-2008 revitalization of entrepreneurialism leads to a world that transparently offers diminishing, not expanding, horizons of human liberty; if it is not first forcibly rejected by the great and growing (populist? (Smith 2016)) army of the quite literally redundant, perhaps leading to the very nemesis of neoliberalism, a post-capitalism (see next chapter; Mason 2015).

From the perspective of a CP/KS analysis, however, we can see how *both* of these arguments are likely not only right in their specifics – rather than contradictory as is often understood to be the case – but also in ways that *feed* each other. The key to this point is that we are not talking about 'innovation' per se but about *neoliberal* innovation: innovation conditioned by and conditioning a neoliberal-dominated system. It is thus a crisis of neoliberal innovation that is the source of the present problem and the specifically destructive form of post-human technological change now emerging. And this because this form of innovation does not just keep neoliberal system government ticking over, solving the problems seemingly 'external' to neoliberalism that it encounters (and raises). But it is also – if not primarily – a key process for the very *constitution* and reproduction of the *specific power/knowledge relations of neoliberalism*, that in turn keep innovation tightly framed within the neoliberal model, thereby compounding the very problems of the crises of innovation, in positive feedback loops. By contrast, this key dynamic and the fact that the crucial product of innovation is a specific *socio-political ordering* within complex power/knowledge systems is entirely missed in the mainstream economic discussions, of both deceleration and acceleration alike.

Consider first deceleration. A CP/KS analysis of neoliberalism readily explains this – and as a specifically rich-world (and especially American) problem, with the tendencies of neoliberalism coming home to roost, having reaped their plundering harvest globally (Pagano and Rossi 2009). Deepening financialization of the political economy, commercialized and marketized destruction of crucial public sources of innovation and sites of productive investment (Block 2008; Mazzucato 2011), overpropertization (especially of intellectual resources and commons) (Heller and Eisenberg 1998), and hypertrophy of dependence on debt-based end-consumer demand (Crouch 2011; Graeber 2014) all contribute to and over-determine the progressive evisceration of the productive economy and the secular fall in productivity in the global North over the neoliberal period (Atkinson 2016).

Connecting to the other Challenges, they also simultaneously undermine both the techno-economic and the political-bureaucratic capacity to mobilize the kind of large-scale public projects, including of infrastructure, that would be needed not only to reboot investment and innovation, but also to deal with the multiple contemporary environmental challenges. Instead, these deepening problems, *amidst the given incumbent power/knowledge relations*, simply lead to further demands for the only way it is accepted that an economy can be stimulated: namely yet more marketization and hi-tech, IPR-intensive labour-substituting innovation – or to sheer paralyzed inaction – hence exacerbating the malaise. So too for acceleration. It is the *Googliberal* model of innovation specifically that prosecutes the heedless acceleration of job-destroying and ever-more concentrated internet-based innovation, while thereby further enabling the forces of Silicon Valley-VC which focus on precisely innovation of that kind.

But these two cycles of innovation-as-politics within and coproducing complex power/knowledge systems also feed each other, as 'stacceleration'. Deceleration and stagnation leads to further corporate and consumer demands for more innovation and productivity – which is understood to mean more labour- and cost-saving innovation *from 'disruptive' digital technology*. But each round of labour-substituting and market/platform-monopolizing innovation simply further erodes the wage-based demand that, in turn, is the foundation of corporate confidence of growing profits and hence productive investment in innovation – thereby *further undermining broad-based, productive growth* ... and compelling yet more labour-substituting innovation. Acceleration meanwhile strengthens the *political* domination of VC/Tech. This weakens both fiscal budgets for public innovation, including through systematic and massive tax evasion (Morozov 2015; Shaxson 2012), and thus prospects for profitable productive innovation investment in a wholesale *systemic* breakthrough to new horizons of profitable innovation (e.g. cleantech); while also strengthening powerful constituencies arguing for further unleashing of the market and specifically Googliberal disruptive 'solutions'.

Finally, together these feed into the general system crises, perpetuating and prolonging inaction regarding global environmental exigencies (since 'disruptive tech' offers nothing on that score (Morozov 2014)) as well as the global crisis of intranational inequality (Piketty 2014; Stiglitz 2013) and its global socio-political repercussions of unrest and rejection; political turbulence and backlash, that is, now with the added intensity of the existential anxiety of post-human redundancy *especially* for those already excluded, afraid and angry.

Far from innovation coming to the rescue, and beyond even it directly *incubating* the global challenges with which it is currently tasked as saviour, we see that innovation – specifically neoliberal innovation-as-politics coproducing that regime of system government – is itself in crisis and even is itself *the* crisis. Situated firmly and inescapably within this crumbling edifice, therefore, we have no option but to focus on a *different* innovation, not just as the purveyor of off-the-peg solutions to other problems, but as itself the crucial node and moment in transforming the *complex power/knowledge system* as a whole. Guiding, shaping and facilitating innovation, however, is no easy matter (e.g. Allenby and Sarewitz 2011; Stirling 2009), and this leads to our final Challenge.

4 The complex government of complex systems

The fourth Great Challenge encompasses and situates the three before it, as is particularly brought out by a CP/KS perspective. For if the three above clearly overlap in complex ways as different aspects of the same challenge of the emergence of a new global system that is simultaneously and inseparably natural, social and technoscientific (respectively) – a new 'techno-science-society' (Maassen *et al.* 2017) – then the CP/KS perspective highlights the necessary corollary of this: that as a *system* of power/knowledge relations, and constituted and reproduced from moment-to-moment as such, both its emergence and (possible) subsequent restabilization entail new processes, practices and institutions for its *self*-government *as* such a global system. The fourth and final (meta-)Challenge is thus the *complex (self-)government of complex systems*, and how this can be forged from within the existing power/knowledge relations and regime of system government (in crisis) without system collapse in the meantime. And the flipside of this perspective is that innovation (-as-politics) is thus one of the – if not *the* – most important but currently neglected arenas of twenty-first century politics.

A CP/KS lens illuminates how complexity *itself* is produced – deliberately cultivated – by neoliberalism, with its Promethean entrepreneurship and disregard for planned system order; and is, therefore, without limit. This is how that regime of system government *works*. Neoliberalism manages the crises it gladly and wilfully whips up by turning any system problems from problems of public, politically-debatable systemic/collective social order – that must then be correctly diagnosed, or known, and then responded to with rationally-crafted appropriate public administration – to opportunities for endless private entrepreneurial risk-taking that, by creating and deepening the reorganization of society and all things *as* (the) market(s), just *is* the best of all possible worlds. Likewise and expressive of its intrinsic limitlessness and radical ontological agnosticism, 'system government' per se is of no concern beyond being the political project of cultivating, and clearing regulatory or socio-political obstacles to, ever-more new opportunities for entrepreneurial exploitation.

As all the above Challenges illustrate, this process tends to the continual growth – in scale, depth, interconnection and complexity – of various aspects of turbulence thus sown, from isolated, exploitable and manageable pockets and niches up to system, and *global* system, level ... culminating in the system crisis of crisis management. In other words, while neoliberalism *as* a regime of system government presupposes (and is parasitic upon pre-existing and established) mechanisms and processes through which the system is held together and reproduced, not just catapulted forwarded and expanded at accelerating pace, it systematically dismantles these while also directly incubating new and profound challenges *at* system level.

This includes the very processes of R&I, as just discussed, thereby hobbling a crucial pillar of neoliberal system renewal – and, indeed, of system renewal itself. The result is the growing gap between contemporary understanding of governance and politics of R&I – perhaps with a view to climate change mitigation and/or equitable, sustainable development (Stirling 2009; Leach *et al* 2012) – as a complex process shaping socio-technical systems that depends upon, and optimally should feed, a broad-based empowerment; and a reality that remains locked into high-carbon and exploitative development, with the barest glimmers of such low-carbon transition at the accelerated global rate actually needed.

This profound Challenge of the production of R&I *as* power/knowledges of government, however, is manifest in, and compounded by, what may, without exaggeration, be called the death of (Enlightenment) knowledge, where this is itself the *deliberate* outcome of the neoliberal MoI and its commercialization of R&I, as discussed above. For claiming to be a capitalist socio-political order based on optimal knowledge, which is then identified with the market, sets up neoliberalism from the *outset* as a project of the radical subversion of Enlightenment epistemology – as representationally objective and normatively neutral, and hence progressively enlightening – that has been the cornerstone of Western-dominated global order for the past two-plus centuries.

Neoliberalism – or rather now its deepening collapse and zombie dominance (Crouch 2011) – is not just stalling R&I capable of the profound *system* transition needed in response to the global system problems, nor even obstructing sufficiently unified and politically-enabled consensus on the ways forward. It is actively frustrating even the unified and politically enabled consensus on the *existence and nature* of these system problems in the first place, since public, objective knowledge itself is systematically undermined by the MoI and its anti-expertise and agnotological common-sense. Again, this is spectacularly evidenced in the successful Trump campaign: sweeping to a shock victory on the basis of outright denial of climate change and direct opposition to globalization per se, on top of all the vitriol and mendacity.[3] But we cannot simply return to the way things were. For in the meantime, the previously effortfully-established bases of 'reasonable', 'objective', 'expert-informed' public debate – which were not without their own major problems – have been profoundly undermined in the very success of neoliberalism; and, recall, where this is the deliberate outcome of that political project, not some unforeseeable and unfortunate side-effect.

It is these system dynamics, thus, that incubate the current *intra*-national legitimacy crises across the world, as the political economic destruction of structures of society and government (especially the industrial working class and public sectors in the Global North) is capped by the double whammy of unshiftable political economic malaise *and* the destabilization of the cultural-political resources of common, reasoned debate. This thus leaves only deeper cultural identities – especially of polarized system winners and losers – and a toxic cycle of fear and mutual loathing and recrimination fed by the post-truth circulation of political knowledge claims (on all sides). Of course, the neoliberalization of states themselves simply compounds the *positive* case for such distrust of political debate and institutions, as in the growing influence of money in elections (e.g. in super-PACs) and administration (via lobbying, revolving doors, expenses gravytrains and the neoliberalization of the civil service (Du Gay 2013)) – all perfectly neoliberal developments in that the state should *also* ideally be a marketplace of ideas, political success going to the highest bidder and deepest pocket. Such is the resulting and cultivated *irr*ationality of the system that a bilious billionaire, epitomizing the hated elite, has become the champion of the angry masses simply by venting and feeding their disoriented anger as a 'political outsider'. As shock vote upon shock vote is delivered, (e.g. Brexit, Trump ...), black swan event after black swan (Taleb 2007), and without the sky falling (yet ...), this also serves further to demean and falsify the 'experts' (of polling, economics, politics ...) who have called things otherwise, seemingly in defence of a corrupt and rotten status quo.

The world thus faces increasingly urgent, massive and intertwined challenges at the same time as the institutions of system government, and Government, are themselves least enabled, in a perfect storm of global system turbulence.[4] Indeed, this simultaneity is no coincidence but precisely the playing out of neoliberalism's system logic (see Figure 2.1). Here we reach the final and crucial point of this analysis of 'where we are now': how these Great Challenges become not just 'great challenges' but the four horsemen of global system *crisis*. The increasingly titanic and all-encompassing storm continues to feed neoliberalism, and thus itself – since neoliberalism *is* system turbulence, the former feeding the latter feeding the former ... without limit ... up to the point at which *limits are indeed and inescapably encountered* (Davies 2016). There are two ways in which such limits are encountered, namely from without and from within, but the totalizing, intra-active and recursive dynamics in play lead to these emerging simultaneously. The three Challenges above are all manifestations of external limits; and, as such, all *necessarily* global, overlapping and existentially threatening. But what is key, transforming even *these* Challenges from 'mere' problems, however grievous, to *system crises*, is the internal limit of neoliberalism of progressive destruction of knowledge and thus of the *epistemic foundations* of neoliberalism itself.

The ultimate challenge, in other words, is how neoliberalism has not only reaped an historic hurricane of new power/knowledge-mediated and innovation-produced destructive creation that is intrinsically threatening to the very integrity of individual societies, human selves and their species being, the human world as a whole and the planet (regarding each of the Four Challenges, in reverse order), while forging all of these (together) *as* interconnected complex global systems. Nor even that it also has in that very process profoundly unsettled the epistemic resources with which we may hope to respond reasonably and efficiently to the challenges, as discussed above. But that destroying knowledge as it is broadly understood and conducted – in public (and civic (Jasanoff 2005)) epistemologies – also destroys its *own* capacities for *crisis management*, its own power/knowledge technologies of

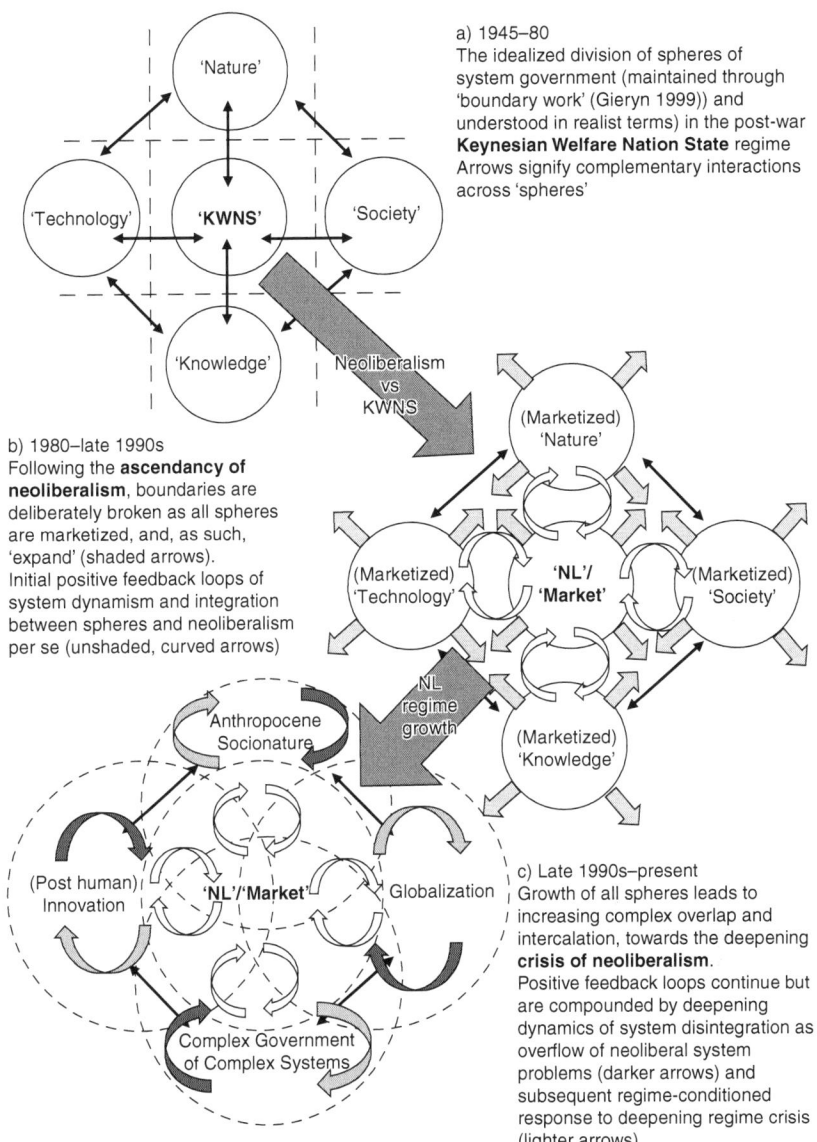

Figure 2.1 Four Great Challenges and the *emergence* of risk-innovation capitalism.

system government, thereby precipitating the systemic inability even to exploit the growing problems for entrepreneurial profit. This is the crux of this age of global turbulence: the crisis of crisis management as crisis of global system government in the simultaneous and intra-active collapse of our relations to knowledge (and thence to knowledgeable, rational government) and the emergence of a new, complex global system, where both of these are the direct and mutually compounding but self-defeating products of neoliberalism. A world system in crisis that is quite literally a *world* gone *mad*.

To repeat, then, the Fourth Great Challenge of the present, characteristic of the terminal crisis of the global regime of neoliberalism, is that of complex government of complex systems per se. Or, to put it differently, it is to forge *new relations to knowledge* adequate to the government of the complex, global systems of today; and, as in the other three Challenges, adequate to their respective overlapping crises and wicked problems; and adequate to moving *beyond* them and the dominant regime of neoliberalism collapsing around us. Yet the CP/KS approach highlights how this predicament inescapably situates us – the world as current global system – *within*, and conditioned by, the dysfunctional cycles of sedimented power/knowledge relations.

This is the problem of the 'emergence' of the complex power/knowledge system of *global risk-innovation capitalism*, and its everyday coproduction and government (for both its initial emergence and subsequent maintenance and reproduction) (Figure 2.1). The present is thus the moment inseparably of both *emergence* of new power/knowledge relations and technologies constitutive of a socio-technical system; AND the *urgence* (Foucault 1980) of how they are as yet ungovernable, exceeding the existing power/knowledge-mediated processes of system government.

This thus spells but one tendential outcome: the collapse and progressive self-destructive crisis of neoliberalism and the innovation model at its core. But what next? How can we avoid this collapse taking us, the world and the planet down with it? It is to this crucial question of the present that we turn in the rest of this book. But in doing so we focus precisely on the privileged lens on the process of unfolding socio-technical futures that is (transition in the system-dominating model of) innovation-as-politics.

Notes

1 The average size of an American house was a staggering 2,164 square foot (201 m²) in 2009 (Wilson N.D.). On showering and energy demand, see Hand *et al.* (2005).
2 Throughout the book, I capitalize 'Government' to distinguish the term when used in the sense of the state or the formal apparatus of 'the' government from the much broader sense of 'government' (as conduct of conduct by dispersed selves and institutions) in which the latter term is usually used here.
3 Meanwhile, embracing Wall Street, Silicon Valley and the displaced American working classes altogether without irony, Clinton preached that the US economy is well on the mend, simply ignoring the challenge of stacceleration. Neither candidate began to address the Challenge(s) of complex system government.
4 Hence, surely, the particular current fascination of *Game of Thrones* (cf. Lanchester 2013).

References

Ali, T. (2011) *The Obama Syndrome*, London: Verso.
Allenby, B. and D. Sarewitz (2011) *The Techno-Human Condition*, Cambridge, MA: MIT Press.
Altenburg, T. and A. Pegels (2012) 'Sustainability-oriented innovation systems: managing the green transformation', *Innovation and Development* 2(1): 5–22.
Arrighi, G. (1994) *The Long Twentieth Century*, London: Verso.

Atkinson, R.D. (2016) *Think Like an Enterprise: Why Nations Need Comprehensive Productivity Strategies*, Washington DC: ITIF.

Bathelt, H., A. Malmberg and P. Maskell (2004) 'Clusters and knowledge: local buzz, global pipelines and the process of knowledge creation', *Progress in Human Geography* 28(1): 31–56.

Beck, U. (2006) *The Cosmopolitan Vision*, Cambridge: Polity.

Beck, U. (2009) *World at risk*, Cambridge: Polity.

Beck, U., A. Blok, D. Tyfield, and J.Y. Zhang (2013) 'Cosmopolitan communities of Climate risk: Conceptual & empirical suggestions for a new research agenda', *Global Networks* 13(1): 1–21.

Bernanke, B. (2004) 'The great moderation', in E.F. Koenig (ed.), *The Taylor Rule and the Transformation of Monetary Policy*, Stanford: Institutions Press Publication Hoover: 145–162.

Birch, K. (2015) *We Have Never Been Neoliberal*, London: Zero Books.

Birch, K. (2016) 'Rethinking value in the bio-economy: Finance, assetization, and the management of value', *Science Technology & Human Values*, 0162243916661633.

Blackburn, S. (2006) 'Finance's fourth dimension', *New Left Review* 39: 39–72.

Block, F. (2008) 'Swimming against the current: The rise of a hidden developmental state in the United States', *Politics & Society* 36(2): 169–206.

Bonneuil, C. and J.-B. Fressoz (2016) *The Shock of the Anthropocene: The Earth, History and Us*, London: Verso.

Brynjolfsson, E. and A. McAfee (2011) *Race Against the Machine*, New York: Digital Frontier Press.

Brynjolfsson, E. and A. McAfee (2016) *The Second Machine Age*, New York: W.W. Norton & Company.

Butchart, S.H.M., M. Walpole, B. Collen *et al.* (2010) 'Global biodiversity: Indicators of recent declines', *Science* 328(5982): 1164–1168.

Carrington, D. (2015) 'Leave fossil fuels buried to prevent climate change, study urges', *Guardian*, 7 January, www.theguardian.com/environment/2015/jan/07/much-worlds-fossil-fuel-reserve-must-stay-buried-prevent-climate-change-study-says.

Carrington, D. (2016) 'Climate change threatens ability of insurers to manage risk', *Guardian*, 7 December, www.theguardian.com/environment/2016/dec/07/climate-change-threatens-ability-insurers-manage-risk.

Carroll, W.K., C. Carson, M. Fennema, E.M. Heemskerk and J.P. Sapinski (2010) *The Making of Transnational Capitalist Class: Corporate Power in the Twenty-First Century*, London: Zed Books.

Corry, O. (2013) *Constructing a Global Polity: Theory, Discourse and Governance*, Basingstoke: Palgrave Macmillan.

Cowan, T. (2011) *The Great Stagnation*, New York: Dutton.

Crouch, C. (2011) *The Strange Non-Death of Neoliberalism*, Cambridge: Polity.

Cudahy, B. (2006) *Box Boats: How Container Ships Changed the World*, New York: Fordham University Press.

Curran, D. (2013) 'Risk society and the distribution of bads: Theorizing class in the risk society', *British Journal of Sociology* 64(1): 44–62.

Darnton, R. (2009) 'Google and the future of books', *New York Review of Books*, 12 February.

Davies, W. (2016) *The Limits of Neoliberalism*, London & Thousand Oaks, CA: Sage.

Delvenne, P. and P. Kreimer (2017) 'World-system analysis 2.0: Globalized science in centers and peripheries', in D. Tyfield, R. Lave, S. Randalls and C. Thorpe (eds), *The Routledge Handbook of the Political Economy of Science*, London: Routledge: 390–404.

Dicken, P. (2014) *Global Shift* (7th edition), London and Thousand Oaks, CA: Sage.

Duara, P. (2014) *The Crisis of Global Modernity*, Cambridge: Cambridge University Press.

Du Gay, P. (2013) 'New spirits of public management ... "Post-Bureaucracy"', in P. Du Gay and G. Morgan (eds), *New Spirits of Capitalism?*, Oxford: Oxford University Press: 274–293.

Duncan, R. (2012) *The New Depression*, Singapore: John Wiley & Sons.

The Economist (2013) 'Unburnable fuel', 4 May, www.economist.com/news/business/21577097-either-governments-are-not-serious-about-climate-change-or-fossil-fuel-firms-are.

The Economist (2016a) Trump's World: The New Nationalism, 19 November www.economist.com/news/leaders/21710249-his-call-put-america-first-donald-trump-latest-recruit-dangerous

The Economist (2016b) 'Why they're wrong', 1 October, www.economist.com/news/leaders/21707926-globalisations-critics-say-it-benefits-only-elite-fact-less-open-world-would-hurt.

Edgerton, D. (2017) 'The political economy of science: Prospects and retrospects', in D. Tyfield, R. Lave, S. Randalls and C. Thorpe (eds), *The Routledge Handbook of the Political Economy of Science*, London: Routledge: 21–31.

Elliott, L. (2016) 'Brexit is a rejection of globalization', *Guardian* 26 June: www.theguardian.com/business/2016/jun/26/brexit-is-the-rejection-of-globalisation.

Erixon, F. and B. Weigel (2016) *The Innovation Illusion*, New Haven, CT: Yale University Press.

Ernst, D. and L. Kim (2002) 'Global production networks, knowledge diffusion, and local capability formation', *Research Policy* 31(8–9): 1417–1429.

Fettweis, G. and E. Zimmermann (2008) 'ICT energy consumption – trends and challenges', paper delivered to The 11th International Symposium on Wireless Personal Multimedia Communications, https://mns.ifn.et.tu-dresden.de/Lists/nPublications/Attachments/559/Fettweis_G_WPMC_08.pdf.

Fisher, M. (2009) *Capitalist Realism*, Ropley: John Hunt Publishing.

Florida, R. (2005) 'The world is spiky: Globalization has changed the economic playing field, but hasn't leveled it', *Atlantic Monthly* 296(3): 48–51.

Ford, M. (2015) *The Rise of the Robots*, New York: Basic Books.

Foucault, M. (1980) *Power/Knowledge. Selected Interviews and Other Writings 1972–1977*, Harlow: Longman.

Foucault, M. (2009) *Security, Territory, Population: Lectures at the Collège de France 1977–1978*. Translated by Graham Burchell, Basingstoke: Palgrave Macmillan.

Gereffi, G., J. Humphrey and T. Sturgeon (2005) 'The governance of global value chains', *Review of International Political Economy* 12(1): 78–104.

Gieryn, T. (1999) *Cultural Boundaries of Science*. Chicago: University of Chicago Press.

Gordon, R. (2012) *Is US Economic Growth Over? Faltering Innovation confronts the Six Headwinds* (No. w18315), National Bureau of Economic Research.

Gordon, R. (2016) *The Rise and Fall of American Growth*, Princeton: Princeton University Press.

Gottdiener, M. (ed.) (2000) *New Forms of Consumption: Consumers, Culture and Commodification*, Lanham (MD): Rowman & Littlefield.

Graeber, D. (2014) *Debt: The First 5000 Years*, London: Melville House Publishing.

Hand, M., Shove, E., and Southerton, D. (2005) 'Explaining showering: a discussion of the material, conventional, and temporal dimensions of practice', *Sociological Research Online* 10(2).

Harvey, D. (1989) *The Condition of Postmodernity*, Oxford and Cambridge (MA): Basil Blackwell.

Held, D. and A. McGrew (eds) *The Global Transformations Reader* (2nd edition), Cambridge: Polity.

Heller, M. and R. Eisenberg (1998) 'Can patents deter innovation? The anti-commons in biomedical research', *Science* 280: 698–701.

Huber, M. (2013) *Lifeblood: Oil, Freedom, and the Forces of Capital*. Minneapolis: University of Minnesota Press.

International Panel on Climate Change (2013) *Climate Change 2013. The Physical Science Basis*, Geneva: IPCC www.climatechange2013.org/.

Janeway, W.H. (2012) *Doing Capitalism in the Innovation Economy*, Cambridge: Cambridge University Press.

Kaufmann, V. (2010) *Rethinking the City*, London: Routledge.

Keen, A. (2015) *The Internet Is Not The Answer*, New York: Atlantic Monthly Press.

Kelly, K. (1997) 'New Rules for the New Economy'. Wired, 9 January, www.wired.com/1997/09/newrules/.

Kosek, J. (2010) 'Ecologies of empire: On the new uses of the honeybee', *Cultural Anthropology* 25(4): 650–678.

Lanchester, J. (2013) 'When did you get hooked?', *London Review of Books* 35(7): 20–22.

Langley, Paul (2008) *The Everyday Life of Global Finance*, Oxford: Oxford University Press.

Lanier, J. (2011) *You Are Not a Gadget*, London: Allen Lane.

Lanier, J. (2013) *Who Owns the Future?*, London: Allen Lane.

Lazonick, W. (2014) 'Profits without prosperity', *Harvard Business Review* 92(9): 46–55.

Lazonick, W., M. Hopkins, K. Jacobson, M.E. Sakinç and O. Tulum (2017) 'US pharma's business model: Why it is broken, and how it can be fixed', in D. Tyfield, R. Lave, S. Randalls and C. Thorpe (eds), *The Routledge Handbook of the Political Economy of Science*, London: Routledge: 83–100.

Leach, M., J. Rockström, P. Raskin, I. Scoones, A. Stirling, A. Smith, J. Thompson, E. Millstone, A. Ely, E. Arond, C. Folke and P. Olsson (2012) 'Transforming innovation for sustainability', *Ecology and Society* 17.2: 11.

Leadbeater, C. (1999) *Living on Thin Air: The New Economy*, New York: Viking.

Leydesdorff, L., C. Wagner, H.W. Park and J. Adams (2013) 'International collaboration in science: The global map and the network', *arXiv* preprint arXiv:1301.0801.

Lohmann, L. (2017) 'Towards a political economy of neoliberal climate science', in D. Tyfield, R. Lave, S. Randalls and C. Thorpe (eds), *The Routledge Handbook of the Political Economy of Science*, London: Routledge: 305–316.

Lynas, M. (2007) *Six Degrees: Our Future on a Hotter Planet*, London: Fourth Estate.

Maassen, S., S. Dickel and C. Schneider (eds) (2017) *Sociology of the Sciences Yearbook: Techno-Science-Society*.

Markusson, N., M. Gjefssen, J. Stephens and D. Tyfield (2017, forthcoming) 'The political economy of technical fixes: the (mis)alignment of clean fossil and political regimes', *Energy Research and Social Science*.

Martin, R. (2002) *Financialization of Daily Life*, Philadelphia: Temple University Press.

Mason, P. (2012) *Why It's Kicking Off Everywhere*, London: Verso.

Mason, P. (2015) *Post-Capitalism*, London: Allen Lane.

Massey, D. (2005) *For Space*. London and Thousand Oaks: SAGE.

Mazzucato, M. (2011) *The Entrepreneurial State*, London: Demos.

McGrath, M. (2016) 'World's poorest countries to aim for 100% green energy', *BBC News*, 18 November www.bbc.co.uk/news/science-environment-38028130.

Mirowski, P. (2014) *Never Let a Serious Crisis Go to Waste: How Neoliberalism Survived the Financial Meltdown*, London: Verso.

Mitchell, T. (2011) *Carbon Democracy*, London: Verso.

Mittra, J., J. Tait and D. Wield (2011) 'From maturity to value-added innovation: Lessons From the pharmaceutical and agro-biotechnology industries', *Trends in Biotechnology* 29: 105–109.

Monbiot, G. (2006) *Heat*, London: Allen Unwin.

Morozov, E. (2014) *To Save Everything, Click Here*, London: Allen Lane.

Morozov, E. (2015) 'A dystopian welfare state funded by clicks'. *Financial Times*, 3 August www.ft.com/cms/s/0/8cc05ef0-37ae-11e5-bdbb-35e55cbae175.htm l.

Nolan, P. (2004) 'China and the global business revolution', in *Transforming China: Globalization, Transition and Development*, London, Anthem: 185–232.

Nonini, D.M. (2008) 'Is China becoming neoliberal?', *Critique of Anthropology* 28(2): 145–176.

Pagano, U. and M.A. Rossi (2009) 'The crash of the knowledge economy', *Cambridge Journal of Economics* 33: 665–683.

Pagano. U, and M.A. Rossi (2017) 'The knowledge economy, the crash and the depression', in D. Tyfield, R. Lave, S. Randalls and C. Thorpe (eds), *The Routledge Handbook of the Political Economy of Science*, London: Routledge: 57–69.

Paterson, M. (2007) *Automobile Politics: Ecology and Cultural Political Economy*. Cambridge: Cambridge University Press.

Piketty, T. (2014) *Capital in the 21st Century*, Cambridge, MA: Harvard University Press.

Rajan, R.G. (2011) *Fault Lines: How Hidden Fractures Still Threaten the World Economy*, Princeton, Princeton University Press.

Rajan, S.C. (2006) 'Automobility and the liberal disposition', *Sociological Review* 54(1): 113–129.

Randalls, S. (2017) 'Commercializing environmental data: seeing like a market', in D. Tyfield, R. Lave, S. Randalls and C. Thorpe (eds), *The Routledge Handbook of the Political Economy of Science*, London: Routledge: 317–328.

Roberts, J.T. and B.C. Parks (2007) *A Climate of Injustice: Global Inequality, North-South Politics and Climate Policy*, Cambridge (MA) & London: MIT Press.

Rockström, J. *et al.* (2009) 'A safe operating space for humanity', *Nature* 461(7263): 472–475.

Saxenian, A. (2005) 'From brain drain to brain circulation: Transnational communities and regional upgrading in India and China', *Studies in Comparative International Development* 40(2): 35–61.

Sayer, A. (2015) *Why We Can't Afford the Rich*, London: Policy Press.

Scholte, J.A. (2000) *Globalization: A Critical Introduction*, Basingstoke: Palgrave.

Shaxson, N. (2012) *Treasure Islands: Tax Havens and the Men who Stole the World*, London: Vintage.

Simpson, T. (2016) 'Tourist utopias: Biopolitics and the genealogy of the post-world tourist city', *Current Issues in Tourism* 19(1): 27–59.

Smith, R. (2016) *The Great Divide: Globalization, Populism and Stumbling towards a Post-Scarcity World*, www.robbsmith.com/the-birth-of-prosperity.

Solow, R.M. (1987) 'We'd better watch out', *New York Times*, 12 July, Book Review, 36.

Steffen, W., K. Richardson, J. Rockström, S.E. Cornell, I. Fetzer, E.M. Bennett, … and C. Folke (2015) 'Planetary boundaries: Guiding human development on a changing planet', *Science*, 347(6223): 1259855.

Stephens, P. (2010) 'Three years on, the markets are masters again', *Financial Times*, 29 July.

Stiglitz, J. (2013) *The Price of Inequality*, London: Penguin.

Stirling, A. (2009) *Direction, Distribution and Diversity! Pluralising Progress in Innovation, Sustainability and Development*, STEPS Working Paper 32, Brighton: STEPS Centre.

Straw, J. and M. Baxter (2014) *iDisrupted*, New York: New Generation Publishing.

Sturgeon, T. (2002) 'Modular production networks: A new American model of industrial organization', *Industrial and Corporate Change* 11(3): 451–496.

Szerszynski, B., M. Kearnes, P. Macnaghten, R. Owen and J. Stilgoe (2013) 'Why solar radiation management geoengineering and democracy won't mix', *Environment and Planning* 45(12): 2809–2816.

Taleb, N.N. (2007) *The Black Swan*, London: Penguin.

Tansey, G. and A. Worsley (2014) *The Food System*, London: Routledge.

Teulings, C. and R. Baldwin (eds) (2014) *Secular Stagnation: Facts, Causes and Cures*, London: CEPR.

Tyfield, D. (2013) 'Transition to science 2.0: 'Remoralizing' the economy of science', *Spontaneous Generations:* Special Issue on 'The Economics of Science', September.

Tyfield, D. (2014) 'King Coal is Dead, Long Live the King! The Coal Renaissance in the Emergence of Low Carbon Societies', *Theory, Culture & Society*, Special Issue on 'Energizing Society' 31(5): 59–81.

Tyfield, D. and J. Urry (2009) 'Cosmopolitan China? Lessons from international collaboration in low –carbon innovation', *The British Journal of Sociology* 60(4): 793–812.

Urry, J. (1995) *Consuming Places*, London: Routledge.

Urry, J. (2007) *Mobilities*, Cambridge: Polity.

Urry, J. and A. Elliott (2010) *Mobile Lives*, Cambridge: Polity.

Urry, J., A. Elliott, D. Radford and N. Pitt (2016) 'Globalisations utopia? On airport atmospheric', *Emotion, Space and Society* 19: 13–20.

Virilio, P. (1986) *Speed and Politics*, New York: Semiotext(e).

Walsh, B. (2013) 'Surprisingly Large Energy Footprint of the Digital Economy' *Time*, 14 August. http://science.time.com/2013/08/14/power-drain-the-digital-cloud-is-using-more-energy-than-you-think/.

Water.org (2015) *Facts about Water & Sanitation*, http://water.org/water-crisis/water-sanitation-facts/.

Wilson, L. (N.D.) 'How big is a house? Average house size by country', *shrinkthatfootprint.com* http://shrinkthatfootprint.com/how-big-is-a-house.

World Bank (2016) *Access to electricity*, http://data.worldbank.org/indicator/EG.ELC.ACCS.ZS.

Zeller, C. (2008) 'From the gene to the Globe: Extracting rents based on intellectual property monopolies', *Review of International Political Economy* 15(1): 86–115.

Zingales, L. (2014) *Capitalism for the People*, New York: Basic Books.

3 The genealogy of the emerging capitalist present

What next? Catastrophe? …

We have explored how the present is a moment – a generation – of global system crisis. This may not be understood, let alone practically intervened in, from the perspective of one, or even several, of the Grand Challenges that confront us. Only a systemic perspective exploring the mutually shaping threads that form the whole will suffice. It is this *system of systems*, its logic and micro manifestations, and its regime of system government that is collapsing, the regime in question today being that of 'neoliberalism'. The challenge of the present, thus, is whether – and, more specifically, *how* – a new system may be constructed out of the active and inertial persistence, exacerbation and collapse of neoliberalism.

How is this possible? And how can we begin to think insightfully about this so as actively to shape it, trapped as we are ourselves within the declining power/knowledge regime? In turning to these questions we must first address some crucial objections regarding the prior question of '*is* this system rejuvenation possible?' In other words, what about the possibility that any such investigation will be entirely overtaken by something much more dramatic. Certainly such scenarios also abound in contemporary discussions. As Morris (2010) notes, this may be either a broader system collapse (e.g. Turner and Alexander 2014; Lovelock 2007; Parenti 2011 cf. Diamond 2007; Tainter 1990) or conversely a utopic/dystopic qualitatively unthinkable transcendence to a trans-human 'singularity' (e.g. Kurzweil 2006; More and Vita-More 2013). In both cases there is an apparently complete blindness beyond a deepening catastrophic or triumphalist climax that can only rule us into dumb silence.

It is impossible to discount such futures, but there are strong reasons not to accept them as the final word (Jackson 2016); or rather to seek understanding with what we do have to hand. First, current evidence still suggests robustness in earth (-cum-social) systems for all the undoubted problems and their urgency (IPCC 2013; Steffen *et al.* 2015) so an imminent total upending of historical continuity remains highly unlikely. Much may indeed profoundly change in the next generation let alone century, therefore, but probably not 'everything'. Second, even granted a force majeur 'Act of God', human society will likely not be anywhere near annihilated so that politics and culture will continue, and both as primary shapers of the resulting socio-natural futures. There is thus both normative reason and analytical possibility for a serious consideration of the contours of the emerging society.

Indeed, from this perspective, the performative effects of catastrophist and eschatological discourses, proffering an historical *deus ex machina* that entirely overwhelms us, themselves deserve critical analysis. Here, then, such discourses can be criticized for not only denying but actively robbing humanity of its agency, while also contributing to the construction of supposedly apolitical futures in which questions of politics and of distribution of goods and bads are, if anything, potentially even more intense and important (Jackson 2015; Bonneuil and Fressoz 2016). Constructing plausible, rigorous and politically engaged accounts of possible social futures – and not just producing critical aetiologies of the present problems that then assume solutions that take familiar forms – is thus a key element of response.

But why suppose there is going to be a *new 'system'*? While we may pragmatically disregard apocalyptic catastrophe, what about the possibility of a long drawn-out winding down (Greer 2008) or conversely the fragmentation into the unstable co-existence of multiple and diverse neo-tribalisms per a '*Mad Max*' new barbarism (Urry 2011)?[1] Why assume there will be – *could* be – a new regime of global system government that (substantially) addresses the problems with the existing system, *especially* today, when these are so profound, far-reaching and historically unprecedented. These are key questions that cannot be brushed off so lightly. Indeed, a rational appraisal of the evidence must begin with these questions remaining fundamentally open and unanswered. So our question must be rephrased in terms of 'is there any evidence of the *actual concrete* and tendential emergence of such a system?'[2]

… Post-capitalism?

This raises questions about a concrete and high-profile answer to precisely this question, namely that of *Post-Capitalism* (Mason 2015; cf. Streeck 2014). We focus on this particular analysis for several reasons. First, because of its significant similarities, overlaps and sympathetic resonances with the approach here, in that it analyses innovation and its broader political economic cycles in attempting to answer questions about the current global system crises; and explicitly with a view to informing a politics of brighter and more equitable futures. Second, because it has garnered much (and well-deserved) attention in the public sphere beyond purely academic discussion. But, third, because it also comes to conclusions starkly different to those presented in this book.

Space prevents the comprehensive response that Mason's eponymous volume merits (though see Tyfield 2015/2016). Instead, we must limit the discussion here to a few key points. '*Post-Capitalism*' is effectively structured around four arguments and/or theories. First, Mason calls upon a variant of the Kondratiev theory of 'long waves' of capitalist growth (or 'K waves'). The variation he inserts is the importance of understanding this rhythmic process in terms that take seriously political (economic) contestation. Resistance to a given phase of capitalist growth is not merely a predictable side-effect that must be historically worked through, perhaps as part of the birth pangs of an 'upswing' (Perez 2002), but is constitutive and irreducible. Without such resistance, capitalists in pursuit of opportunities for competitive profit are not compelled to pursue the radical socio-technical and socio-political innovation producing the new 'upswing' and subsequent 'Golden Age'. Instead they can continue to pursue the 'easy' option of tightening the screws of existing mechanisms of labour exploitation in order to eke out a profit, even as the system tends to stagnation.

Through this lens, Mason argues that a new 'wave' should now have emerged, given the demise of the 'last' wave and the empirical periodicity of the process as a whole. Yet, through a series of interesting graphs ((Mason 2015: p. 94 et seq.), on which more later), Mason argues instead we are amidst an historical anomaly – the exceptional extension of a downswing that has endured now some 20 years beyond its proper 'date of death'. To Mason, this shows that the cycle is broken, and with it the dynamic of the capitalist growth engine that is the cycle of long waves; hence the turbulence, stagnation and general disorientation of the present. Capitalism needs a new upswing for it to survive. Yet none has emerged for some two decades now. The reason for this, according to Mason, is ironically precisely the global triumph of capitalism under neoliberalism. The destruction of the working class (of the Global North), which has been this 'victory's' primary means and end, was also the destruction of the most powerful socio-political mechanism forcing the renewal of capitalism, as just described.

Second, not only has capitalism not yet changed, but nor is it pregnant with a new upswing. To the contrary, the currently dominant domain of socio-technical innovation that neoliberalism's 'long down-swing' has actually generated – i.e. the ICT revolution – is systematically incompatible with capitalist relations of production. To make this argument, Mason calls on two further economic theories. First, he explores theories of information economies and information capitalism. 'Information' is an ever-increasing part of economic activity and value, including as the catch-all accountancy term of 'goodwill' on corporate balance sheets. Following many other (high-profile) scholars (e.g. Foray 2004; Benkler 2006; David and Dasgupta 1994; Arrow 1962), Mason conceptualizes 'information' as intangible, non-appropriable, non-rival and hence – now enabled by ICTs and the internet – 'free' in the double sense of freely circulating (as in 'free speech') and with a marginal cost of (re)production (hence price) that tends to zero (as in 'free beer'). In short, the very technological productivity of capitalism, still constantly revolutionizing the means of production, has now produced cutting-edge transformations in those means of production that also systematically undermine the price mechanism on which competitive profit through the market utterly depends. Developments in use value directly *undermine*, rather than renew and expand, exchange value. In these circumstances, then, we can see how the harder capitalism tries to renew itself, the more it simply deepens its crisis. No wonder no new 'long wave' has emerged.

For Mason, the booming of information and information-based production, including software, big data, interconnected materialities (the 'internet of things') and learning machines, is not just the latest revolutionizing of the means of production, the next step in a familiar process now several hundred years old through cycles of K waves. Rather, the information revolution's tendency to zero marginal costs combined with the fundamental capitalist mechanism of harnessing and appropriating the value of labour in production processes tends to the progressive and relentless destruction of the capacity of the market to coordinate political economic activity.

First, the increasing use-value importance of information to production processes translates into falling prices for those commodities, as competitive advantages and monopoly pricing are undermined by information's free circulation. The reduction in commodity prices, however, in turn reduces the cost of the basket of goods and services that adds up to the wage. The exchange value of labour thus also falls. This is then combined with the progressive replacement of labour with high-(information)-technology that is itself increasingly cheap, and often cheaper (hence 'cost-saving' in austere times) than workers. Put this together and capitalism enters a death spiral, for the combination of ever-reducing inputs by labour AND ever-falling value of that labour destroys the central motor of the expansion of capitalism, namely *increasing* production of value. If the value input of labour is declining and attempts to rectify this (i.e. more 'innovation') simply *accelerate* the devaluation of labour, there is no escape from the stagnation and demise of a system that is systematically dependent upon endless growth.

There is a crucial twist here, though, for Mason. Capitalist enterprises, in their individual pursuit of short-term competitive gain, are meanwhile busily constructing the technologies of an information economy that are increasingly affording spontaneous political economic coordination *outside* the market. In short, the inexorable and accelerating movement is towards a political economy that prioritizes use-value over exchange value. This is an 'economy' – more accurately a social formation, the idea of a standalone 'economy' itself a peculiarly capitalist notion – that is increasingly not only capable of being organized by way of voluntary knowledge input and labour, but also is one that is, conversely, impossible to coordinate and run on the basis of *capitalist* employment for production of commodities to be sold for a profit. So arises post-capitalism.

Altogether, this makes for a compelling theoretical edifice offering a profound and hopeful political vision. Against many commentators who take offence at the wide-ranging and imaginative synthesis of unapologetically heterodox theories, I find Mason's choice and use of theory for his argument informed, informative and sensible. It seems by far the greater error in this moment of unquestionable political economic turbulence to remain timidly locked within an orthodoxy that admits only a timeless economy of markets of physical goods tending to equilibrium, rather than to stride out and experiment with bolder, more political and more historical theories of the evolution of capitalism. The problems, however, arise in the particular theoretical synthesis thus constructed, its use to explain the present predicament and its fit with empirical evidence, including that which Mason himself presents.

First, regarding informationalization and its imputed destruction of the price mechanism, this argument draws upon on overly abstract and idealized concept of 'information' (and markets thereof). Admittedly, this concept is common in much economics of information, as mentioned above, yet it remains highly problematic. For information is *not* a purely abstract and intangible thing that floats free of a materiality against which it is defined. Rather, valuable information (i.e. forms of knowledge, but also semiotic or signifying phenomena more broadly, as in brands, 'look-and-feel', genre or cultural value etc....) is not only always materialized in some way, on some physical substrate, but also – and more importantly – founded on a much deeper ontology of knowledge-saturated things including, crucially, forms of tacit and embodied knowledge in specific human beings.

In all these ways, therefore, knowledge and information is not only appropri*able* in many (or most) instances – and so controllable by capitalist enterprise – but often appropri*ated*; born not 'free' but owned. This is especially the case for the key form of information for an emergent 'knowledge capitalism', namely markets of knowledge-intensive labour power. The expert surgeon, lawyer, engineer, musician, hair stylist, carpenter etc.... are all capable of appropriating quasi-monopoly rent gains from sales of their services, even as these are competitive and dynamic markets. The challenge of informationalization for capitalism, therefore, is not so much the collapse of the price mechanism as prices fall uncontrollably to zero. Rather it is two-fold: first, the acceleration of competition in knowledge-skilled industries, and the rate of innovation implicit in that; and second, the division of labour markets into high-knowledge skilled and low-knowledge skilled, where the former can build competitive CVs of unique expertise and experience to their personal advantage from which the latter are locked out, while the latter are locked *into* careers of increasing competition with robot intelligence and automation.

In both cases, however, these are challenges that are, first, socio-political challenges regarding the capacity of the broader complex power/knowledge system of 'society' to shape and/or accommodate these changes of accelerated innovation and/or deepening polarization of the workforce; challenges to which it is an open and empirical question whether or not, and how, capitalist society can adapt. And, second, both are transparently challenges that are not completely new, but evidently manifest already: in the acceleration of knowledge-based competition and innovation (Straw and Baxter 2014) and the sleepless, hyper-mobile lives of the 'working rich' (*The Economist* 2014; Crary 2014; Birtchnell and Calétrio 2013), on the one hand; and the clear polarization of the workforce into 'high' and 'low' skilled by the knowledge economy (Acemoglu and Autor 2011; Frey and Osborne 2013), on the other.

To be sure, these are still profound challenges. But they are challenges for capitalism about which we have to *see* whether or not they can be managed, not abstract dynamics that in themselves spell the sure collapse of capitalism. And, indeed, here we must also recall not just the exceptional resilience and flexibility of capitalism, but also that these dynamics only have to work *for capitalism* and its system winners in order to have a viable future. They do not need to promise *good* jobs and satisfying lives and bright futures for everyone – no matter how morally offensive one (hopefully) finds that. Indeed, only by admitting this do we begin to see the political issues and dangers currently in play.

Second, though, Mason's use of the K wave dynamic is also problematic. The K wave is here used to argue that a new wave should have emerged around 1990; that it did not and *still* has not while the last wave has continued zombie-like; and that it *cannot*, because of both the impossibility in principle of informational capitalism and how the key mechanism of such K wave upswing, namely working class protest, was destroyed by neoliberalism. First, regarding the K wave itself, while Mason's attempt to bring crucial political economic dynamics, constitutive of capitalism, to bear on these rhythms is admirable, it does not and cannot work since it attempts to do too much with a single dynamic, and that being one of techno-economic change. Capitalism, however, is a system that certainly does elevate techno-economic forces to a new level of importance but is *also* irreducibly a globalizing and ever-expanding complex system of power/knowledge relations, with all this entails (Figure 3.1).

Absent these considerations, K wave theory, including Mason's laudable and ambitious twist on it, gets wrapped up in insoluble problems of trying to explain more than it possibly can and/or the flipside of failing to explain even what it wants to. For instance, the timing and periodicity of the K wave *alone* is confronted with the absurdity of having to treat the entire period across the two World Wars (1914–1945) including the Great Depression of the 1930s – a period of extraordinary and hugely significant techno-economic change, let alone socio-political change – as an extrinsic accident and anomaly (Korotayev and Tsirel 2010; Perez 2002). This is to tear a huge hole in the relatively short history it is supposed to be able to illuminate, straining its credibility to breaking point. These are problems that Mason's use of the theory does not escape.

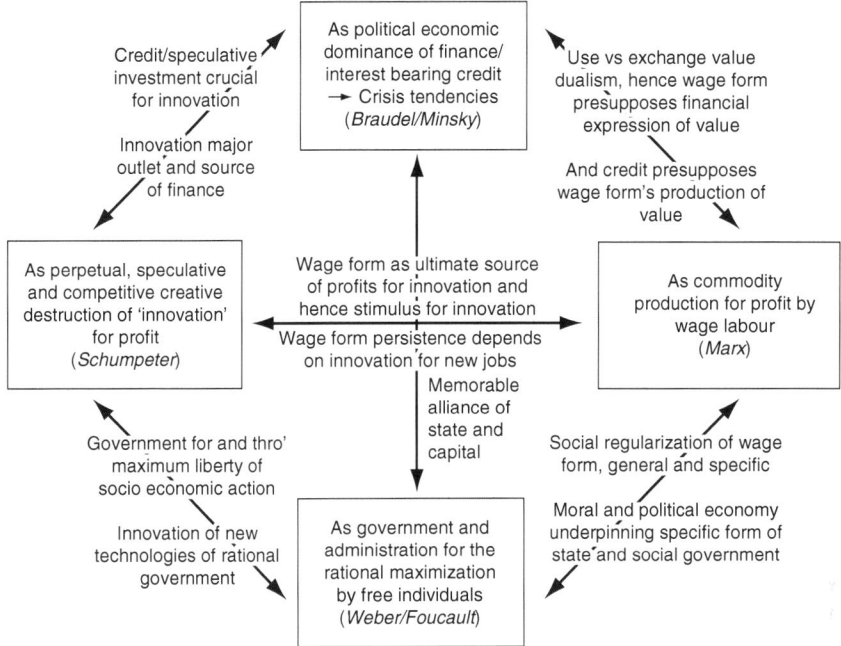

Figure 3.1 Capitalism.

Second, Mason (2015: 94 et seq.) himself presents compelling evidence that the strength of the working class in the global North at the end of the prior K wave (from the late 1960s through the mid-1980s) was at historic highs – as it unarguably was. The key mechanism to kickstart a new K wave was therefore, *ex hypothesi*, at its strongest ever, raising the dilemma that either he has identified the crucial mechanism of the renewal of capitalism and it worked, or it failed and therefore the importance of this mechanism is in doubt. Siding with the former, here, it seems rather that the very strength of the Northern working classes led to a titanic capitalist backlash (Harvey 2005), in the form of neoliberalism and its specific model of financialized globalization that, in turn, sponsored precisely a new K wave, founded on ICT innovation, and the massive global roll-out of oil-based (and debt-fuelled) consumerism. On this account, however, it is clear not that the mechanism of capitalist renewal that Mason insightfully identifies *failed*, but that it *succeeded*, spectacularly! And it did so in ways that did indeed seed both a new K wave and a change of political economic model underpinning a massive secular upswing in the global economy. The terrible irony was just that the overwhelming losers of that process of successful working class resistance were those very Northern working classes.

And, indeed, Mason's graphs (mentioned above, see pp. 94 et seq.) illustrate precisely the periodicity we would expect from a new K wave starting around 1990 (i.e. concomitant with the collapse of real-existing communism in the USSR – and, arguably, China) in a boost to capitalist economic growth over the next 15–20 years (see especially pp. 100, 101, 104), up to the GFC of 2007/2008, and beyond in the crucial case of the global South, now the engine of system growth more generally. As his Figure on p. 102 shows, however, there is one clear and absolute loser to this process of global growth and realignment of distribution, namely the working and middle classes of the global North. Moreover, this spectacular socioeconomic development in the global South points to a second key problem with Mason's account; namely its parochial focus on the unquestionable destruction of that *global North* working class as if it spells the end of the working class (or a capitalist subaltern as political constituency) per se. To the contrary, though, in China alone, the working class will by 2020 likely be bigger than that of the EU and US combined, at some 533 million people (Jacques 2009: 186). Far from establishing a conclusive, self-defeating victory of capital over labour, in other words, the neoliberal period has dismantled that of the global North only to incubate a much larger and global working class.

In all these ways, therefore, it seems that the evidence points not to an imminent collapse or implosion of global capitalism but rather to its robust health, albeit while currently undergoing one of its periodic paroxysms of transformation and resettlement highlighted by the 'Arrighi/Braudel' (A/B) cycles of global capitalist hegemony (Arrighi 1994, 2007). Pairing expanding territorial and capitalist logics, Arrighi reveals a neo-Gramscian geopolitical dynamic of cycles through phases of system dominance of productive capital (MC), financial capital (CM') and then system turbulence and 'reset', centred on core hegemonic polities of increasing scale, power and reach (Italian city states → United Provinces of the Netherlands → Great Britain → United States) stretching back as much as 800 years. Combined with that key dynamic – missing and, indeed, explicitly eschewed in Mason's analysis – it is clear that far from stalled and winding down, the present fits perfectly within an *unbroken* cycle of K waves (cf. Mason 2015, see Figure 3.2). For it is the A/B cycle that explains the system turbulence not a break in the K wave dynamic, which, rather, is strongly evidenced as a new upswing from around 1990 (as just described). This would place the present firmly in a peaking K wave based on the initial emergence of ICTs, turning to its deepening consolidation into the broader industrial economy; the beginning of a phase of maturity and decline of the last K wave of fossil-fuelled consumerist growth and the internal combustion engine; and all in the middle of a periodic inter-regnum between a declining financialized hegemon and the embryonic emergence of a new hegemon on expanded scale founded on a new regime of productive capital accumulation and non-zero-sum growth (see Tyfield 2015/2016).

Of course, these dynamics themselves permit no easy extrapolation to a certain future. Rather, they demand again attention to the concrete, unfolding and possibly embryonic evidence of such a new emergent regime of accumulation and hegemony. But they also give us sufficient reason to treat the continuation, resurgence and mutation of capitalism as our starting or default hypothesis; the *prima facie* case on which the burden of proof lies almost, but not quite, on the side of having to *disprove* such a trajectory. To be sure, a positive case for strategic alignment to this emerging future is needed, and may seem hard to furnish, even implausible, at present. Yet it would serve us well to recall just how implausible the prosperity and peace (in the global North) and decolonization (in the global South) of the post-war period was in the darkest days of the 1930s. Nonetheless, the question still remains: how can we test this default hypothesis?

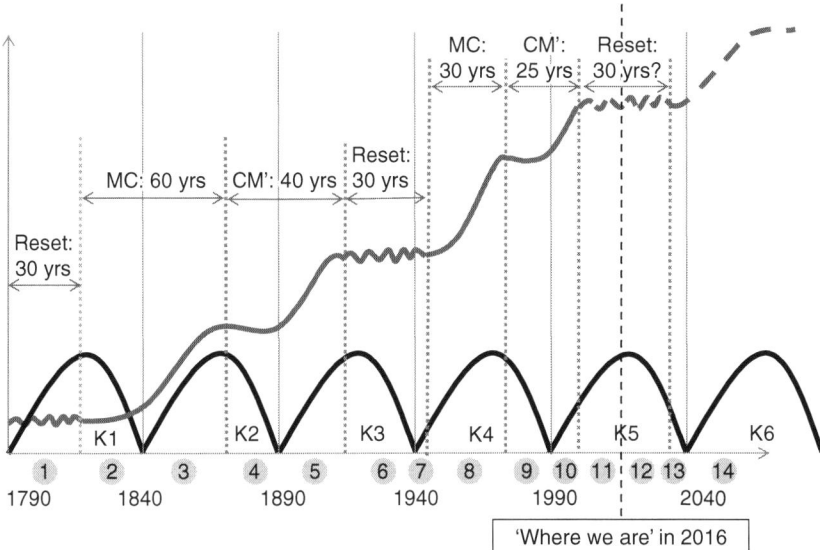

* For more details, see Tyfield (2015/16):
http://www.lancaster.ac.uk/staff/tyfield/On_Postcapitalism_3.pdf

1 1790–1815   Napoleonic Wars and 1st Industrial Revolution

2 1815–1840   Post-Napoleonic depression under restabilized British
hegemony

3 1840–1875   2nd K wave upswing, 'Age of Capital'

4 1875–1890   (First) Great Depression, Financialization, 'Age of Empire'

5 1890–1914   *Belle Époque*, 3rd K wave upswing, twilight of British hegemony
with interimperialist rivalry

6 1914–1940   WW1, peak and fall of K wave in Great Depression, collapse of
British hegemony, US not yet hegemonic

7 1940–1945   WW2, Allied win, 4th K wave upswing begins, US ascendant to
hegemony

8 1945–1975   *Trente Glorieuses*, secular coincidence of new hegemony and K
wave upswing in 'MC' cycle.

9 1975–1990   Signal crisis with 4th K wave downswing, US from hegemonic to
dominant, ascendancy of neoliberalism (inc. globalization and
financialisation).

10 1990–2001   'Roaring 90s', 5th K wave upswing, growth of Global South
'emerging markets', US unipolar superpower.

11 2001–early '20s?   Collapse of US hegemony and neoliberalism with low growth in
Global North and declining, volatile global growth and no new
emergent hegemon

12 Early 2020s–2030?   Global economic downswing exacerbating global turbulence,
ascendant new hegemon

13 2030–2040?   New hegemon established but post-geopolitical-'reset' recession
as K wave downswing ends (Cf the post-Napoleonic war
depression)

14 2040–?   New 'Golden Age' centred on new hegemon with 6th K wave
upswing

Figure 3.2 The rhythms of the capitalist global system from the *combination* of K waves
and A/B cycles*.

Looking forward: a phronetic synthesis of Gramsci and Foucault

The key methodological/theoretical issues in tackling this question hinge on how we may go about exploring the unfolding future in a way that manages the epistemic challenges of both novelty *and* continuity in the context of *systemic* change where what may and may not be taken as the latter – and hence a given premise (of which *some* are necessary conditions of intelligibility for any analysis) – is radically unclear and uncertain. The actual working through existing evidence is necessarily a messy process – what in an epistemologically neater and more confident (and seemingly less complex, uncertain and 'liquid' (Urry 2002; Gross 2010; Bauman 2000, respectively)) age may have been called 'dialectical'. But in retrospect we can observe that the process involves the constant shuttling back-and-forth and cross-checking between modes of thought of differing qualities: imaginative and speculative vs critical and contrastive; constructive vs deconstructive; interpretive vs explanatory; synthetic vs analytical; engaged in the practical and political present vs abstract and scientific etc....

More specifically, the analysis in this book is both critical realist and post-structuralist in orientation, reflecting the productive synthesis of approaches constitutive of a cultural political economy (cf. Sum and Jessop 2013) but with greater focus on strategic intervention in the present and the key issue of innovation of power/knowledge technologies (Tyfield 2015). By 'critical realist' we mean analysis that is fallibly realist in epistemology, involving the conjunction of an epistemic social relativism (i.e. knowledge is always irreducibly socio-historically situated) with the possibility nevertheless of judgementally rational (i.e. reasonably conclusive and objective, if fallible) conclusions regarding representationally true statements about a non-mind-dependent reality. This takes the form of explanation in terms of major causal powers and tendencies that are real but not necessarily actual (Sayer 2000; Bhaskar 1998). Critical realism thus emphasizes the identification of forms of (conditional and contingently emergent) 'natural necessity' regarding 'what is the case' from the messy confusion of the contingent course of actual events; together with the inescapability of ontological commitment to *some* such realities and contours of reality itself in the day-to-day process of practical judgement.

Conversely by 'poststructuralism', we mean here primarily a genealogical form of analysis exemplified in the (later) work of Michel Foucault (2004, 2009, 2010). This emphasizes both the practical nature of knowledge claims and their saturation thus with strategic projects of specific living persons and groups. This goes together with a conceptualization of power – as power-knowledge – described above (Chapter 1). Recall, power is productive and constitutive of social formations, relational, dispersed and normatively ambivalent and complex; in each respect of which it contrasts with the standard (modern) juridico-discursive conception of power respectively as coercive, a property or asset, concentrated and held (by definition) by 'the powerful' and presumptively bad until legitimated and tamed by institutions of consent of the 'ruled'. Hence the emphasis is on the *deconstructive* genealogy of supposed 'necessities' of a given and supposedly universalistic 'common sense' into the (normatively ambiguous if not always, per Nietzsche, scandalous) contingencies from which they were actively founded.

This seemingly awkward pairing of perspectives, though, is secured on a fundamental reorientation in the social sciences towards phronesis. This is the epistemological and methodological corollary of the CP/KS approach introduced above, given that the researchers are always *themselves* situated within such particular and substantive systems of power/knowledge relations. Following the work of Flyvbjerg and a growing movement or 'real social science' (Flyvbjerg *et al.* 2012; Flyvbjerg 2001), '*phronesis*' refers to the primary Aristotelian epistemic form and virtue of a situated practical wisdom. It is thus to be contrasted with the more familiar knowledge forms of *episteme* (objective scientific reasoning of universalistic laws) and *techne* (concrete know-how).

In the modern period, characterized by a strict Cartesian dualism of subject/object and growing scientific mastery, phronesis has been forgotten if not actively denigrated. Yet as Flyvbjerg shows, it is not only presupposed in all cases of episteme and techne, rendering it primary, but also its explicit rehabilitation is a matter of utmost urgency today. This imperative is even clearer in the context of the Four Challenges, and especially that of complex government of complex systems. Only diverse and practically-engaged knowledges offer any prospect of meaningful responses, while the complexity, 'wickedness' and systemic profundity of these problems means both episteme and techne – universalistic prescriptions and detailed practical tinkering alike – are largely impotent and often self-defeating, at least without their broader contextualization in projects of phronesis.

Such phronetic knowledge is also both fallibly realist and, in a modern twist explicitly following Foucault (Flyvbjerg *et al.* 2012), power-attentive, strategic (not purely epistemic) and restlessly non-foundational. A phronetic reorientation thus affords a productive if necessarily messy dialogue (rather than a dialectical synthesis) of these seemingly opposed ontological-cum-epistemological positions insofar as *both* are also framed within a phronetic project. Pairing critical realism's Gramscian concern for fallible definition of real tendencies and (conditional) necessities and a Foucauldian attention to the strategic, performative and power-relational nature of knowledge claims – hence to openness and contingency and the active work put into their closure – a productive phronetic methodology may be formulated that can study 'transition' to or 'emergence' of a new social system in ways that attend to the openness and uncertainty confronting such analysis while not letting this rule us into dumb silence.

Key here is the reconceptualization of the process of system transition (and of innovation per se) through the CP/KS lens, in particular incorporating just such a strategic, productive, relational and ambivalent conception of power (Tyfield 2014). To recap, reframed in CP/KS terms, transition becomes a process of power relation-mediated strategic agency that is qualitatively shaping *new* power-relations (and hence power/knowledge systems). With innovation itself conceived as socio-technical power process, it also becomes a privileged window into this process, precisely as the key reflexive moment of *power/knowledge acting on itself*, i.e. *the paramount arena of this power process of transforming and (re)constructing society*. This is especially the case regarding the power/knowledge technologies that are themselves being innovated in the context of *existing* power relations, systems crisis, zeitgeist etc....

The aim of this process of analysis is to trace two things, using and iteratively reinterpreting concrete evidence (especially regarding contemporary processes of innovation in particular): on the one hand, the glimmers of both the relational structure of such a system and its internally-related and constitutive power 'logics'; *and inseparably*, on the other, the (power-saturated) process of *how it could emerge from where we are*, given the power relations and power/knowledge technologies of the present including, crucially, the dynamics of the 'crises' themselves.

This clearly involves the gathering, organization and analysis of evidence in the *present* of an (if not unimaginable, then) empirically inaccessible and hence speculative future. This is a shift in temporality that has significant implications for the standard forms of both a critical realist and a genealogical analysis. Usually both of these approaches explore the present and open it up to informed political response using empirically available evidence from the past. Yet if the present was formed in the past, we have no reason to suppose the future is not being formed in the present, with contingencies being transformed *right now* into apparent necessities and seemingly unarguable truths. Precisely the awareness of this contingency and ongoing construction thus raises the demand – especially in moments of system crises, normative epistemic disorientation and systemic strategic opportunity – that we seek to understand the *present* processes from which *future* 'common senses' are currently emerging.

Accordingly, the familiar *post hoc* forms of both critical realist and genealogical analysis must be altered. This shift in temporal gaze – from the past-viewed-from-the-critically-engaged-present to the present-viewed-from-the-critically-engaged-emerging-future – demands that both work together in complement and over an expanded analytical process of three steps. The goal of that expansion is twofold: to triangulate in the construction of a credible future that responds to the responsibility of exploring the emerging regime, its logics and 'common senses' – especially in a moment of systemic inter-regnum, but also more generally with complex system government as now an emerging and durable, but currently profoundly problematic, predicament – while also seeking to remain faithful to the ontological and political openness of the future thus described *and* the *strategic* impact of such analysis itself on the future thereby *actually* constructed. In other words, it is to construct a *genealogy of the emerging present as a project of phronesis.*

The three steps consist of three questions that respectively look forward, explore that emergent future and then look backward *from there* to the present (cf. Flyvbjerg *et al.* 2012):

1   Where are we going (tendentially) in terms of emerging, possibly embryonic, systemic logics and common senses, and how?
2   Is there a 'there' there? Does this system have an internal strategic-systemic coherence, hence a possibly self-sustaining *power momentum, and* an external coherence, particularly vis-à-vis the deepening systemic crises engendered by the disintegrating incumbent? What is the logic of the system (or regime thereof) and how does it 'work' in terms of its self-sustaining system dynamics? *Who* does it primarily serve and empower?
3   If 2) is affirmative, what does a critical analysis of this system reveal in terms of *its* limits, contradictions, problems, social exclusions and the naturalized and legitimated contingencies of its construction? And hence, what can be done in the present to shape it?

Step 1 draws on existing evidence and trends to construct a possible emergent system. Step 2 accepts this provisional characterization and explores the resilience and dynamism of that system on its own terms and especially in the current context of systemic breakdown. To the extent that a picture emerges from this analysis, therefore, then and only then can one *conclude* with any degree of confidence (but certainly not complacency or 'rational optimism' (cf. Ridley 2010)) that there is

likely to be a new system (i.e. *in this case*); the abstract conclusion *follows* the concretely characterized one. To this point, such an analysis *cannot help but be*, in the
first instance, itself both an implicit apology (if not lauding) for, as well as a performative intervention in the construction of, that system. Indeed, the more compelling
the analysis the greater its fatalistic acceptance and naturalizing effect will appear.
This could lead to an agential and normative disabling. At least, this is so to the
extent that the project as a whole is framed as a purely epistemic and 'disinterested'
scientific enquiry. Certainly, there are no clean hands – whether of the 'disinterested'
scientist or a 'critical' praxis.

As a phronetic project, however, having assumed this theoretical perspective and
constructed such a credible picture, the question that *immediately* presents itself is:
*what is this new system like – socially and politically – and what can/should be done
about this?* This leads to the crucial third and final step in which both the more
familiar tools of critical realist critique and genealogical criticism may once again be
deployed, but here on – and situated *within* – a speculatively projected emerging
future. This emerging future is thus both *constructed* and then *deconstructively re-
opened*. From a purely epistemic perspective this may seem a crazy waste of considerable effort: one step forward, one step back. But back where? Back to the *present*, of
course, where shaping the future *must* happen and which is the only place it *can*
happen. From a phronetic perspective, in other words, the 'journey' is both crucial
and unquestionably worthwhile for one's strategic orientation to an emerging
present – to an emerging *system* that will likely dominate 'common sense' and social
change for a generation or more and, today, with seemingly the very 'world' at stake
– is now thoroughly (in)formed for action in real-time in that present.

Testing our default hypothesis in this way, however, we are further assisted by a
substantive insight regarding the particular meso-level dynamics of incumbent
system change today. Specifically, the twinned critical realist/Gramscian and
Foucauldian analysis of a CP/KS perspective highlights a key substantive system-
constitutive power dynamic of modern capitalist societies that must take centre-stage
in a self-consciously strategic but critical explanatory analysis of real-time system
transition: the essentially contested dynamism of liberty-security constitutive of liberalism (Foucault 2009, 2010).

The complex system of liberalism

Through the modern era, to the present day, the power regime that has grown
increasingly ecologically dominant (Jessop 2014) may be called 'liberalism'. Today,
however, 'liberalism' is a much abused and confused term even as, amidst the crises
of global capitalism and its possible refounding, it is also absolutely key, connoting
the CP/KS that is in crisis. To understand this crisis, therefore, we must understand
liberalism. This task is significantly complicated by contemporary common-sense use
of the term.

A main culprit in this process is the American sense of the word, which has
become particularly dominant in recent years, increasingly even in the UK. Here
'liberal' is contrasted with 'conservative', which are then placed on top of US party
lines, Democrat and Republican respectively (e.g. Krugman 2009; Frank 2016). The
term is thus supposed to capture a settled political spectrum, readily intelligible to
all, usefully reduced to the short-hand of these single terms. Through the neoliberal
period and now today into the age of Trump, 'liberal' has become a 'conservative'

swearword, meaning the 'bleeding heart' and 'politically correct' supporter of big and activist government. This has recently coalesced into a particular hate figure of the new populism (Müller 2016), the 'liberal establishment' of the *bien pensant* metropolitan cultural and media – but also legal, political and financial – elite.

This understanding of 'liberal' has long been more confusing than illuminating. For instance, it is no longer clear what is 'liberal' – i.e. regarding a foundational orientation to liberty – about state provision; or, vice versa, what is 'conservative' about letting the market destroy precious national institutions? Moreover, how do this 'liberalism' and the dominant regime of the day, *neo*liberalism, fit together? Indeed, the confusion here is a mark of the conceptual hegemony and strategic upper-hand of neoliberalism against any 'progressive' political programmes, denied even a meaningful language of their own differentiation and self-definition (of course, especially in the US).

In the age of Trump, and hence profound Government-sponsored *il*-liberalism (at the centre of the global capitalist system), though, this dominant understanding of 'liberalism' is transparently self-defeating and problematic. For thus defined, 'liberalism' is the compromised property of a particular mid-twentieth century political project that has transparently failed – in the very election of Trump vs Clinton – to counter that resurgent illiberal nastiness and threat to system integration. A major part of the problem is that different aspects of 'liberalism', as political orientation, are conflated here, as the more nuanced Anglo-European sense reveals, distinguishing between the social liberal and economic liberal. As the Clinton campaign exemplifies, to its profound electoral cost, 'liberal' today apparently encompasses both the anti-WTO activist and the Wall Street CEO committed to global economic liberalization; the advertiser pushing a global brand and homogenized culture (cf. Ritzer 2014) and the local stalwart stewarding local tradition or the counter-cultural *avant garde*; the gig economy employer paying below the minimum wage and the struggling working classes of the precariat (Standing 2011). No wonder the entire spectrum of metropolitan political opinion can now be dismissed as all-of-a-piece, the 'liberal establishment'.

But this distinction of social vs economic liberal does not itself fare much better in terms of illuminating matters today. The banker may be obviously economically liberal, but is probably also socially liberal in terms of openness to hiring the best 'talent', be they women, LGBTQ, people of colour, or immigrants. Yet s/he remains an altogether and obviously different shade of political opinion to the radical feminist or the campaigner for global environmental justice. Against those who argue that these are profoundly *illiberal* times, since *their* chosen definition of liberalism is currently under attack, it seems that the problem is rather that the common-sense understanding of 'liberal' itself has been evacuated of meaningful and discriminating substance. 'Liberalism', in other words, is under threat – thereby *affording* the strategic advantage to a rough, messy coalition of the illiberal aggrieved – because its own house is so profoundly in disarray; both discursively, regarding the meaning of 'liberalism', and politically, regarding what status quo 'liberalism' offers the majority.[3]

Yet, insofar as we still live in thoroughly *capitalist* times – indeed, arguably as never before – 'liberalism' remains a key term, if defined appropriately in CP/KS terms. We must be absolutely clear that 'liberalism' (as used here) refers to a family of regimes of complex power/knowledge system government and not an explicit political philosophy or orientation (which are, rather, among its strategic power/knowledge technologies). Nor does it mean liberal *democracy*.[4] In particular, by 'liberalism' we mean a regime primarily characterized by socio-political and personal self-ordering through the production and consumption of new (primarily individual) freedoms – the regime of 'living dangerously' (Foucault 2010: 66). Liberalism is a power regime characterized by systems of liberal *individualizing* selves, actively expanded *rational* technoscience, *secular materialist* exploitation of reality, associated institutional and organizational forms and the promise or dogma of *rational progress*, all in self-advancing pursuit of and through their growing *liberty* (hence precisely 'liberalism').

It is also, of course, characterized by (ever-expanding) capitalist relations of production. Liberalism is thus the regime of *capitalist* system government, and it has taken multiple specific forms over the past 200–300 years. This, of course, includes neoliberalism, but also the social liberalism of the post-war welfare state, from which the still-incumbent but now utterly out-dated and otiose American common-sense of 'liberal' hails. But note also how, from this complex power/knowledge systems perspective, the ontologically primary aspect of these systems is their strategic-relational power logic, or rather the complex assemblage of power/knowledges, technologies, institutions and selves, *not* (just) the capitalist relations of production. Though, to be sure, the latter forms a crucial aspect of the whole, mediating the intense and historically unprecedented dynamism of (distinctively liberal!) power/knowledge 'innovation', as well as thereby creating socio-technologies capable of harnessing the massive and endlessly growing energic costs of the system's growth (Biel 2012).

Liberty-security

Of greatest importance for our present task, however, is that liberalism may be characterized by a specific dynamic that is the source of its (presently unrivalled) world-producing strategic advantages. This key dynamic (of the present), which underpins *both* the reproduction and sedimentation of specific forms of liberal regime *and* their transition and emergence, is that of liberty-security.[5]

Liberalism, to repeat, is government through production and consumption of new freedoms (Foucault 2010: 63) while also in the process conditioning and enabling specific, concrete subjectivities that continue actively to press and mobilize for their increase. This is thus an intrinsically dynamic and expanding power regime. But the innovation of new forms of individually-enabled action necessarily raises a perpetually new set of challenges for the preservation of system integrity. In other words, the very source of liberalism's system dynamism and productivity also and necessarily constructs phenomena that are (rightly) interpreted by actors within that time-space as existential threats to the collective integrity of their 'society', or 'security threats' (Foucault 2010: 64–65).

This process involves 'overflow' (cf. Callon 1998) in two senses:

- First, the necessarily limited definitions and conceptualizations of 'common-sense' power/knowledge relations-technologies of government exclude, penalize and burden with the system's costs contingent groups that are *not* the group thus *enabled* by this system, thereby generating ever greater sources of objectively 'legitimate' and irrepressible anti-system grievance.
- Second, the continual expansion of new liberties enables new forms of individual(ist) action and practice that are not yet 'governed' by the sedimented systems of power/knowledge technologies and relations.

The build-up of both of these – growing 'security threats' and system destabilizing liberties respectively, both of which are socio-technically-mediated – increasingly threatens to exceed the current power/knowledge technologies of government in their conceptualization and/or capabilities. But, of course, the latter too are constantly being innovated. Insofar as their trajectories of innovation are relatively resonant or parallel, therefore, the 'normal' way of managing these system-productive-but-disruptive challenges may continue with familiar forms of 'crisis management'. Hence the growing forms of anti-system grievance and new innovations may both be managed to the extent that the system is *also* growing both the asymmetrically-distributed enablement of system supporters and beneficiaries – deepening system integrity – and its corollary, the capacities of power/knowledge technologies of government to effect the (self-)policing systemically needed.

At its heart, liberalism is thus characterized and constituted by the hugely social productive positive feedback loop amongst new sociotechnically-mediated 'liberties' for specific groups, enabling the further innovation of new socio-technologies to their specific strategic advantage. This is thus a profoundly Dionysian process, harnessing deep-seated human aspiration and greed, hope and despair, celebration and envy, to the formation of seemingly Apollonian structures of power/knowledge technologies and common-sense rationalities that tame and shape those animal appetites in turn. In short, against the 'liberalism' of contemporary befuddled common-sense, there is nothing cuddly or even necessarily compassionate about liberalism. Rather it is a pitiless power regime of consuming and expanding liberties of specific humans in their pursuit of 'rational' and 'legitimate' self-advancement.

However, the very dynamism of liberalism and the proliferation of innovations (of power/knowledge technologies i.e. as *political* process) necessarily tends to the production of new forms of liberty that qualitatively exceed even the incremental innovation of governmental technologies, especially as this must take place within limits to the (contemporaneously socio-technically accessible) exergy (Biel 2012) available for the needs of the growing system. The result, therefore, is the deepening emergence of existential systemic 'security threats' – which are real! – not dealt with adequately by the current regime and its processes and logics of government and hence emergent precisely in the form of a system '*urgence*' (Foucault 1980). In short, the emergence of deepening crises *of crisis management* (Jessop 2013).

Moreover, where the twin challenges of unmanageable overspill have accumulated to the points of crisis of crisis management, liberal systems are confronted with a fundamental but profoundly dynamizing challenge. On the one hand, the ceaseless innovation and reproduction of liberties must continue as this is the life-blood of systemic dynamism. But, on the other, there must also be innovation in power/knowledge technologies of government – so that the former may be newly 'governed' – that mark a distinctive *break* from the existing ecologically dominant relations-technologies of government, i.e. exactly the conditions that need to be overcome. In short, this context of 'new' and 'old' plays out as a turbulent antagonism between system *dynamism* – as survival vs *stagnation and breakdown*, given the intrinsic growth dynamic of liberal(-capitalist) systems – and system *integrity* – as survival vs *disintegration*.[6]

Security (threats), then, may be understood as the necessary flipside and product of (also ever-self-proliferating) liberties. Bringing security threats explicitly into the picture, however, also further clarifies the dynamics of liberal systems. The deepening challenges of system *success* elicit action attempting to manage both new liberties and *old* security threats, where these are likely to be overlapping and mutually compounding processes (e.g. how ISIS is both enabled by globalization and new social media and a familiar neo-traditionalist movement of anti-American imperialism, respectively). However, in both cases, given incumbent systems – processes and logics – this is done in ways that are framed by these 'old' systems and hence in ways that tend to *fail actively* in both regards; hence merely exacerbating both challenges.

Focusing on security threats also helps us recall the essentially arational and non-cognitive but power/knowledge-mediated process at work here. In other words, security threats are viscerally encountered as such: not just as new unmanageable problems that are confounding for the cool, rational, liberal intellect, but utterly baffling and disorientating challenges that present potentially existential dangers (especially to those least enabled to respond). In these circumstances, then, a key aspect to the dynamism of liberty-security dynamics is that they tap into these deep-rooted, intuitive motivational drivers (cf. Haidt 2013) of human agency. This includes such powerful affective registers as fear and hope, hatred and love, but also the non-rational, practical and cultural richness of human selves and identities, personal and collective (cf. Fischer 2009). Crucially, though, this is particularly enabled when done in ways that are specifically oriented to the novel formulation of new 'universal' power/knowledge technologies of 'reason', 'rational action' and (individual) 'liberty', given the fundamental mechanisms of incumbent system government through consumption and expansion of liberties. The Apollonian face of a 'reasonable' and 'liberal' regime thus is forged anew in the fiery productive chaos of the Dionysian contestation of liberty-security.

This is thus an essentially productive process from the perspective of complex power/knowledge systems of liberalism, with innovation (today especially) as key. We can consider this process in two analytically distinct, but ontologically inseparable and co-productive, aspects, namely the material and discursive/ideational dimensions. In both cases, these are liberty/security dynamics. First, regarding material social change, rather than just eliciting a dynamic of deepening system disintegration, the emergence of the new liberties (together with old liberties still 'active') and of the security threats as systemic challenges *elicit further 'innovation'* (as socio-political-technical change).

This innovation takes two forms:

- 'security measures', i.e. new forms and technologies of (self-) government *against* the newly-perceived security threats emergent from the old and new liberties alike that now exceed system management (e.g. measures to combat the negative corollaries of the internet, and digital social media, in new horizons of intellectual property piracy, trolling, bullying and misogyny, access to pornography reaching new levels of sexual dysfunction in young men and sexual objectification of young women, child pornography and sexual abuse, challenges to privacy and data security, cyber-snooping technologies, fake news etc....); AND crucially
- actions in turn *by* the new (and old (cf. Edgerton 2011)) liberties to preserve themselves *against* the reciprocal overspill of those same 'security measures', perceived as *security threats to these liberties* (e.g. all the digital technologies and socio-technical innovations that attempt to preserve online anonymity (e.g. Ghostery, Startpage, Tor ...) or erase the permanent record of interaction (e.g. Snapchat) or strikeback against state/corporate surveillance (e.g. hactivism, Anonymous, Wikileaks etc....)).

Hence a dynamic emerges of active innovation and counter-innovation, and where these are increasingly *explicitly* framed around issues of 'freedom' and 'security' even as (or rather because) the substantive *definitions* of these two terms are themselves essentially contested through that very process. Dynamics of innovation-as-politics are thus both propelled by *and* harnessed to the active construction of new worlds – and, indeed, specifically *liberal* systems – constituted by ever-new horizons of empowered liberties (Figure 3.3).

The flipside of this 'material' process of constructing 'new worlds', however, is equally crucial; namely that as a human (and not an 'automatic' or 'natural') process the power involved, and the power relational system conditioning the present and emerging from the innovation, is irreducibly conceptual and intellective. It thus concerns, builds and uses new ('universal', 'rational') power/*knowledges* and concepts, both technical and legitimatory, both descriptive/scientific and normative. This matters profoundly, not least in understanding the arational, power/knowledge dynamics of construction and *re*-construction of liberal socio-political order.

The *power*-knowledge innovation dynamism of liberalism literally creates new worlds (or complex power-knowledge systems) for the government of which existing resources and technologies of understanding are *necessarily* inadequate. Moments of system breakdown, of given 'normal' common-senses and forms of world-sense-making and crisis management, boost the appeal to meaning-seeking beings like ourselves of highly idealistic and/or populist ideologies promising to restore a 'lost'

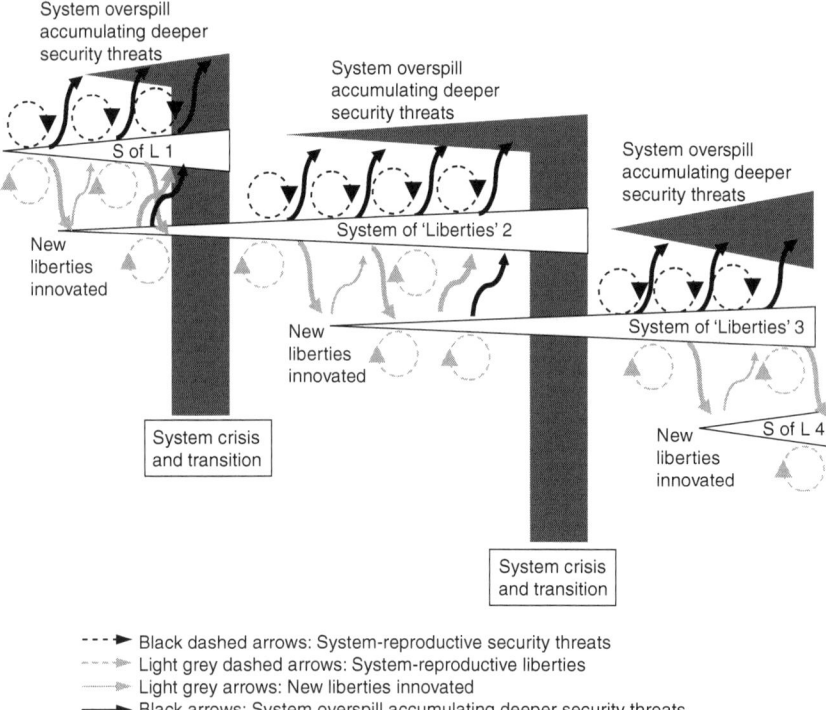

Figure 3.3 Cycles of liberty-security in complex power/knowledge systems.[7]

order. Given systemic crisis and openness, this can in turn feed their seeming world-productive self-legitimation, empowering them further. Yet this very process is also divisive and an affront to established liberties, and so tends simply to contribute to system crisis as the performative precondition of broad-based ideational consensus (or hegemonic common-sense) becomes ever-more transparently groundless. In short, a crucial mediation of system crisis is the breakdown of a largely unreflective, 'pragmatic' but system-performative trust and 'common sense' in the workings of that power-knowledge regime of system government and its 'predictable' (by its members) trajectory of development.[8]

But the opposite is also true: system renewal essentially involves the *re*-building of the power-knowledge 'normalities' and new, rebased common-senses. This process, however, is one that is intrinsically more enabled as a process of pragmatic accommodation of the actual changing power-knowledge system produced by and through (recursive, piecemeal transformation of the dominant incumbent model of) 'innovation' than as a process conducted specifically, and in the first instance, at the level of ideational-political programme formulation – even as tempers are high and 'radicalism' of all types is apparently ascendant. For political movements of the latter sort are primarily engaged in deploying given (and hence always retrograde and retrospective) conceptual resources in a (necessarily, more-or-less authoritarian) attempt to fit the world to their chosen preferences; almost as a necessary condition and corollary of what essentially contested popular appeal they manage to garner.

Meanwhile, though, those successfully pursuing the pragmatic expansion of their own liberties are enabled by a flexible responsiveness to the actual turbulent changing of power-knowledge relations, and hence also by their pragmatic not zealous (whether idealistic, hate-filled or aimlessly angry) demands and aspirations. This affords a surfing of the waves of not just contending socio-technical innovations and counter-innovations, but also of increasingly polarized public *discourse* about issues of 'liberty' and 'security'. Instead, these agents, by staying single-mindedly focused on continual pragmatic recasting of their own common-sense so as to optimize their personal (strategic and empowered) liberty and *understanding* thereof (e.g. as per the changing *meaning* of 'enlightened self-interest'), actively shape the actual locus of both personal enablement *and* collective system resettlement. The intrinsically pragmatic process of such agents thus, ironically, achieves obliquely what the direct approach of radicals cannot; namely the construction of a new system-integrating 'common sense', in both 'knowledge' and socio-technical materiality. Moreover, the very radicalism of the explicitly ideological or politically radical approach tends to alienate this self-advancing pragmatic constituency, the former *itself* becoming labelled as a security threat to the latter's liberty. It thus receives their deepening, and increasingly explicit, rejection in the formulation of new liberal common-sense.

While ostensibly opposites, therefore, discourses of and political movements identified with – and innovation practices of – 'liberty' and 'security' are mutually constitutive, including *through* their very antagonism to the former's recurring advantage. Crucially, though, liberal regimes and agents are also systematically blind to that mutual conditioning, generating the particular form of 'depoliticized' and naturalized self-righteousness on which a liberal power system rests. The broader process of *liberal system* emergence thus tends to be the beneficiary, whichever side of a given debate has the upper hand for the time being – hence generating an accelerating and ontologically-deepening dynamic, a system-constitutive power momentum of turbulent but highly productive positive feedback loops amounting to system transition *as* power-knowledge transition. In short, where a new *urgence* emerges, and this comes to be framed – as is likely the case, given the quintessentially liberal-capitalist character of *existing* dominant power-knowledge relations – as the antagonism of liberty vs security, demanding a new set of political-institutional-knowledge 'answers', a liberal system emergence is *tendentially privileged* (but not, of course, guaranteed) to be the outcome.

Conclusion

I have presented here an intelligible, and moreover traceable, *political* dynamic of system emergence of freedom/security for phronetic analysis of the genealogy of the emerging present in this moment of incumbent liberal system crisis. Novel freedoms are actively generated, developed, adopted and consumed – not least through processes of 'innovation' – that in turn elicit new ontological anxieties regarding newly or more intensively perceived 'security threats', leading to calls for new security measures – themselves involving innovations – that are, in turn, *themselves* experienced as 'security threats' to the new freedoms, and so on … This generates a new thrust of socio-technical change and political-cultural strategic action, starting the cycle again, all the while transforming in the background the socio-technical 'facts on the ground' and the 'common-sense' bases of public debate.

Hence, working again from our starting hypothesis of a new *capitalism* translated into an investigation of an emerging *power/knowledge regime*, our question may now be even more accurately specified as: '*is* a liberal capitalist resettlement possible through cycles of liberty-security? And if so, taking what form, or how?' where the answer to the *latter question determines the former, in that order not vice versa*.

The primary tool – or power/knowledge technology, itself an innovation of this work – at our disposal in this venture, then, is strategic, phronetic analysis of the dynamics of (specifically liberal capitalist) complex systems and their (immanent) system crises through the lens of their key source of world-producing dynamism, namely the massive and global enterprises of research and innovation that they instigate and impel. But, clearly, 'innovation' per se is too big a subject matter to analyse in requisite detail. Selection of specific fields of innovation are thus needed, focusing on domains and locations that the structural, critical realist moment of our analysis above would suggest are likely to be particularly significant within an emerging global capitalist system, to the extent there are indeed embryonic indications of its resettlement. Today, then, this means innovation in response to the Four Challenges, and as it is actually taking shape in a political community of sufficient territorial and geopolitical heft to take over and expand capitalism beyond the capacities of the incumbent hegemon of the United States. There is just one place that has the slightest chance of fulfilling these basic criteria in the early twenty-first century: China.

Notes

1 Indeed, we consider something similar in Chapter 11.
2 We note that a new regime is also potentially complementary in terms of both system dynamics and temporalities with a long winding down of capitalist industrial society. Just as peak oil does not spell the sudden collapse of fossil-fuel-based civilization, so too 'peak capitalism' (arguably reached on some measurements with neoliberalism) could involve what seems to be significant capitalist renaissance 'on the way down', and

where this temporal perspective matters profoundly since 'in the long-run we are all dead', as Keynes reminds us.

3 The corollary of this is that the ubiquity of the term 'liberalism' in November 2016, in the aftermath of Trump's election, and its equally widespread conceptual confusion may come to be seen in retrospect as the beginning of the process of reassessing and reclaiming a *specific substantive*, and hence possibly powerful and empowering, meaning to the term, as a key step in the rebuilding of global liberal capitalism from its current nadir.

4 On the contested relationship of liberalism and democracy see e.g. Dean 2003; Losurdo 2010.

5 Note that this is liberty-security *not* liberty *vs* security. The distinction is important. The latter would give the misleading impression that 'liberty' and 'security' are predefined and/or really existing things with given natures, with the 'essential contestation' being *between* them, as if dynamically adducing a balance or equilibrium. Such a reading would be a gross misunderstanding. Rather both 'liberty' and 'security' are materialized-discourses *themselves* subject to essential contestation and strategic jockeying in their definition and concrete/experienced forms, but in a process of constant and complex cross-definition, mutual presupposition and contra-distinction with the other term. Crucially, this whole process must also be understood as mediated by and mediating dynamic *power/knowledge relations* that are constitutive of actual social forms.

6 Cf. Gramsci (1971) on the 'old dying but the new is not ready to be born'.

7 Note how this dynamic resonates both with that mapped out above in Figure 3.2 regarding the overlapping interaction of K waves and Arrighi/Braudel cycles; and with insights from complex system science. For, as Lane *et al.* (2009) have described the process of complex systems development dynamics, the *acceleration* of growth of the system ultimately leads either to collapse or novel system emergence.

8 Witness the collapse in meaning of 'liberalism' discussed above.

References

Acemoglu, D. and D. Autor (2011) Skills, tasks and technologies: Implications for employment and earnings. *Handbook of Labor Economics*, 4: 1043–1171.

Arrighi, G. (1994) *The Long Twentieth Century*, London: Verso.

Arrighi, G. (2007) *Adam Smith in Beijing: Lineages of the Twenty-First Century*, London: Verso.

Arrow, K. (1962) 'Economic welfare and the allocation of resources for invention', in *National Bureau of Economic Research, The Rate and Direction of Inventive Activity*, Princeton: Princeton University Press: 609–626.

Bauman, Z. (2000) *Liquid Modernity*, Cambridge: Polity.

Benkler, Y. (2006) *The Wealth of Networks: How Social Production Transforms Markets and Freedom*, London and New Haven, CT: Yale University Press.

Bhaskar, R. (1998) *The Possibility of Naturalism* (3rd edition), London and New York: Routledge.

Biel, R. (2012) *The Entropy of Capitalism*, Boston and Leiden: Brill.

Birtchnell, T. and J. Caletrío (eds) (2013) *Elite Mobilities*, London: Routledge.

Bonneuil, C. and J.-B. Fressoz (2016) *The Shock of the Anthropocene: The Earth, History and Us*, London: Verso.

Callon, M. (1998) 'An essay on framing and overflowing: economic externalities revisited by sociology', *The Sociological Review* 46(S1): 244–269.

Crary, J. (2014) *24/7: Late Capitalism and the Ends of Sleep*, London: Verso.

David, P. and P. Dasgupta (1994) 'Toward a New Economics of Science', *Research Policy* 23: 487–521.

Dean, K. (2003) *Capitalism and Citizenship: The Impossible Partnership*, London: Routledge.

Diamond, J. (2007) *Collapse*, London: Penguin.

The Economist (2014) 'Nice work if you can get out', 19 April.

Edgerton, D. (2011) *Shock of the Old: Technology and Global History since 1900*, Profile books.

Fischer, F. (2009) *Democracy and Expertise: Reorienting Policy Inquiry*, Oxford: Oxford University Press.

Flyvbjerg, B. (2001) *Making Social Science Matter*, Cambridge: Cambridge University Press.

Flyvbjerg, B., T. Landman and S. Schram (2012) *Real Social Science – Applied Phronesis*, Cambridge: Cambridge University Press.

Foray, D. (2004) *The Economics of Knowledge*, Cambridge (MA): MIT Press.

Foucault, M. (1980) *Power/Knowledge: Selected Interviews and Other Writings 1971–1977*, Harlow: Longman.

Foucault, M. (2004) *Society Must be Defended: Lectures at the Collège de France 1975–1976*. Translated by David Macey. London: Penguin.

Foucault, M. (2009) *Security, Territory, Population: Lectures at the Collège de France 1977–1978*. Translated by Graham Burchell. Basingstoke: Palgrave Macmillan.

Foucault, M. (2010) *The Birth of Biopolitics: Lectures at the Collège de France 1978–1979*. Translated by Graham Burchell. Basingstoke: Palgrave Macmillan.

Frank, T. (2016) *Listen, Liberal*, Brunswick and London: Scribe Publications.

Frey, C.B. and M.A. Osborne (2013) *The Future of Employment: How Susceptible are Jobs to Computerisation?*, Oxford Martin School Paper, www.oxfordmartin.ox.ac.uk/downloads/academic/The_Future_of_Employment.pdf.

Gramsci, A. (1971) *Selections from the Prison Notebooks*, London: Lawrence and Wishart.

Greer, J.M. (2008) *The Long Descent*, Gabriola, BC: New Society Publishers.

Gross, M. (2010) *Ignorance and Surprise*, Cambridge, MA: MIT Press.

Haidt, J. (2013) *The Righteous Mind*, London: Penguin.

Harvey, D. (2005) *A Brief History of Neoliberalism*, Oxford: Oxford University Press.

Ingham, G. (2008) *Capitalism*, Cambridge: Polity.

International Panel on Climate Change (IPCC) (2013) *Climate Change 2013. The Physical Science Basis*, Geneva: IPCC, www.climatechange2013.org/.

Jackson, S. (2015) Climate Change, Rhetoric and Catastrophe, PhD Thesis, Lancaster University.

Jackson, S. (2016) 'Catastrophism is as much an obstacle to addressing climate change as denial', OpenDemocracy 6 September, www.opendemocracy.net/transformation/stephen-jackson/catastrophism-is-as-much-obstacle-to-addressing-climate-change-as-den.

Jacques, M. (2009) *When China Rules the World*, London: Allen Lane.

Jessop, B. (2013) 'Revisiting the regulation approach: Critical reflections on the contradictions, dilemmas, fixes and crisis dynamics of growth regimes', *Capital & Class*, 37(1): 5–24.

Jessop, B. (2014) 'Capitalist diversity and variety: Variegation, the world market, compossibility and ecological dominance', *Capital & Class* 38(1): 45–58.

Korotayev, A.V. and S.V. Tsirel (2010) 'A spectral analysis of world GDP dynamics: Kondratieff waves, Kuznets swings, Juglar and Kitchin cycles in global economic development, and the 2008–2009 economic crisis', *Structure & Dynamics* 4(1), http://escholarship.org/uc/item/9jv108xp.

Krugman, P. (2009) *The Conscience of a Liberal*, London: Penguin.

Kurzweil, R. (2006) *The Singularity is Near*, New York: Penguin.

Lane, D., D. Pumain, S.E. van der Leeuw, and G. West (eds) (2009) *Complexity Perspectives in Innovation and Social Change*, Dordrecht: Springer.

Losurdo, D. (2010) *Liberalism: A Counter-History*, London: Verso.

Lovelock, J. (2007) *The Revenge of Gaia*, London: Penguin.

Mason, P. (2015) *PostCapitalism*, London: Allen Lane.

More, M. and N. Vita-More (eds) (2013) *The Transhumanist Reader*, Chichester: Wiley-Blackwell.

Morris, I. (2010) *Why the West Rules – For Now*, London: Profile Books.

Müller, J.-W. (2016) 'Capitalism in one family', *London Review of Books* 38(23): 10–14.

Parenti, C. (2011) *The Tropic of Chaos*, New York: Nation Books.

Perez, C. (2002) *Technological Revolutions and Financial Capital – The Dynamics of Bubbles and Golden Ages*, Cheltenham and Northampton (MA): Edward Elgar.

Ridley, M. (2010) *The Rational Optimist*, London: Fourth Estate.

Ritzer, G. (2014) *The McDonaldization of Society* (8th edition), London and Thousand Oaks, CA: Sage.

Sayer, A. (2000) *Realism and Social Science*, London and Thousand Oaks (CA): Sage.

Standing, G. (2011) *The Precariat: The New Dangerous Class*, London: A&C Black.

Steffen, W., K. Richardson, J. Rockström, S.E. Cornell, I. Fetzer, E.M. Bennett, … and C. Folke (2015) 'Planetary boundaries: Guiding human development on a changing planet', *Science* 347(6223): 736–747.

Straw, J. and M. Baxter (2014) *iDisrupted*, New York: New Generation Publishing.

Streeck, W. (2014) 'How will capitalism end?', *New Left Review* 87: 35–64.

Sum, N.-L. and B. Jessop (2013) *Towards a Cultural Political Economy*, Cheltenham: Edward Elgar.

Tainter, J. (1990) *The Collapse of Complex Societies*, Cambridge: Cambridge University Press.

Turner, G. and C. Alexander (2014) '*Limits to Growth* was right. New research shows we're nearing collapse', *Guardian*, 2 September.

Tyfield, D. (2014) 'Putting the power in 'socio-technical regimes' – E-mobility transition in China as political process', *Mobilities* 9(4): 585–603.

Tyfield, D. (2015) 'What is to be done? Insights and blindspots from cultural political economy(s)', *Journal of Critical Realism* 14(5): 530–548.

Tyfield, D. (2015/2016) 'An extended review of Mason's PostCapitalism', Lancaster University, available at: www.lancaster.ac.uk/staff/tyfield/On_Postcapitalism_1.pdf; www.lancaster.ac.uk/staff/tyfield/On_Postcapitalism_2.pdf; www.lancaster.ac.uk/staff/tyfield/On_Postcapitalism_3.pdf.

Urry, J. (2002) *Global Complexity*, Cambridge: Polity.

Urry, J. (2011) *Climate Change and Society*, Cambridge: Polity.

Part II

Where are we?

Innovation in China

4 Will China rule the world?

The emergence of Chinese capitalism

The China–world problems of the early twenty-first century

Our central questions are 'what is coming next, as a strategic landscape, in the context of the Four Challenges and innovation? And how?' so that we may think and act strategically in response. In the early twenty-first century, still amidst the interregnum of the terminal crises of neoliberalism, there is one country in particular to which we must attend in thinking about this: China. For across the Four Challenges, in China we encounter the society that is not only clearly of pivotal global significance in terms of the size and scope of the manifestation of these issues and, possibly, of the 'fixes' and responses (if not necessarily 'solutions') that emerge from there – and noticeably of relevance for both developed, stagnating global North and developing global South. But China also, and inseparably, stands out in the intensity of all Four Challenges, since this is precisely what imparts such world-significant dynamism to the socio-technical and knowledge-political innovation taking place in China.

Regarding the Challenge of cosmopolitized globalism, we must start with the spectacular growth of the Chinese economy over a generation, with an average 8 per cent per annum GDP growth for over 3 decades since the post-Mao "Reform and Opening Up" (*gaige kaifang*) of Deng Xiaoping starting in 1978 (Naughton 2006; World Bank 2015). This presents not only the most striking political economic and geopolitical development of this period, nor even the most extraordinary capitalist economic take-off in history (in the largest country in the world of 1.3 billion people or approximately 20–25 per cent of the world's population), including lifting over 600 million people (or nearly 10 per cent of the *global* population) above the $2-a-day poverty line between 1981 and 2004 (World Bank 2010). But contemporary China also thereby presents *prima facie* the most arresting trajectory of contemporary history, with its (re-)insertion back to a central role in the global system, after two-plus centuries of internal turmoil and external 'humiliation', an apparently unstoppable supertanker.

Yet for all this, there remains a lively and insightful debate, both outside China and within it, regarding whether or not the seemingly 'logical' culmination of this process, rising to global dominance, will happen *at all*, not just when and how (e.g. Fenby 2014; Jacques 2009; Hung 2016; Shambaugh 2016; Halper 2010; Zhang 2012). Indeed, far from being a peripheral and academic debate, entertaining important but recondite issues, counterfactuals and hypotheticals, this is the dominant register for the amplified contemporary interest in China. The continued ascendancy of China is thus *fundamentally* in question, signalling both profound analytical objections that highlight significant challenges ahead, and, even more so, an essentially contested landscape, of hope and fear, regarding what it means for various powerful groups and constituencies.

There are many reasons why this is so. At a cultural level, the rise of China spells a profound and unprecedented challenge to the global dominance of Western modernity over the past few centuries: both as the first ethnically non-Western European (dominated) nation state to be a viable contender as hegemon of the unified and unifying 'globe' that has itself been the product of this period; and one, to boot, that is unquestionably amongst the most enduring and impressive of global civilizations and is profoundly self-aware of this – however much or little credence we give to hyperbolic statements of '5,000 years' of unbroken history – *as well as* one of the least familiar and similar to existing Western modernity. Indeed, if we compare a broadly Sino-Confucian cultural heritage with contemporary manifestations of what have been described as the other three 'Axial' civilizations (Armstrong 2007), we find that China's position – as civilization not nation-state (Pye 1992; Zhang 2012) – in the early twenty-first century globalizing world is actually marked by its relative *smallness*, isolation and unfamiliarity vis-à-vis the rest of the world, not its unrivalled size and ubiquity (see Table 4.1).[1]

Add in the deliberate educational programmes of a specifically chauvinistic nationalism (Hughes 2007; Zhao 2013) and the profound sense of civilizational superiority this cultivates (Jacques 2009), a massive population that is actually extraordinarily ethnically homogeneous for its size (compared to say, India, Indonesia, Brazil, Russia or the United States), and the fact that, for all the growth in foreign travel, Chinese (i.e. PRC-domiciled) interaction with the world (and vice versa) is still proportionally so small, and you have what may be described, without exaggeration, as a 'China-world' problem; that is, the problem *for* (contemporary) China (and its much wished-for rise to global centrality) of an increasingly interconnected, cosmopolitized world, on the one hand, and the problem *of* that China and its increasing global importance for that world, on the other. From this perspective, it seems that amongst the most pressing aspects of the challenge of cosmopolitized globalism is precisely how 'world' and 'China' can learn to accommodate each other, as surely they must.

Of course, though, the most obvious, if often unspoken, issue raising doubts about the continued global rise of China concerns how *this present* China can possibly become the political and economic core of *this present* global order: how can a nominally *Communist* and staunchly illiberal one-party state become the centre of global capitalism, as we currently understand both of these and their intrinsic opposition? *This* is the world historical conundrum – the meeting of an 'immoveable object' and an 'unstoppable force' respectively – actually underlying the profound concern and discombobulation amongst the powerful on all sides. And this has, in turn, fomented the high-profile debate that dominates current discussion about and interest in the future of China … though it is often not in the interest of parties to be so candid about this, for reasons discussed below. Once spelt out thus, it is crystal clear that this is indeed a question that can only point to profound, uncertain, and likely highly unsettling, *qualitative* change in the world vis-à-vis our current common-sense understandings of both the world and ourselves within it. In short, somewhere something profound will have to give.

We find similarly enigmatic puzzles also regarding each of the other three Challenges. Take for instance, the issue of post-human innovation. Manifest as the issue

Table 4.1 The global presence and familiarity of the 4 Axial Civilizations in the early twenty-first century: a back-of-the-envelope comparison

'Axial' civilization	Modern manifestation	Global total		Outside the respective 'core'[1]	
		Population (2016)[2]	% global population[3]	Population	% global population
'North' – post-ancient	Incorporated into global markets	• 7.43 billion[4]	• 99	• 6.4 billion	• 85
Greek European cultures	• Living in a liberal democracy	• 4 billion[5]	• 53.5	• 3.4 billion	• 46
'West' – West Asian Monotheism	• Speakers of a European language	• 2.55 billion[6]	• 34	• 1.8 billion	• 24
	• Monotheist self-described believers	• 3.79 billion	• 51	• 3.5 billion	• 47
'South' – South Asian, Indo-Aryan cultures	• South Asian country population[7] and diaspora	• 1.75 billion	• 23.5	• N/A	• N/A
	• Global Hindu, Jain, Sikh and Buddhist communities (and/or practising mindfulness or yoga)	• 1.53 billion Both together • 2.23 billion[8]	• 20.5 • 30	• 480 million[9] • N/A	• 6.5 • N/A
'East' – Sino-Confucian cultures	• Citizen and/or diaspora from Confucian-based country[10]	• 1.65 billion	• 22	• 50 million[12]	• 0.67
	• Practitioners of a martial art[11]	• N/A	• N/A	• 20 million[13]	• 0.27
	• Speaking north east Asian language	• 1.36 billion	• 18	• N/A	• N/A

Notes

1 Respectively, outside Europe, the Middle East, the Indian subcontinent and north east Asia.
2 Figures from www.worldometers.info/world-population/unless otherwise specified, 20 December 2016.
3 Total global population taken as 7,472,500,000.
4 Excluding uncontacted peoples (approx. 20,000 at most, https://en.wikipedia.org/wiki/Uncontacted_peoples), nomads (30 to 40 million www.newworldencyclopedia.org/entry/Nomad) and North Korea (25 million www.worldometers.info/world-population/north-korea-population/) but not Cuba (or any other Bolivarian socialist country). Estimates of numbers of subsistence peasant farmers are difficult to find and many would be integrated into the monetary market economy in any case. We take here the contested thesis that the global peasantry are thus, in any case, incorporated into global capitalism.
5 https://ourworldindata.org/democracy/.
6 https://en.wikipedia.org/wiki/List_of_languages_by_total_number_of_speakers.
7 Includes India, Pakistan, Bangladesh, Sri Lanka, Nepal, Bhutan.
8 https://en.wikipedia.org/wiki/List_of_religious_populations.
9 https://en.wikipedia.org/wiki/List_of_religious_populations.
10 Includes People's Republic of China (all nationalities/minzu), Japan, North Korea, South Korea, Taiwan, Hong Kong, Mongolia and diaspora.
11 This category is chosen in preference to global numbers of Confucianism, Taoism and Shintoism since these are so small outside the 'core'.
12 https://en.wikipedia.org/wiki/Overseas_Chinese.
13 Taking a generous 1% (https://www.quora.com/What-percentage-of-people-practice-martial-arts) of populations of the Americas, Europe and Oceania.

of Chinese innovation competitiveness in a global capitalism that is increasingly innovation-centric (see next chapters) we find again precisely the combination of extraordinary, globally-significant ascendency across multiple metrics, and under a project of increasingly focused state support, but alongside continuing and intense problems. Regarding the specific issue of future technological un(der)-employment, we also find in contemporary China potentially the most combustible conjunction anywhere in the world. Here a government, business community and consumer public displaying unrivalled interest in automation and AI sits alongside a polity characterized by the perennial governmental challenge of what to do with a surplus of labour (not a relative scarcity, as in the immigrant-built United States) and with a history, still officially celebrated, of revolutionary worker agitation. In this context, an officially Communist Party-state supposedly primarily allied to, backed by, and acting for the largest national industrial working class the world has ever seen yet actually committed to a project of automation, renders the tension between contemporary job-destroying capitalist innovation and workers all the more intense, since it is refracted through an existential challenge to the incumbent state form.

Regarding the environment too, in China we find both problems of the environmental consequences of late twentieth century carbon capitalism of exceptional scale and intensity and, increasingly, massive initiatives with some embryonic successes in terms of incubating 'green' energy industries and innovations that could come to present a meaningful challenge at system level (ClimateGroup 2009, 2014, 2015; Green and Stern 2015). Again, we also see how it is in the potentially explosive consequences of simmering dissatisfaction about the environment – i.e. of the environmental challenges experienced as *political* ones – to the existing Party-state constitution, and not just to the tenure of a particular leader or party in office, that gives the environment such political purchase and intensity in China vis-à-vis other countries, in both global North and South. But, on the flipside, this is compounded further by the very complexity and novelty of China's contemporary environmental woes, making them so much more challenging than simply following the example of a London or Pittsburgh in sorting out their mid-twentieth century smog, despite the ubiquity of this misleading comparison in both Chinese and Western literature (e.g. Kahn and Zheng 2016). For these demand more dispersed and inclusive models of government, information collection, transparency and accountability that pose equal and opposite challenges to that existing party-state order.

This thus points to the final challenge, of the complex government of complex systems, which brings together the three before. On the one hand, the very determination of the CCP regime to preserve itself instils a non-negotiable inflexibility towards preference for top-down control that militates directly against effective experiment to resolve, and tends to exacerbate, the proliferating problems of complex systems, irreducible difference and moves towards their enabled *self*-government. Looking for implementable solutions that can be copied from elsewhere, this also conditions a preference amongst government towards narratives of 'catch-up' with the dominant global North along established linear trajectories.

Moreover, these are challenges that, like the environment, are particularly intense in China, as it undergoes a 'compressed modernity' (Chang 2010) in which its encounter with – in the language of Ulrich Beck – both the problems of industrial first modernity and those of knowledge-capitalist second modernity are happening simultaneously, considerably complicating and compounding both (Han and Shim 2010). It is precisely in this way, for instance, that addressing Beijing's smog in 2016 is exactly *not* equivalent to London's in 1956. But so too with legion other issues of profound, often global, risk in China, including finance, health, mobility, employment.... Determination to follow in the footsteps of the mid-twentieth century US and Western Europe to solve any of these issues leads only to their deepening crisis – as is increasingly acknowledged even by the central government. Contemporary CCP China thus, on the one hand, faces starker challenges of complexity with, on the other, permanently hobbled resources due to its systematic bent to top-down solutions.

Finally, in returning to the challenges of cosmopolitized globalism, we see here too one final big unanswered question and objection to the continued rise of China: that contemporary China is quite evidently still miles from being an acceptable capitalist hegemon. China today does not embody the most progressive and enabled frontiers of socio-economic development, profitable enterprise and individual and cultural flourishing such that it could win the willing acceptance of its geopolitical dominance by a global 'power majority' (if not necessarily demographic, democratic majority) of the contemporary world; *much less* the emerging acme of sophisticated, enlightened, democratically empowering complex self-government. In other words, measured against what a genuinely 'progressive', if still capitalist (see Chapter 3), form of government fit for the novel challenges of the twenty-first century would look like, the (titanic?) supertanker seems but a boastful imposter, certainly no awesome 'dragon' as the Western business press *cliché du jour* would have it.

The foregoing thus all adds up to rendering one question amongst the most important for our understanding of the genealogy of the emerging present: 'Will China rule the world?' (Fenby 2014; Hung 2016; Jacques 2009). But where qualitative and strategic analysis of this ubiquitous question through a CP/KS lens of innovation-as-politics profoundly reframes it; indeed does so regarding each of its elements in ways that almost all of the literature dominating this discussion does not attempt, much less complete. Our question, in other words, is now: '"rule" in *what sense* and how? *Which* "China" and *which* "world"? And even in what sense "will"?' – the last of these a question that is then explicitly pragmatic and strategic, eliciting the phronetic approach adopted in this book. To get a better understanding of how these enormous macro trends are actually interacting, and hence answers to *these* questions, we can turn to study innovation – the creation and adoption of power/ knowledge technologies and relations and the remaking of the 'world' that is innovation-as-politics – and *in China itself*, where this means 'located at specific places in the massive and diverse territory of the People's Republic' while also tracing their irreducible interconnections with the ongoing constitution of 'China' itself as a nation-state and with a cosmopolitizing 'global' world. In this way, we can begin to get at the *qualitative* changes concealed by questions posed using seemingly familiar terms – such as 'China', 'rule', 'the world' – which is where the action of socio-technical and political change that will ultimately 'answer' the question is actually unfolding.

Immoveable object vs unstoppable force

By this deliberately allegorical and imprecise but vivid dualism, used to stimulate strategic insight rather than a careful analytical truth, it must be understood that we mean *China* as the immoveable object and *global capitalism* as the unstoppable force.[2] What is unequivocally *not* meant, therefore, is any mistaken conflation with a supposed dichotomy of 'state vs market'. This point bears emphasis for several reasons. First, because the usual presumed use and meaning of the 'state vs market' dichotomy (or 'state capitalism vs free-market capitalism' etc....) is not only profoundly misleading in its own terms, alluding to two ideal types that are of limited applicability, especially outside the Euro-American sphere, as well as pitting them against each other as qualitatively incommensurable. But it also thereby effectively sets up a theoretical framework that makes empirical understanding of the indissolubly state-market hybrids characteristic of the Chinese political economy in the late twentieth/early twenty-first century all but impossible, occluding more than it illuminates.

Framed thus, in other words, we are already firmly positioned within a specifically US-centric approach that sets the (supposedly irrational, backward, authoritarian) 'state' against the (rational, progressive, free and apolitical) 'market', and effectively asks with deepening consternation "how the hell has contemporary Chinese capitalism not collapsed?!" (cf. Chang 2003) even as it has been the most significant global exception for a generation, persistently ignoring the 'free market' Washington Consensus (Boyer 2016). Indeed, adding to the affront China poses to this incumbent dominant common-sense, Chinese authoritarian state capitalism is surely the *most successful* example of capitalist development ever; and certainly much more impressive than the contemporaneous and woeful record of 'free market' capitalism of neoliberal globalization, now culminating in global economic malaise even in the countries it has served to benefit the most (see Chapter 2). The unspoken horror underlying such an approach, in other words, is the unthinkable possibility to its proponents that the 'state' could ultimately 'beat' 'the market', the mere possibility of which would rock the foundations of contemporary US capitalist order even more than did the collapse of Lehman Brothers (cf. Greenspan 2008).

Instead of the analytically superficial and tacitly ideological dualism of 'state' vs 'market', therefore, both Immoveable Object (IO) and Unstoppable Force (UF) are here conceptualized as complex systems of power/knowledge relations. The IO is thus precisely *China*, while the UF is *global capitalism*, and both in all their systemic richness and complexity and over time. As they are two incarnations of essentially the same kind of thing (i.e. a CP/KS) we can then proceed to explore empirically and in detail *how* they interact and perhaps become mutually constitutive, and so perhaps qualitatively inter-penetrating, at the meso- and micro-levels.

Here, global capitalism (see Chapter 3) is an essentially expanding (and hence immanently 'global') system of power/knowledge relations driven by and for the purpose of maximizing the accumulation of capital. Granted this is inextricable from the 'rule of the market' since the law of value can only be realized through markets (Marx 1999; Fleetwood 2001). But it is also, and no less importantly, an arena of forces of state, the corporation, the financial network, individualized popular manoeuvring, class etc.... (Ingham 2008) that are equally systemically essential as all of these are so many institutional condensations of dynamic and essentially jockeying power/knowledge relations – *including* 'the market' (Fligstein 1996) (or simply 'markets' (Aspers 2011)) itself.

Focusing on the CP/KS of capitalism, as opposed to the 'free market', thus not only allows us to dig much deeper in our understanding of *how* the political economy is organized, and how 'state' and 'market' interact *together* to co-produce the dynamic global system of capitalism (and be produced by it in turn) – as already mentioned, a particularly crucial task when studying contemporary China. But it also does so in ways that, since the whole ontology is thoroughly (power/knowledge) relational, affords an analysis that can investigate how this system of capitalism can itself be profoundly, constitutively and qualitatively *shaped* by its encounter with other complex power/knowledge systems (and *vice versa*). This affords rich qualitative exploration and fecund conceptual innovation for the understanding of new, emerging 'worlds', not just setting out a more-or-less stable hybrid political economic model of 'state' and 'market', with both of those taken as fixed concepts and/or phenomena.[3]

This latter point really matters here. Not just because of the profound qualitative change we cannot but witness in the coming decades given the Four Challenges and crises of neoliberalism. But also because – turning to the 'immoveable object' – of what may be called the 'conceptual challenge of China' for the incumbent forms of societal self-knowledge, in terms of both the (Western-dominated) professional, academic social sciences and lay understanding that together in part constitute the contemporary world ... and its current inter-regnum (Tyfield 2017). For China really is profoundly unfamiliar and different to incumbent, dominant Euro-American understanding but in ways that themselves push towards a more pragmatic, strategic and relational systems perspective that strongly resonates with the theoretical shift adopted in this analysis. Indeed, a CP/KS perspective also yields highly productive meso-level insights into the specific dynamics and challenges of government that are constitutive of the system we call 'China'.

China is, to be sure, currently constituted as a nation-state, and one that currently stands out for the intransigent persistence of its explicitly illiberal model of strong, central authoritarian government despite rapid capitalist economic development, and with apparently great success. But China is also, of course, very much more than that. As a nation-state China is actually extremely young, founded only in 1949, yet one would have to be blind to miss also its (much played upon) extraordinary longevity as a *civilization* state (Pye 1992). Indeed, in China, and in China alone, we must grapple to understand a civilization that is extraordinary and exceptional in several key regards: its unification into an imperial state extremely early in (its) history as well as its extraordinary size, both geographically and demographically; and then with unrivalled continuity, notwithstanding a history punctuated with periods of exceptionally bloody internal collapse, up to the present, such that the modern Chinese can read ancient texts with a comparative (but not seamless) ease inconceivable across most of the rest of the world.

Our starting point for thinking about China, thus, is an abstract CP/KS historically composed of two key inter-related and mutually constitutive elements: an imperial state of concentric rings in a strict hierarchy converging on the single person of the Emperor, whose primary concern is only with the integrity of the imperial constitution; and conversely, a massive population, overwhelmingly of agricultural, village-based peasants but also from early on of market-urban dwellers and tradespeople engaged in essentially pragmatic practices of pursuing the interest of oneself and one's family within other concentric rings of inter-personal trust-based networks (Fei *et al.* 1992). The strict top-down hierarchy of government, unified by a single script across multiple languages, from emperor down through local dignitaries to the massive and diverse population, thus pulls the 'whole' that is 'China' together into an emergent power/knowledge system that is more than the sum of its parts. While, conversely, the relative autonomy and profound self-governing pragmatism of the Chinese people, responsibly tending their own families' gardens, serves as the bedrock of Chinese civilizational greatness and its sheer heft; a size simply governmentally unthinkable if based on strong, let alone authoritarian, institutions before the modern age.

As many scholars have noted, therefore, China is extraordinary for the enduring *lack* of institutionalized government that characterizes it across history, with the 'rule by man', rather than 'rule of law', and dependence on personal and familial connections of *guanxi* rather than arms-length rules and institutions the norm. From an orthodox Western social theory perspective, then, this paints China as the very acme of oriental despotism, and thus fatally flawed and inadequate as regards realizing the supposedly 'natural' and 'rational' will of the 'human individual' (itself a modern Western confection).

But from the pragmatic, relational (and quintessentially amodern) perspective of a CP/KS analysis that recognizes also the essentially relational and strategic pragmatism of both the Chinese imperial constitution and the Chinese people, the very opposite conclusion arises. It is precisely the conjunction of this top-down imperial hierarchy and the bottom-up resilience and pragmatic personal-relational manoeuvring, and their mutual acknowledgement *and disregard or neglect*, that afforded the integration of such a massive group into a single territorial civilization and its robust durability down the millennia. The persistent presence of a strong imperial authority, if possibly distant from day-to-day life, afforded a unified socio-political order and social peace that, so long as it did not interfere too much, underpinned the continuation of everyday life and participation in a great civilizational project that *made* everyday life in China *Chinese*; i.e. in Chinese language, 'civilized' (*Han)* from the 'Middle Kingdom' at the centre of the world (*Zhongguo*). Conversely, with such a massive empire beneath it, unified by connections of tribute and cultural integration (including the meritocracy of the imperial examinations), 'China' could emerge personified in the unrivalled splendour of the Emperor, so long as there was reasonably competent and unexploitative government. Each is also then the check and counterbalance of the other, engendering a power/knowledge dynamic that tends towards the long-term maintenance of socio-political stability once it has been effortfully and contingently attained.

Feeding off each other in pragmatic, strategic feedback loops, the relation of Imperial state and populace to each other cultivated the specifically pragmatic and systemic approach in that counterpart that then, in turn, emerges as the key manifestations of Chinese culture (Duara 2014), 'high' and 'low' – of text, 'philosophy'/ 'religion', ritual and magic, calendar and practice, including a worship of ancestors and respect for immemorial practical tradition itself. In other words, subjectivity and forms of state are co-produced, bottom-up and top-down, and as substantively Chinese.

For instance, one's awareness of the omnipresence of the strong top-down state and its potential capriciousness and under-institutionalization incubates precisely a turn to reliance on networks of personal trust and acquaintance, and then their formalization in forms of Confucian ethics of the family and filial piety, and of the state as family (*guojia*), etc.… (Hsu 1998). While conversely, ultimately dependent upon the maintenance of sufficient social order and peace in a systematically underinstitutionalized polity, the Emperor's rule is practiced and formalized as being dependent upon the Mandate of Heaven, manifest in the acceptance of the regime by that massive and pragmatically-ruled peasantry. This counsels formalization into specifically strategic and pragmatic forms of top-down government and knowledge thereof – giving Chinese 'philosophy' and 'religion' their characteristically practical and naturalistic flavour, as multiple contending schools oriented primarily to the question of how to conduct good top-down maintenance of harmonious social order, and hence of the (*Chinese*) state. This culminates in the enduring cultural form of the virtuous Confucian ruler whose character is the final guarantor of good government, as opposed to formal constitutions aiming productively and justly to balance contending but institutionalized powers, as in the modern West and much of the world it colonized.

A brief history of China and capitalism

To be sure, modern China has undergone considerable, wrenching socio-political change in the past two centuries (Mitter 2005; Yu 2012). Yet, while there is no reified cultural Chinese 'essence' that has been preserved, there remains an unquestionable continuity across the period, and extending further back, between the 'China' today and that captured in the description above. It is in this respect, therefore, that we may think of China as an immoveable object. For 'China' is a complex power/knowledge system that continues to exhibit exceptional durability and scale, and with these two characteristics inseparable given their common genesis in the specifically Chinese dynamics of deep-seated strategic pragmatism that is precisely attentive to the reproduction and development of China-as-system and state.

Yet that recent history illustrates perfectly how this complex power/knowledge system is capable not just of conditioning the emergence and maintenance of a socio-cultural order of unique longevity. The flipside of that pragmatic tendency towards comparatively harmonious social stability is that this system logic is also largely devoid of immanent sources of system dynamism and renewal. No matter how large and successful a specific form of Chinese order is, it remains only a relatively closed system. A time will come when it must confront new sources of disorder and system productivity, including new power/knowledges, that may well exceed the capacity of the existing regime to restabilize around. At this point, the very dynamics of stabilization unravel into a positive feedback loop of instability and social breakdown that will manifest in part in the growing *dysfunctional* interrelation between the summit of state and the base of the population, between high and low culture (cf. Duara 2014). As the crisis deepens, then, the Mandate of Heaven really *is* quite literally revoked!

Of course, this conjuncture precisely describes the decline and fall of the Chinese Empire in late nineteenth and early twentieth centuries. And while no doubt a crude simplification, it is pragmatically enlightening to view this process as the long arc of the deepening encounter, since the sixteenth century and its very emergence as a global force (Hung 2016), with perhaps the most powerful of such disruptive forces in history: capitalism. Here we have an immanently global force that instantiates precisely those characteristics, in its essential turbulence and creative destruction, that intrinsically oppose the maintenance of stable, harmonious social order by the Chinese CP/KS. Insofar as, on the one hand, capitalism is indeed presumptively and insatiably global and, on the other, today China can reattain its civilization greatness only by remaining and reinventing its distinctive 'Chineseness' in the context of global capitalism, therefore, this is a world historical clash that has been several centuries in the making.

Hung (2016) presents an insightful account of this history from just such a *longue durée* perspective. China was already being pulled into the system-constitutive cycles of global capital through its demand for the growing volumes of silver coming from the Spanish and Portuguese conquests of the New World in the 1500s. But the tale begins in earnest in the mid/late-eighteenth century when trading delegations from the new, proto-industrial north western European powers started to knock on China's door for access to its thriving markets and commerce. Hung corrects widespread contemporary misunderstanding that China's was already a stagnating, if not backward, economy at this point in time, having shut itself off from world trade in the late fifteenth century. To the contrary, under the early Qing dynasty, China's economy and society was attaining new heights of greatness, and it certainly outshone in the marks of its civilization the belligerent upstarts of the Europeans from their tiny, cold, remote outcrops at the other end of the world.

Moreover, the Chinese eighteenth century saw a commercial and even 'industrious' (as opposed to 'industrial' (De Vries 1994)) revolution that saw the emergence of significant industrial capacity and the rise of fortunes built upon market-based revenues (Hung 2016). This was a China already showing many of the key features of a supposedly modern market economy. Indeed, in many ways this strong market economy instantiated the writings of seminal Western political economists like Adam Smith more closely than did the West itself (Arrighi 2007). Yet it was precisely in doing so *without* building the power- and exploitation-saturated *capitalist* political economy – that these early Liberal economists judiciously and self-servingly neglected to theorize in their pictures of individuals in rational market exchange – that proved to be its (strategic) weakness, not its strength, in the medium-term.

The response to this shift in economic gear from the impetus of a growing global capitalist system precisely captures the difference in immanent system dynamics between that 'unstoppable force' and the 'immoveable object' of China. For the very size, stability and stature of the latter meant that this first profound encounter of these two systems took the form of the attempted digestion of capitalism by the existing imperial order. Growing concentrations of newly accumulated capitalist wealth in Europe fed the growth of insatiably acquisitive bourgeois and capitalist classes in compounding feedback loops underpinning a deepening capitalist revolution of social and political structures – generating the 'memorable alliance' of capital and state power (Ingham 2008). In China, however, success in business was quickly cashed in for the much more valued power/knowledge resource of imperial favour and political power, in the form of powerful bureaucratic sinecures, perhaps even at the cost of collapse of their original commercial fortune (Hung 2016). The only clear exceptions to this rule were among the growing and systematically insecure Chinese diaspora of familial capitalism (Yeung 2006), spread across south east Asia in particular; a division of 'Chinese capitalism' into thwarted mainland and resilient and distinctive diaspora that has continued to date and is of continuing supreme consequence (see below).

In similar vein, industrial advance was not concentrated in the hands of a few hyper-competitive private and state institutions, thereby empowering them further and driving ever-more industrial innovation. Instead, fortunes remained dispersed and diffuse, and so with no specifically capitalist agency underpinning the violent concentration of capital from the existing agricultural surplus that is the 'primitive accumulation' of capital (Hung 2016), nor the specifically capitalist state powers necessary to regularize and channel its inescapably violent transformation of society to self-sustaining ends. As such, like being big and strong enough to swallow a ticking bomb, late imperial China successfully digested the ever-expanding and sundering pill of capitalism only to be destroyed by it, as the twin pressures of internal differentiation and unequal development and external European (and latterly Japanese) capitalist imperialism grew far beyond the limits of even its unique powers of assimilation.

As Hung shows, it ultimately took a national Communist Party (CCP) to reestablish China following this collapse of its age-old agricultural-imperial form into a modern nation-state. For this Leninist party-state (with the Party pre-existing the state, not *vice versa* as with Western political parties (Saich 2004)) achieved, with the given power/knowledge relational legacies of mid-twentieth century China, the institutional capacity to oversee and drive through the primitive accumulation of capital and systematic pro-urban bias necessary to secure China's existence as a powerful sovereign nation-state in the modern capitalist world system; and to do this in the name of, and arguably for the benefit of, the vast majority of China's still then highly rural population. But in perhaps one of recent history's greatest ironies, it was also therefore the Communist Party-state that has provided the foundations for China's subsequent and spectacular economic development through its deepening reinsertion back into global capitalism since 1978.

This, however, also qualitatively transformed further the relation between the immoveable object and the unstoppable force. For whereas eighteenth century imperial China felt only orthogonally or obliquely challenged by global capitalism, at the cost of shaking China to its very foundations, its re-establishment after and in the context of that internal turmoil came with a self-definition of their essential opposition. At first, of course, this took the form of China's complicated role (given the Sino-Soviet split of 1960) in the Cold War under Mao. But even in the post-Mao Deng era and since, China's progressive opening to capitalist global markets has still brought with every step of opening at least an equal and opposite transformation of the Chinese Party-state regime to secure its continued monopoly of power, understood (and not entirely without reason, as we have seen) as the ultimate guarantor of Chinese sovereignty and its path back to self-determination. As such, the very success of the reestablishment of 'China' as a global force per se, let alone one to be reckoned with, in the past 80 years has further crystallized the confrontation of China and global capitalism precisely into one of increasingly mutually implacable interdefinition *as* immoveable object *to* capitalism and unstoppable force *vis-à-vis* China.

This has been compounded further in recent years through a parallel dynamic. For the imperial-agricultural state was founded on, both *de facto* and *de jure*, and co-produced a broadly Confucian-Legalist-Daoist culture that weaved together the twin poles of imperial hierarchy and pragmatist population into a self-consciously Great Civilization that specifically celebrated its ancientness and respect for ancestral tradition stretching back to mythical origins. As such, and as a crucial element of its very longevity, this was a complex power/knowledge system that was explicitly aimed at upholding and instantiating an order of other – i.e. quasi-transcendent, heavenly if naturalistic – provenance (Han and Park 2014).

By contrast, the Maoist Party-state was founded explicitly on the radical repudiation of this traditionalism, in an essentially riven and unstable order seeking to reinvent China without all that was understood to be – and indeed living and in practice as – specifically Chinese. Its achievement of a new overweening authority, necessary for China's modernization through the concentration of capital in the hands of the Party-state, was thus based upon discourses and practices oriented to the new transcendent promise of an essentially alien secular, modernist doctrine of Marxism-Leninism (Xu

2012), translated and inflected with Chinese characteristics into Maoism. This was thus an essentially unstable – indeed, revolutionary to the point of self-consumption (Yu 2012) – CP/KS in which the state was re-established and re-empowered, and society entirely reorganized and subjectivities reshaped, in reference to the authoritarian top-down realization of the goal of Chinese-centric global socialism. Maoism thus took on the essential task of reimagining and modernizing the Chinese power/knowledge system – a system lacking powerful immanent forces of renewal, as we saw above. But in the process it also, and inevitably, attempted, at the cost of enormous social upheaval and suffering, to realign Chineseness with a new transcendent ideal that was not only essentially foreign but also, and more importantly, incapable of supporting the burden of revitalizing China that was placed upon it (Mitter 2005; Brown 2010).

As such, in the aftermath of the forlorn florescence of desire for popular sovereignty and the shattering of the dream of socialist democracy in 1989, as climax to the early reform period renaissance (Gittings 2005), the reassertion of the CCP regime and its renewed kickstarting of reform and opening in 1992 brought with it an even starker power/knowledge system: no longer even apparently committed to its self-preservation for the greater goal of global socialism, the CCP Party-state was now transparently wedded only to its own self-preservation, diluted with the *quid pro quo* of popular acceptance on the basis of unbroken, rollicking economic growth and a newly strident and chauvinistic nationalism (Zhao 2013). In other words, the very success of the CCP Party-state through the Mao years, the early 'dual-track' reform period and the post-'92 period of ever-deeper insertion into global capitalism has brought with it an ever-deepening self-referential 'immoveability' and inflexibility to 'China' *as* Party-state and *specifically* vis-à-vis *global capitalism*, notwithstanding the CCP's impressive capacity for adaptation alongside its inescapable atrophy (Shambaugh 2009; Zheng 2009).

Yet, it is this very deepening 'immoveability' that has been so crucial in cultivating the conditions that have afforded the potential for the specifically massive, pragmatic, under-institutionalized polity that is *China* to survive – and indeed thrive this time around – off that deepening encounter with global capitalism. And this, moreover, in a period of particular 'unstoppability', or uncontrollability, of global capitalism, namely the rise of neoliberalism (see Chapter 1) with its rampant form of financialized globalization and its 'business revolution' (Nolan 2004) of fragmented supply chains dominated by Global North-domiciled giant transnational corporations (Dicken 2014).[4] Yet far from being destroyed by neoliberal globalization and the historic heights to which it has driven the untrammelled growth of capitalist exploitation around the world, growing into a system and power momentum incalculably greater than that which confronted and undid China in the eighteenth and nineteenth centuries, China – and *Communist* China – has managed to harness that dynamism to its continuing and spectacular advantage. Indeed, it seems arguable, given the current system crisis, that China has benefitted more from neoliberal globalization than neoliberalism has itself.

What is clear, regardless, is how the long arc of China's encounter with global capitalism over several centuries has involved both the deepening opposition of these two complex power-knowledge systems *and*, inseparably, their deepening intercalation and inter-dependence. Indeed, the former dynamic has fed the latter, by preventing the ultimate (but premature) 'triumph' of one over the other or the dissolution of that opposition into some messy compromise. From this perspective, then, the climactic heights to which this essential opposition has risen – with the ascendancy of CCP-led China to the cusp of global hegemony, on the one hand, and terminal but turbulent crises of neoliberalism, on the other – simply highlights what is perhaps the most awkward geopolitical fact of the present conjuncture for both sides: that, while evidence for the collapse of neoliberalism in its declining core is everywhere (e.g. Trump and the ensuing 'liberal' disorientation in the US, Brexit and its chaotic aftermath and party political cross-dressing in the UK, continuing Eurozone malaise and political polarization, the growing calls for unorthodox monetary policy now combined with fiscal stimulus from such bastions of neoliberal orthodoxy as central banks and global financial institutions etc....) the single most robust remaining pillar of the global neoliberal power/knowledge system is the Chinese Communist Party.

There are multiple ways in which this is the case, but all hinge on how the growth and continuation of neoliberal globalization and its privatized Keynesianism in the global North (Crouch 2011), of debt-fuelled consumption compensating for stagnating wages, would not have been able to last anywhere near as long as it has done without the parallel transformation of China: into the low-cost workshop of the world; as inexhaustible sink for surplus capital as FDI into pitilessly disciplined factories, also serving to discipline Western workers with credible and persistent threats of offshoring; and as global creditor of last resort, with its effectively insatiable appetite for US Treasury debt. In other words, without a country of the size of China systematically ignoring the diktats of the Washington Consensus except to the extent that observing them served essentially internal, domestic political purposes (e.g. Lardy 2004 on the WTO).

Approaching the riddle of Chinese innovation

With the nature and inter-relations of the immovable object and unstoppable force thus clarified, we can now return to our original and central question of "will China rule the world?", and give to it some more concrete, if provisional, substance (concluded in Chapter 11). The foregoing shows that it is specifically regarding the complex and hugely dynamic power/knowledge systems of China as the *CCP Party-state*, now on the essentially contested and uncertain cusp of global hegemony, and *neoliberal global capitalism*, now in terminal crisis, that the clash of 'immoveable object' (IO) and 'unstoppable force' (UF) has attained new intensity and importance in our tracing of emerging futures. The urgent question that emerges, therefore, is 'what happens next in this encounter?', where that can only be asked on the expectation that it will involve significant qualitative novelty; just as this encounter presents an essential paradox to our current understanding, thereby demanding conceptual innovation. This is a question that can only be investigated from a perspective capable of affording such qualitative insights, rather than one that analytically examines empirical trends on the basis of theory that (has no option but to) take(s) definitions and the objects of study as fixed and/or structural.[5]

But while asking 'what next?' may appear to dump a great imponderable on our lap – especially as we acknowledge the essentially open and uncertain *qualitative system* change implicit in this question, and particularly amidst the Four Challenges – the CP/KS work already done above affords a way forward. Instead of simply asking 'will China rule the world?', we can now ask instead:

- How will China and global capitalism *continue* to shape each other; and especially now, at the moment of crises and/or apex of the specifically uncontrollable incarnation of the latter that is neoliberalism *and* a seemingly historical climax of the chaotic but ostensibly unstoppable rise of the CCP Party-state *within* that geopolitical order?

Here, in other words, is opened a window through which we can explore, and in meso-level analysis and micro-level detail, what now appears as the most credible future given this great historical clash, deserving at least our initial attention (while taking no future as written and certain): that China and capitalism do indeed transform each other, precisely under the profound, alchemical pressure of each in its present incarnation on the other, into the breakthrough of a new flexibility in the immoveable object and a new discipline and control over the unstoppable force that would be a new relatively settled regime of global capitalism as complex power/knowledge system. And where the emergent qualitative transformation of both is essentially *surprising* not only to our common-senses, situated in the present, but especially to the *IO and UF themselves*, being also what *neither* of these incumbent hugely powerful systems presently conceives to be what they want to happen.

We cannot expect, in other words, to find out or comment informatively on the trajectory of China's (geo)political (economic) future – and/or that of capitalism, reciprocally – through a straightforward audit of its existing strengths and weaknesses, perhaps in comparison with the incumbent hegemon of the United States, where both 'strengths' and 'weaknesses' are defined in terms of the *current* global regime of capitalism and China's place in it. This theoretical problem – 'where, then, can we start?' – may seem to rule us into dumb silence but, to the contrary, its solution lies close-at-hand, emergent from precisely the same theoretical reorientation that allowed us to formulate this question in the first place: i.e. in the study of innovation-as-politics in coproduction of complex power/knowledge systems as a genealogy of the emerging present (Chapter 3), yielding strategically productive (if not necessarily – and necessarily *not* – objectively correct and certain) tracings of emerging systemic futures that are exactly such qualitative novelty.

As already discussed, innovation is a key and privileged window into present processes of socio-political (re-)ordering and (re-)production; and this undoubtedly would include this world historical encounter of China and capitalism. On both sides, innovation has assumed increasing and explicitly celebrated system centrality to both the polity that is China and global capitalist hegemony – the broad-based acceptance of a global socio-political order emerging centred on one (nation-state) government and the domination of its territorial ordering (Arrighi 1994) in conditions of continued global capitalist (now increasingly *knowledge* capitalist) system growth. In innovation (and especially innovation directly targeted at issues of the Four Challenges) we thus also find in China issues, case studies and developments that are simultaneously less fraught and politically sensitive than the usual focus of the encounter of China and capitalism at its most tense (e.g. diplomatic tensions regarding economic or military issues, or clampdowns on civic freedoms and 'Western' ideas), and arguably, precisely as such, more productive and informative regarding their twinned trajectories.

Furthermore, as soon as we turn in this way to innovation in China as our empirical field, we also find another reason to have adopted this approach. The headline character of contemporary Chinese innovation is its strikingly haphazard, chaotic, non-linear nature that has proven itself nonetheless exceptionally productive – but in both respects, regarding the rhythm and the outcome, in ways that defy the expectations of understanding of innovation of both orthodox (largely Western) economic and the supposed overlords of this process itself, the CCP Party-state.

Chinese innovation thus raises another set of questions and conundra that themselves cry out for explanation. But addressing this challenge – another instance of the conceptual challenge of China, mentioned above – calls for an appreciation of how the socio-economic and cultural-political process of innovation in China takes place inextricably in the shadow of and mediated by, and so is unintelligible absent, exactly the key meso-level dynamic of IO vs UF. As such, even to understand the trajectory of contemporary Chinese innovation as a purely techno-economic matter, this demands looking 'under the hood', beyond given dichotomies of state vs market, domestic vs international, to the broader parallel and intra-active development of innovation and complex power/knowledge system. And this demands a form of analysis that systematically attends not just to what is directly produced (let alone expressly intended) by innovation policy and business strategy in China, but also what is being inadvertently produced in response, and both directly at the level of agency and indirectly at the level of systems.[6]

But, conversely, it follows that such an analysis also thereby illuminates that broader question of socio-political evolution that is the mutual qualitative shaping of IO and UF ... possibly to the constitution of a new capitalist settlement of spatio-temporal fixes that would constitute a new Chinese global capitalist hegemonic regime in the medium term, insofar as evidence for this does emerge when viewed through this lens.

Conclusion

In what follows, therefore, we conduct just such an analysis, considering both the key strengths and weaknesses of contemporary China not just against incumbent system definitions but also regarding the possibly embryonically-observable emergence of a new hegemonic global order. Hence we are not interested so much in extrapolatable actual trends, let alone given time-slice static characterization. Instead we focus on insights from innovation-as-politics into key dynamics and real (if possibly abstract) tensions regarding the current tendential evolution of 'China' and 'capitalism', and cognisant of how profoundly China's political order can – and has already – changed in the long arc of that deepening inter-relation.

We do this through a CP/KS assessment of the co-evolution of Chinese innovation and capitalism from the 'supply' and 'demand' sides amidst the Four Challenges – regarding the world significance, or otherwise, of its innovation competitiveness (Chapters 5 and 6), and a profile of consumer demand in the form of the emerging Chinese 'middle class' (Chapters 7 and 8) respectively – before exploring their immanent (and possibly imminent) tendential convergence in a qualitatively new and distinctively Chinese liberal hegemony of green, knowledge, networked global capitalism in a specific field of complexity-attentive innovation, low-carbon urban mobility (Chapters 9 and 10).

Notes

1 Note also that, absent growth of Sino-cultural familiarity, the trend is *downward* given the shrinking and aging populations of the whole East Asian region.
2 We refer here to global capitalism not liberalism to focus on the evident historical conjunction of China and global capitalist political economy without prejudging their deepening synthesis in Chinese liberalism and/or liberal China.
3 Of equal importance, if in the other direction, is how studying global capitalism as a complex system of power/knowledge relations as opposed to as a political economic model of 'capitalism' understood to be already well-characterized and understood, as in Marxian political economy of industrial capitalism, allows analysis that is less structural and so more open to insights about *qualitative* transformation or unfolding in the very 'kind' of capitalism, not just as 'tokens' of the same thing. This allows an analysis, for instance, that can conclude that China is indeed heading for forms of capitalist hegemony and will only achieve that ascendancy to the extent it is indeed capitalist (as in Hung 2016; Panitch and Gindin 2013 or Starrs 2014), but without this leading inexorably to the conclusion that 'nothing will change' and Chinese hegemony will simply be an extension of US domination or even neoliberalism (see Chapter 11).
4 As many scholars have now noted, therefore, the parallels between the rise of China since 1978 and of other East Asian economies in the post-war period are overwhelmed by their differences, regarding both the institutions and processes within these countries and the external, geopolitical climate without (e.g. McNally 2012).
5 This, for instance, is precisely the step that Hung's (2016) otherwise highly informative analysis fails to take.
6 This thus sets up the illustrative device of a 2×2 grid – of changes intended and those unintended on the one hand and those achieved directly via/to agency and indirectly via/to systems – that we will deploy throughout the following discussion.

References

Armstrong, K. (2007) *The Great Transformation*, New York: Atlantic Books.
Arrighi, G. (1994) *The Long Twentieth Century*, London: Verso.
Arrighi, G. (2007) *Adam Smith in Beijing: Lineages of the Twenty-First Century*. London: Verso.
Aspers, P. (2011) *Markets*, Cambridge: Polity.
Boyer, R. (2016) 'How the specificity of Chinese capitalism explains its position in the world economy', *Working Paper* http://robertboyer.org/download/How%20the%20specificity%20of%20Chinese%20capitalism%20explains%20its%20position.pdf.
Brown, A. (2010) *The Rise and Fall of Communism*, London: Vintage.
Chang, G. (2003) *The Coming Collapse of China*, London: Arrow.
Chang, K.S. (2010) 'The second modern condition? Compressed modernity as internalized reflexive cosmopolitization' *British Journal of Sociology* 61(3): 444–464.
ClimateGroup (2009) *China's Clean Revolution*. Beijing & London: ClimateGroup.
ClimateGroup (2014) *Inside China*. Beijing & London: ClimateGroup.
ClimateGroup (2015) *Eco-Civilization: China's Blueprint for a New Era*. Beijing & London: ClimateGroup.
Crouch, C. (2011) *The Strange Non-Death of Neoliberalism*, Cambridge: Polity.
De Vries, J. (1994) 'The industrial revolution and the industrious revolution', *The Journal of Economic History* 54(2): 249–270.
Dicken, P. (2014) *Global Shift* (7th edition), London and Thousand Oaks, CA: Sage.
Duara, P. (2014) *The Crisis of Global Modernity*, Cambridge: Cambridge University Press.
Fei, X., G.G. Hamilton and Z. Wang (1992) *From the Soil: The Foundations of Chinese Society*, Berkeley: University of California Press.

Fenby, J. (2014) *Will China Dominate the 21st Century?*, Cambridge: Polity.

Fleetwood, S. (2001) 'What kind of *theory* is Marx's Labour *Theory* of Value? A critical realist inquiry', *Capital & Class* 73: 41–77.

Fligstein, N. (1996) 'Markets as politics: A political-cultural approach to market institutions', *American Sociological Review* 61(4): 656–673.

Gittings, J. (2005) *The Changing Face of China: From Mao to Market*, Oxford: Oxford University Press.

Green, F., and N. Stern (2015) *China's 'New Normal': Better Growth, Better Climate*. Policy Paper, London: Grantham Research Institute on Climate Change.

Greenspan, A. (2008) 'Evidence given on 23 October 2008', Washington DC: House Committee on Oversight and Government Reform.

Halper, S. (2010) *The Beijing Consensus*, New York: Basic Books.

Han, S.J. and Y.D. Park (2014) 'Another cosmopolitanism: a critical reconstruction of the neo-Confucian conception of *Tianxiaweigong* in the age of global risks', *Development and Society* 43(2): 185–206.

Han, S.J. and Y.H. Shim (2010) 'Redefining second modernity for East Asia: a critical assessment', *British Journal of Sociology* 61(3): 465–488.

Hsu, F. (1998) 'Confucianism in comparative context', in W.H. Slote and G.A. De Vos (eds) *Confucianism and the Family*, New York: SUNY Press.

Hughes, C. (2007) *Chinese Nationalism in the Global Era*, London: Routledge.

Hung, H.F. (2016) *The China Boom*, New York: Columbia University Press.

Ingham, G. (2008) *Capitalism*, Cambridge: Polity.

Jacques, M. (2009) *When China Rules the World*, London: Allen Lane.

Kahn, M.E. and S. Zheng (2016) *Blue Skies over Beijing*, Oxford and Princeton, NJ: Princeton University Press.

Lardy, N.R. (2004) *Integrating China into the Global Economy*, New York: Brookings Institution Press.

Marx, K. (1999) *Capital*, abridged by D. McLellan, Oxford: Oxford University Press.

McNally, C.A. (2012) 'Sino-Capitalism: China's re-emergence and the international political economy', *World Politics* 64(4): 741–76.

Mitter, R. (2005) *A Bitter Revolution*, Oxford: Oxford University Press.

Naughton, B. (2006) *The Chinese Economy: Transitions and Growth*, Cambridge (MA): MIT Press.

Nolan, P. (2004) 'China and the global business revolution', in *Transforming China: Globalization, Transition and Development*, London: Anthem. Ch.6: 185–232.

Panitch, L. and S. Gindin (2013) 'The integration of China into global capitalism', *International Critical Thought* 3(2): 146–158.

Pye, L. (1992) *The Spirit of Chinese Politics*, Cambridge, MA: Harvard University Press.

Saich, A. (2004) *Governance and Politics of China*, Basingstoke: Palgrave Macmillan.

Shambaugh, D. (2009) *China's Communist Party: Atrophy and Adaptation*, Berkeley, CA: University of California Press.

Shambaugh, D. (2016) *China's Future*, Cambridge: Polity.

Starrs, S. (2014) 'The chimera of global convergence', *New Left Review* 87: 81–96.

Tyfield, D. (2017) 'Realizing the Beckian vision: Cosmopolitan cosmopolitanism and low-carbon China as political Education', *Theory, Culture & Society* online 20 October 2016, doi: 10.1177/0263276416669413.

World Bank (2010) Results Profile: China Poverty Alleviation www.worldbank.org/en/news/feature/2010/03/19/results-profile-china-poverty-reduction.

World Bank (2015) GDP growth (annual percent). http://data.worldbank.org/indicator/NY.GDP.MKTP.KD.ZG/.

Xu, C.F. (2012) 'The incomplete transformation of Sinicized Marxism', *Socialism and Democracy* 26(1): 1–17.

Yeung, H.W.-C. (2006) *Chinese Capitalism in a Global Era*, London: Routledge.

Yu, H. (2012) *China in Ten Words*, translated by A.H. Barr, London: Duckworth Overlook.

Zhang, W.W. (2012) *The China Wave: Rise of a Civilizational State*, Shanghai: World Century Publishing.

Zhao, S. (2013) 'Foreign policy implications of Chinese nationalism revisited: The strident turn,' *Journal of Contemporary China* 22(82): 535–553.

Zheng, Y. (2009) *The Chinese Communist Party as Organizational Emperor*, London: Routledge.

5 The supply side
Debates and paradoxes regarding Chinese innovation upgrade

Introduction

There seems scant chance any time soon of a broad consensus regarding the significance and future of Chinese innovation. Like the debate about whether China will 'rule the world', discussed above (Chapter 4), and in part a key element of it (Segal 2012; Jakobson 2007; Fu 2016; Lazonick *et al.* 2016; Lewin *et al.* 2016), this issue today is essentially contested. Underpinning this essential disagreement, though, is a broad-based agreement that assessing China's capacity for innovation is indeed a key lens onto this issue; that China will succeed or fail in its self-styled historic mission of claiming global centrality in large part depending upon its capacity for innovation.

We open our discussion here with a balance sheet of arguments for the China optimists and pessimists, since this serves the dual purpose of providing an introductory outline to the Chinese innovation system and debate thereon *and*, as importantly, its paradoxes and conceptual challenges. The latter are rarely picked up in most of the 'will China rule the world' ('with innovation') literature, just as it tends to compare existing evidence of Chinese innovation strengths and weaknesses set against a definition of 'innovation' and a presumed system of innovation that is taken as given, ahistorical and unchanging in precisely the way we have been challenging in the last chapter.

By optimists, we mean those convinced that Chinese innovation already has (and will continue to) improve(d) significantly, imminently reaching heights of global leadership in some, if not many, techno-scientific and/or socio-technical fields. Conversely, pessimists offer counter-arguments regarding the continuing essential weaknesses of Chinese innovation, both in itself and comparatively with existing global leaders, notably the US. They are 'optimists' and 'pessimists' here, therefore, regardless of whether or not they welcome or fear the outcome they describe; though, as one would expect given the high political stakes involved, optimists tend to cheer China's rise (e.g. Lin 2016; Jacques 2009) and pessimists fret about it in order to dismiss it (e.g. Abrami *et al.* 2014).

The optimist's case

Optimists have a huge and growing body of evidence, often quantitative, on which to draw in arguing for China's imminent global strength in innovation. While not strictly an issue of innovation per se, unsurprisingly the first fact raised is often the GDP growth 'miracle' of China over the past three decades, during which it has grown at an unprecedented 8 per cent per annum on average (World Bank 2015), increasing its GDP approximately 14-fold between 1998 and 2013. This has brought with it a rise in average per capita annual income from $3,800 in 2000 to $11,500 in 2013 – higher than this in its developed coastal cities and provinces. China's economy will thus likely soon overtake the United States as the world's biggest, at least in purchasing power parity (PPP) terms, if it has not already done so (IMF 2014).

From here, the next headline figure is invariably the parallel growth of China's generalized expenditure on R&D (GERD), a key (if not altogether insightful) metric of the size of innovation activity in a country. Here again, China's rise is clearly the story of the age, witnessing astonishing growth. The Cultural Revolution (1966–1976) saw the dismantling of China's R&D institutions, but research and innovation (R&I) – or 'sci-tech', *keji* (the abbreviated conjunction of *kexue-jishu*) – has been at the heart of the Chinese project of reform and opening up, ever since 1978, in Deng Xiaoping's contemporaneous high-level government slogan of the 'Four Modernizations'.[1] This has translated, especially in the 'second' post-'89 reform period (Naughton 2006, Huang 2008), into striking growth in investment in R&D, rising from 0.7 per cent of GDP (or RMB15.1 billion, US$2.83 billion) in 1991 to 2.0 per cent of GDP (RMB1.185 trillion, US$191 billion) in 2013, a rise of nearly 100-fold, and with a compound annual growth rate of 20 per cent since 2005 (Lewin *et al.* 2016: 3, 6). As a result, at current comparative growth rates China's GERD is set to overtake the United States in global pole position by the mid-2020s, while it is already global no. 2 in PPP terms (OECD 2014; Van Noorden 2016).

Moreover, the profile of this R&D spending has changed profoundly in ways signalling growing innovation competitiveness, 70 per cent now accounted for by industrial enterprise rather than the public purse (Cao 2015; OECD 2014: 292). China is also home to more than 1,300 R&D centres for transnational corporations (Suttmeier 2017), conducting R&D that is increasingly at the cutting-edge of that business's innovation rather than downstream adaptation to local tastes or data processing.

Another favoured metric for optimists is the growth in Chinese patents, both within China and overseas in key markets. For instance, the Chinese ICT firm Huawei ranked first in the world in numbers of patents filed in 2014 and 2015, overtaking SonyEricsson and Qualcomm (Yu 2016). This performance is arguably indicative of a broader national transformation. Regarding filings at the US Patent and Trademark Office (USPTO), for instance, China's patents have seen an exponential growth (just ahead of a similar trend from India) that has catapulted China from the low ranks of developing countries up towards developed countries, following a pattern similar to that historically traced by Japan, Taiwan and South Korea (Lewin *et al.* 2016: 6). University patents have also increased significantly in recent years (Cheng and Huang 2016), following legislation specifically aiming to mimic the supposed stimulus in university-business connections and innovation of the US's Bayh–Dole Act in 1981 at the very start of the neoliberal era (see Mowery *et al.* 2004 for a critical assessment of that prevailing axiom). Indeed, in terms of quality as well as quantity, it is arguable that the pace of the improvement in China's innovation capacity is unprecedented (Fu 2016).

The result of this ongoing process is the progressive emergence of a national innovation system not just of extraordinary size but also increasingly confident in leadership or global parity of key technological sectors. This would include not just

ICTs and the internet, with a growing set of global giant firms and the world's largest online population, overtaking the US (now with an unbridgeable lead) in 2007 (Tse 2016: 82), but also various cleantech and renewable energy industries and firms that are now amongst the world's biggest (notably wind power, solar PV and hydro (Zhang and He 2013; Li, H. 2015; Lewis 2013; Urban *et al.* 2016; Geall 2016)). Also, given China's emergence as industrial workshop of the world, Chinese leadership is also visible in multiple fields of heavy industry, such as 'construction, high-speed rail, heavy engineering, shipbuilding and steel' (Lewin *et al.* 2016: 2). This heavy industry also matters as it continues to play a crucial role in the ongoing urbanization project which is now arguably China's most important site of economic activity, while conversely, the factory-light model of neoliberal globalized trans-national corporations has systematically gutted the heavy manufacturing capacity of many globally-leading hi-tech firms from the global North.

And both upstream and downstream, in research and development, deployment and diffusion (RDD&D), there is abundant evidence of China's fast improvement. Downstream, China's venture capital sector, focused on the cities of Beijing, Shang-hai and Shenzhen, has boomed in recent years to become the world's second largest and a site of frenzied opportunity. Government too is increasingly signalling not just the reaffirmed and intensified central importance of (hi-tech) innovation to the national project (e.g. Suttmeier 2017; Lewin *et al.* 2016; Butollo and Lüthje 2016; regarding Xi 2015; Li, K.Q. 2015; Xi 2016), but also important signals regarding changes in the policy framework towards enabling a greater (indeed, 'decisive' according to Xi Jinping (*The Economist* 2016d)) role for market forces (cf. DRC/World Bank 2013). This is also being prosecuted alongside an unrelenting anti-corruption campaign, and plans to reform further the still massive state-owned enter-prise sector, both aiming to iron out continuing major challenges.

Meanwhile, regarding the 'upstream' of science and engineering research, invest-ment in this sector has grown significantly since the 1990s, especially to top univer-sities identified and then generously funded through the national 985 and 211 projects (Suttmeier 2017). To improve the allocation of funding with competitive peer review, the National Science Foundation of China has also been successfully established, dispensing RMB18.35 billion in 2015 (NSFC 2016). University and research institution R&D funding increased at approximately 20 per cent per year, or a total of five-fold, between 2004–2013. This has yielded significant output, with the number and international quality (for instance, as measured using various cita-tion metrics) of scientific peer-reviewed journal articles increasing significantly (Bound *et al.* 2013; Chen *et al.* 2015). This includes global leadership in some fields, such as nanotechnology, supercomputing and space. As Freeman and Huang (2015, cited in Kahn and Zheng 2016: 48) put it, therefore, China has gone in under two decades 'from bit player in global science and engineering to become the world's second largest source of S&E graduates … and the second largest producer of scientific papers.' By 2020, China's graduate workforce, stacked towards the 'hard' sciences and business/economics, many of whom will have overseas degrees, will be approximately 200 million people, or larger than the entire workforce of the United States (Tse 2016: 107).

Together then, this flood of impressive statistics detailing extraordinarily rapid growth and at unrivalled scale reaffirms the appositeness of the metaphorical supertanker alluded to above. Except that now we see that this colossal vehicle is also being re-engineered as it goes and with increasing skill and technological sophistication, such that it seems highly plausible that it cannot *but* overtake the United States (and all other developed economies besides) as the very acme of hi-tech innovation.

The pessimist's case

The approach of the pessimist is generally significantly different, as it surely must be when seeking to rebut all this hard evidence of an awe-inspiring historical force on the move. Rather than directly rebutting all these quantitative trends, therefore, the pessimist can instead draw attention to a different and more structural set of concerns that look behind the headlines, examine the essential qualitative aspects of the innovation being discussed – an issue that becomes all the more important as China approaches the global cutting-edge – and set the heights of Chinese innovation in the context of the broader Chinese system.

Perhaps the simplest objection here is that the sheer size of China necessarily translates into figures that will necessarily be absolutely massive in comparison to any other country (except India), even one as large as the United States. For instance, starting again with GDP and GDP *per capita*, while both measurements have certainly grown significantly in the last few decades and China's GDP (whether in PPP terms or not) may overtake the US in the short-to-medium term, it is equally all-but-certain that national *per capita* income (again in PPP or not) will remain significantly smaller than most developed countries for very much longer, and this is even more so once the level of income inequality is taken into account. While there are very likely to be many tens or even hundreds of millions of comparatively prosperous urban Chinese over the next few decades (Chapter 8), it is equally likely that there will remain a great many more who do not enjoy this prosperity for some time to come.

Turning then to innovation itself, the pessimist accepts the definition of the optimist regarding what 'innovation' means, and how Chinese innovation fares against that measurement; namely as new-to-the-world, high-technology innovation that is owned and primarily developed by, and so profits, institutions domiciled in that country. But, turning the argument on its head, it argues that such world-leading innovation is impossible absent globally-attractive urban spaces and unique institutionalized cultures of open sharing of ideas, competitive and accessible funding for the best of these ideas and reliable constitutional and/or legal guarantees that, if successful, the entrepreneur will be able to enjoy her spoils rather than face the risk of dispossession by a capricious or disapproving state. And that China's experience not only offers little evidence against this understanding of innovation competitiveness but plenty of evidence to confirm it, to China's cost. In other words, China, as a Communist one-party state, will remain permanently hobbled, unable to take the all-important final but qualitatively hardest step, or quantum leap, from fast-follower (however fast) to global leader.

There are five elements to this argument. First, the surging forefronts of Chinese innovation are set against the much bigger and broader picture of China's technology sector. Here we find, on the one hand, a huge number of firms that – servicing primarily the massive domestic market but in a country with only embryonic and ineffective legal redress and under intense pressure to produce and sell at all costs – are not only copycats but manufacture poor quality, even dangerous, products for as much short-term profit as possible (Midler 2011). This is hardly, therefore, the domain of patient and massively concentrated investment in R&D that is necessary for global cutting-edge innovation leadership. Moreover, the mass of dynamic private firms that do produce goods of quality high enough for the global market, clustered into infrastructurally accessible and possibly low-tariff zones, are equally low-margin businesses, engaged in cut-throat competition for business overwhelmingly from high-technology corporations based overseas and/or largely dependent upon them for technological upgrade. Again, therefore, these are businesses that struggle to invest in a business strategy of R&D investment (Steinfeld 2004). While China's hi-tech sector is indeed massive, much of it is simply engaged in the lowest value-added step of assembly for global markets, hence the parallel rise in both imports and subsequent export of hi-tech goods (Xing 2011).

Second, then, is the industrial structure of China. Chinese private firms, revenues and profits have grown significantly, and now far outweigh the state-owned sector in such purely quantitative terms. The state-owned enterprises (SOEs) – held at various levels, including central and local government, and by various ministries – have undergone rounds of reform, especially in the late 1990s (in preparation for and driven through on the pretext of China's membership of the WTO). These reforms have rationalized their number and reorganized their management and ownership (for instance, for the biggest and most nationally important firms, into share ownership via a central state body, SASAC (Naughton 2011)) so that they are more subject to the hard-budget constraints, elements of (perhaps oligopolistic) competition and expectations of profitability of 'normal' businesses. Yet the Party-state continues to hold the reins of control over the industrial summits of most of the Chinese ('knowledge') economy, including across the heavy industries, energy, telecoms, banking, education and media. And as a Party-state, not merely a state apparatus, with the parallel chain of command across these firms of the CCP, the essential shaping of management decision-making by concerns other than maximized business competitiveness – not least the cadre-managers likely promotion and circulation through other business or government institutions – remains deeply entrenched and institutionalized. Moreover, spreading the influence of this system beyond the strictly state-owned sector is the complicated hybrid political economy of China's firms (e.g. Tse 2016), in which strict legal ownership does not necessarily delineate independence from the party-state, for its favour or command (Fuller 2016).

On the one hand, the unintended but inevitable distortions of institutionalized incentives for innovation-intensive upgrade are thus legion. For instance, in control of both the banking system and multiple systemically important industries, the allocation of debt capital by the state-owned banks is systematically biased towards the latter, on more-than-commercial grounds of national policy, generating a massive misallocation of finance. The results are a build-up of corporate and local government debt to levels that are patently unsustainable (Chu 2013) – at over 260 per cent of GDP (*The Economist* 2016a), or over twice the level of debt in the US at the time of the Lehman Brothers credit crunch – thereby resurrecting a massive problem of non-performing loans from the late 1990s that was temporarily 'fixed' in the early 2000s. Such misallocation has also led to massive over-investment in surplus industrial capacity, further exacerbating the economic challenges of repaying those debts. Meanwhile, private businesses struggle to access finance from the banks, thereby simultaneously starving the most productive sectors of the Chinese economy or pushing them towards poorly regulated shadow or online P2P financing (Rabinovitch 2016; *The Economist* 2016b).

Yet, on the other hand, still systematically under-institutionalized and far from a single and unified chain of command issuing from Beijing, China's governmental structure is a patchwork of mutually contending local governments, each nurturing their own local developmental state (Oi 1992) and with multiple contending ministries at every level in a system of extreme complexity but also strict hierarchy. This fragmented authoritarianism (Lieberthal 1992; Mertha 2009) thus not only effectively rules out the nationally-integrated programmes of industrial policy that underpinned the successful development of the East Asian economies in the post-war period. It also presents business with an institutionally baffling governance environment littered with systemically irrational incentives and disincentives, including disincentives for long-term investment in innovation upgrade (e.g. Chen 2008). Instead, at best, Chinese firms – both state and private – confront a strategic landscape of 'structured uncertainty' (Breznitz and Murphree 2011) and 'un-decision' (Segal 2003), that structures or builds in profound uncertainty and complexity for management as a matter of course; at worst, they are confronted with straightforwardly perverse, if unintended, incentives.

For instance, in environmental governance, a local/provincial-level Environmental Protection Bureau (EPB) may be charged with implementing a newly rigorous anti-pollution law on the factories in its territory. Yet the most polluting SOEs may be owned by a central government ministry and so, as such, rank higher than the EPB and so pull rank. Moreover, with the leading cadres in the provincial government personally assessed primarily for the GDP growth stimulated in their region (though this is changing (Kahn and Zheng 2016)), strict enforcement by the EPB may be further discouraged by their own direct seniors.

Under this combined pressure of both 'too much' (Party-)state and 'too little', therefore, China is almost the very antithesis of the 'entrepreneurial state' (Mazzucato 2011) now being theorized and lionized in heterodox Western innovation economics. Rather, with both a bloated and difficult-to-reform SOE (or 'state-favoured' as Fuller 2016 nicely puts it (see also Fuller *et al.* 2015)) sector and stuctured uncertainty disincentivizing long-term innovation investment, which is premised on a relatively stable and predictable business climate, Chinese innovation faces deep-seated systemic problems. For all the success to date, therefore, pessimists argue that such catch-up can only face a rude awakening in the medium-term to the stark inadequacy of this structure for the incubation of genuinely globally-competitive and world-leading firms.

This leads to the third objection, regarding the clear weakness or simple absence of the institutional preconditions of attracting and cultivating such firms and entrepreneurs. Implicitly, if not explicitly, the presumed comparison here is with the conditions that have made Silicon Valley – and other leading hi-tech clusters in the US and elsewhere – the acme of innovation competitiveness. This includes the qualitative weakness, for all its size, of the Chinese venture capital sector, which too is overwhelmingly either directly government-funded or part-managed by government institutions. Again, looking behind the numbers thus reveals a sector that is most successful when there is some, but not total, government involvement – since purely government VC is poorly managed while totally private and/or foreign VC has poor connections and market understanding (Fuller 2009; Cao X. 2016). But this comparative performance simply highlights the importance of the management of *political* or governmental uncertainty by those connections in Chinese VC – and all the time and effort that entrepreneurs must waste on the cultivation of the necessary personal connections or *guanxi* (Tse 2016). While undoubtedly a booming investment segment, to the extent it is a success Chinese VC is also overwhelmingly late-stage private equity (with 'private' itself in scare quotes), rather than early-stage 'venture' capital incubating an ecosystem of cutting-edge innovation, and has very few new-to-the-world innovation success stories to its name.

Similarly, other important aspects of a broader national, or even local (e.g. in Shanghai and the Yangtze River Delta, or Shenzhen and the Pearl River Delta (PRD)), culture of buzzing (cf. Bathelt *et al.* 2004) innovation across multiple firms, rather than a few stand-out exceptions, are also apparently missing. The flipside of all the effort needed to cultivate the crucial and scarce business resource of powerful *guanxi*, for instance, is the lack of strong national/local lobbies for small and medium-sized enterprises (SMEs) and other start-ups. In a global market for hi-tech innovation with circulation of 'talent' and with cross-border partnerships and collaborations into global innovation networks (GINs) (Ernst 2006), crucial elements of innovation competitiveness, the multiple barriers and deterrents to such movement into and out of China are also important structural impediments, not least of which is the sheer uncertainty but potentially high stakes of Party-state attention and disapproval.

Even the systematic weakness of Chinese trade unions – ironically in a Communist state, being nationally organized governmental institutions that act primarily as bulwarks of stability against labour unrest, not independent institutional representatives of workers' grievances – harms its innovation upgrade and its corollary of 'social upgrading', creating highly-skilled knowledge workers (Butollo 2014). This is especially so in hi-tech manufacturing sectors, where highly-skilled and company-loyal workers on the shop floor are crucial to the knowledge and information collection that feeds the constant incremental improvement that maintains a competitive advantage (Herrigel *et al.* 2013). And, of course, at the very fundamentals of (current dominant understanding of) a vibrant innovation system, China systematically lacks secure private property rights, including of intellectual property and its enforcement (however much that system is improving (Liang and Xue 2010)), that supposedly provide the strong, guaranteed incentives necessary for personal entrepreneurship (cf. Boldrin and Levine 2010; Jaffe and Lerner 2007).

It is no surprise, therefore, that the outcome of all this is a system that is, if anything considering its size, growing wealth and national power, characterized by the striking weakness of its cutting-edge innovation and the paucity of its new-to-the-world innovations and global hi-tech brands. Within 20–30 years of Japan's or South Korea's post-war take-off, Western markets were already ready consumers of their cars and electronics, creating global champions such as Toyota, Sony or Samsung. Today, comparable Chinese brands continue to rank few and low.

And so to the fourth objection, namely a direct assault, with groundwork now laid, on the statistics of success themselves. This goes beyond simple doubt about Chinese statistics, though including this perennial objection. And it goes beyond setting the statistics in comparative context, whether regarding competition with existing innovation 'superpowers', important as this undoubtedly is. For example, for all the titanic growth in Chinese USPTO patents, the entire mainland still remains behind the comparatively minuscule Taiwan (Lewin *et al.* 2016: 7), and an order of magnitude behind the US and Japan; there is clearly a long way for China still to go.

More importantly, though, the pessimist case here depends on the compelling refrain that quantity does not equal quality ... and it is quality that matters. For instance, even a figure as seemingly unarguable as China's ballooning GERD (in both absolute monetary figures and relative to its GDP), affords considerable scepticism regarding what it actually means. As Suttmeier (2017) notes, comparing national GERD as like-with-like assumes that one dollar of R&D expenditure achieves the same amount of actual R&D in both countries. This only needs to be stated to be seen as clearly false in principle. It is also obviously false in fact regarding China.

> The R&D spending surge over the past 15 years has clearly outpaced institutional design, leaving a legacy of problems including considerable derivative research, scientific misconduct, widespread filing of low-quality patents, waste and misuse of R&D funding, and the development of a technical talent pool, a large portion of which seemed to lack the training and socialization needed for original research.
>
> (Suttmeier 2017: 370)

To this we can add similar questions regarding other elements of the innovation system occluded and possibly tacitly presumed as equivalent in other figures. For instance, while under intense pressure to cultivate connections with business for translation of research into innovation (following similar trends in the West), including a pressure to patent, universities and research institutions continue to have poor links to business as a legacy of the structure and incentives for jealously protected institutional isolation since their (re-) establishment under a USSR-inspired model in the Mao era. Certainly, incentives are now changing fast, not least at the nationally leading institutions, where pressures to publish in top international journals, participate in global research networks and win research funding, from government and business, are now even more intense than in Western institutions (Xu and Ye 2017). But again, the very success of such pressure in the context of the continuing Party-state meta-institutional governance structure simply compounds, rather than alleviates, the perverse incentives: e.g. in a surge in the number of university patents that thereby meet the metric of their personal assessment but are of poor quality (Cheng and Huang 2016). Moreover, caught between the demand to connect with business and the ongoing merciless anti-corruption drive, which very much applies to university and research institution staff, who are considered civil servants, these researchers are also paralyzed by uncertainty regarding what is permitted and what could land them in very hot water. Here, therefore, the broader argument regarding terminal systemic weakness, rooted in the Party-state, comes home to roost in undermining the very foundations of the optimists' argument.

The final straw, however, perhaps comes from adding an objection of our own, in pointing out that this entire argument so far has been premised upon assessing China's innovation capacity as against the global neoliberal innovation system of the 1990s to the present. For this innovation model, the pinnacle of innovation is proprietary hi-tech consumer goods, then dismantled into multiple steps parcelled out across global production networks and supply chains of modularized manufacturing (Breznitz 2007). Innovation today, however, is increasingly tasked with much greater challenges that call for a transformation of innovation practice, the development of genuinely entrepreneurial state apparatuses and the management of much more complex issues, namely the Fur Challenges. While much of the pessimists' argument does indeed accept the neoliberal status quo and implicitly asks the question "can innovation thrive without democracy?", albeit largely as a rhetorical question, this argument strikes one as *all the more* compelling once we acknowledge the profound *change* currently taking place in the tasks with which we are entrusting innovation.

For if China has managed a startling quantitative, and a modicum of qualitative, improvement in its innovation over the past 2–3 decades without any parallel opening up of its politics and civic public sphere, it is hard indeed to see how it can continue such achievement in this new context, where stakeholder involvement, open information, data (and possibly hard asset) sharing and levelled, predictable state support are so important (Stirling 2009). This objection is simply compounded by recent trends in China, where the new leadership of Xi Jinping seems, if anything, to be moving these crucial socio-political aspects of a vibrant innovation culture in precisely the opposite direction. For all the expressions of a deepening economic liberalism, these are at least matched, and possibly undermined, by an equal and opposite push in the other direction regarding freedom of speech, assembly and human rights. The personal accumulation of power in the person of Xi Jinping, not seen since Deng (Brown 2016; Mitchell 2016), and the crackdown on liberal voices, 'Western ideas' and international NGOs (*The Economist* 2015a), outspoken entrepreneurs and celebrities (*The Economist* 2015b), and lawyers or labour activists (*The Economist* 2016c), all previously tolerated, marks a definitive shift away from, not towards, the opening and inclusion that, elsewhere, government pronouncements themselves acknowledge to be necessary to tackle socio-environmental problems. Even the anti-corruption campaign, for all its popularity and necessity, is just as credibly interpreted, by Chinese citizens and entrepreneurs, as a capricious fire to which one should not get too close.

In short, the pessimist can show how many of the so-called strengths of Chinese innovation are no such thing, but are largely puff and show, massively inflated by the sheer size and unique political structure of China; mere Potemkin villages, albeit gigantic and impressive ones, crudely masking what remain as profound underlying weaknesses that are systemically entrenched and, if anything, facing worsening trends.

Innovation competitiveness – not what the CCP wants

We have thus encountered two diametrically opposed positions, both able to mobilize significant evidence. Attempting to adjudicate between the optimist and pessimist cases, however, is a fool's errand. For both are right. And both are wrong. And it is precisely this feature of China's innovation system that is its most important and defining characteristic and, in the greatest irony of all, its greatest strength. We have referred above to the conceptual challenge of China. But here we are confronted with it in all its head-spinning glory. For how can both cases be right? How can we even conceptualize a system in which that is the case?

Our starting point here (and our dialectical conclusion, to come) is that in China, and also regarding its innovation and political economic ascendancy, nothing is just as it seems. 'Strength' is not just strength, nor 'weakness' weakness. Rather we must adopt a perspective that can explore the *constitutive and dynamic inter-relations* of supposed 'strengths' and 'weaknesses', and thereby illuminate how they are shaping each other's emerging parallel trajectories. This approach is the exploration of innovation-as-politics co-produced with complex power/knowledge systems.

Neither the optimist nor the pessimist case is the final word, though each is a crucial step along the way, and one that remains in place, highlighting conditions that remain and continue actively to shape the ongoing evolution of the Chinese innovation system and the competitiveness of its broader political economy within global capitalism. Pulling us onward to a third perspective, however, is a growing body of literature that highlights a 'qualified-optimist' response to the pessimist case. This argument, to which we now turn, highlights a host of inter-related and crucial insights missed by both optimist and pessimist cases, effectively turning our attention, and our perspective, to seeing strengths in China's innovation system that are unexpected, both in nature and in the process of their provenance.

In doing so, this argument picks up the baton of China's multiple sub-optimal political-governmental and socio-cultural conditions for new-to-the-world, globally-leading hi-tech innovation. But turning these on their head, and looking through the Chinese looking glass (cf. Breznitz and Murphree 2011), they are shown to be incubators for forms of innovation that are the secret, and resilient, strengths of Chinese innovation.

Crucially, at the heart of this topsy-turvy institutional logic is precisely the key dynamic of the immoveable object (IO) vs the unstoppable force (UF), described above. As such, we may summarize the trajectory of the Chinese innovation capacity under the slogan of 'not what the CCP wants', the subtitle of this section. This captures both essential and profound shaping of Chinese innovation by what the Party-state *does* want and powerfully support, and the equally irreducible counter-logic that produces something that is quite different (if not necessarily what the CCP does *not* want). As described above, therefore, understanding Chinese innovation demands that we go beyond *both* the headlines of propaganda and PR monologue regarding what policy intends and has actually been achieved (with Government claiming credit) – all regarding what supposedly is a uniquely-enabled top-down authoritarian state, as in the (often approving!) mantra of Western chambers of commerce that 'the Chinese (government) gets things done' – *and* the counter-arguments of how Communist China is intrinsically incapable of innovation, as direct riposte.

Rather, we must explore how, mediated through and intensifying the productive tension of IO vs UF, Chinese innovation makes sense as a dynamic system *only* once we have examined both the direct and intended outcomes and those that are unintended and/or indirectly produced, perhaps in the evolution and emergence of systems and system capacities (see Table 5.1). This is precisely the case regarding the Chinese 'disruptors' (Tse 2016; Breznitz and Murphree 2011; Zeng and Williamson 2007; Tyfield *et al.* 2010), moving to the third box in the top right-hand corner of Table 5.1. But beyond this step, we must also examine where this innovation is *itself* leading, taking us to the final box – a step usually not taken to date by the 'disruptors' literature, but compelled and facilitated by the CP/KS approach.

In what sense is the Chinese innovation system and capacity 'not what the CCP wants'? We have already seen key aspects of this above in the pessimists' case. The CCP's current project, since 1992 and with ever-increasing emphasis on innovation in recent years, is to oversee and orchestrate – and then peremptorily take credit for – the rise to global leadership of Chinese innovation and enterprise, thereby securing the twin existential imperatives of the People's Republic of China, namely: the PRC's relative independence and sovereignty in its foreign affairs; and the continuing monopoly of legitimate organization and state power of the CCP at home. As

Table 5.1 The quadrant of Chinese innovation-as-politics

	Direct effects (at agent level)	*Indirect effects (at system level)*
Intended (immoveable object)	What the CCP Party-state wants to have happened and has, indeed, happened	What has emerged as the case in a seeming vindication of Party-state policy but entirely separately from, or even in direct opposition to, express governmental intentions and levers
Responded (unstoppable force)	What the CCP Party-state directly produces in its deepening encounter with global capitalism, thereby thwarting its own goals	What is in turn emerging from or immanent within these system-functional effects

regards innovation, and the upgrade of its global competitiveness and the broader structure of its political economy, this takes shape as:

1   a strong, globally competitive techno-nationalist state (Hughes 2007; Zhao 2010), characterized by high capacity for 'indigenous innovation' (*zizhu chuangxin*, perhaps better translated as 'self-directed innovation' (Jakobson 2007; Schwaag-Serger and Breidne 2007; OECD 2008; Cao, Suttmeier and Simon 2006; Suttmeier 2017; Lazonick *et al.* 2016)) that secures innovation independence from foreign transnational corporations and builds national defence capacity (Liu 2006)....

2   a political economy under, and supporting, the continued control by the (central) Government of the key pillars of the industrial/financial structure through state-owned and state-favoured enterprises....

3   generation of specifically high-technology global leadership in key industries, securing an enduring competitive advantage together with transformation and upgrade in the model of development, investment and employment....

4   the underpinning of the emergence of a moderately prosperous (*xiaokang*) and harmonious society, with good and high-quality jobs and services, delivering the crucial political *quid pro quo* of unbroken (if no longer now, runaway) economic growth and continually rising standards of living.

As the optimists illustrate, there is a strong case that this has indeed been developed over the past couple of decades. But as the pessimists also counter, this is not just very far from the whole story but a highly misleading half (or quarter?) truth. For alongside and sustaining any stories of success against these metrics are four very different respective stories.

1   The fragmented authoritarianism of local developmental state corporatism and horizontal inter-ministerial contestation. These systematically prevent effective national industrial policy; condition massive industrial overcapacity and excessive

infrastructural investment at an aggregate national level, as well as perverse local incentives and outright corruption; present enduring hurdles to implementation of central government laws and regulations (with "national plans … [remaining] mostly on paper" (Breznitz and Murphree 2011: 26)); and underpin the systemic division and inequality between regions (Solinger 1996; Thun 2006). The combination of this comparative state weakness with the equally undeniable and unrestrained power of the Party-state makes the *authoritarian* aspect of this system no less significant than its *fragmentation*, acting as a kind of permanent but capricious, illegible and threatening veto power that makes both policy and business decision-making exceptionally complicated and, hence, short-termist (Breznitz and Murphree 2011: 22–23, 75).

2   Second, while the state is still in control of key economic sectors, the overwhelming source of Chinese economic dynamism is elsewhere in the 'private' sector, of domestic and overseas firms, creating a situation of 'one country, two economies'. That is only sustained by the systematic favour and neglect of the Party-state towards its two parts respectively, even as it is the *neglected* part that is the real driver of growth and development, across a host of measurements. There were still 2.3 million SOEs in 2013, despite concerted drives at reform and consolidation, but this is now dwarfed by the 12 million private firms and 42 million proprietorships (Tse 2016: 13). Similarly, while SOE revenues (often in protected and 'natural' monopolies) rose to RMB25 trillion in 2013, private sector revenues were over three times larger, at over RMB76 trillion, and private sector profits were rising over three times more rapidly (Tse 2016: 13). Privately-run businesses thus account for 75–80 per cent of GDP (Tse 2016: xii) and rising, whereas SOE profits/GDP peaked in 2007. Private firms also account for 60 per cent of taxes, and have risen to account for more than two-thirds of total capital investment, from just one-quarter in the late 1990s (Tse 2016: 49). Such is the growth of the private sector that a best estimate concludes that it now accounts for over two-thirds of urban employment, or effectively all the growth in this measurement since 1978 (when it was effectively non-existent) – in other words, for the entire Chinese economic miracle (Tse 2016).

Moreover, investment of overseas capital, and especially ethnic Chinese investment (from Hong Kong and Taiwan and, to a lesser extent, south east Asia), is also a crucial aspect of this 'one country, two economies' system, and its undeniable growth success; formally so in the case of Hong Kong under the 'one country, two systems' settlement since it rejoined China in 1997. Overseas Chinese investment in the mainland is thus identified as the most important ingredient in what innovative capacity improvement has in fact taken place (Fuller 2016). This diaspora pairs a cultural understanding and personal networks (especially from Hong Kong into the PRD, and Taiwan into Fujian province) that bring access to the domestic market and business opportunities with the hard budgetary constraints imposed by dependence on financing with full exposure to global capital flows and demands for return on investment.

3   Third, China has thereby become in recent decades the unquestionable workshop of the world, at a scale and pace without precedent. Chinese manufacturing accounted for nearly 71 per cent of its energy use in 2011, whereas that peaked at 41 per cent in the US in 1951 (Kahn and Zheng 2016: 21). Yet the very proliferation of the private, overseas-funded firms locks many of them into

cut-throat competition over tiny profit margins (Steinfeld 2004), which thereby affords little investment in R&D and innovation advantage that could break them out of that cycle. Meanwhile, despite decades of effort and government restrictions and *quid pro quos* for access to the Chinese market, in terms of joint ventures, technological transfer and Chinese competitiveness in key industries, Chinese-developed and dominated technologies and standards have remained obstinately slow to emerge (e.g. Breznitz and Murphree 2011: 71–75). Instead, this dynamic has generated the imbalanced dependence of China's political economy on its low-cost and low-wage export sector, and hence on robust global demand on the one hand, alongside the parallel dependence on its hypertrophied industrial and infrastructural state sector on the other.

4   Balanced unstably and unsustainably, therefore, between dependence on a low-cost export sector and an uncompetitive and overcapitalized state sector, the 'China Dream' of the *xiaokang* society of 'socialism with Chinese characteristics' (or, 'capitalism with Chinese characteristics' (Huang 2008)) is also threadbare. There are now a ballooning number of Chinese dollar billionaires, but the broader picture displays massive inequality that has taken China from amongst the most equal in 1978 to amongst the most unequal distributions of income in 2012 as measured by the Gini coefficient above 0.5 (Xie and Zhou 2014). This has continued even as recent policies of the New Socialist Countryside have made some steps towards mitigating, but not reversing, that trend (Lin and Wong 2012). This is inequality across rural–urban and geographical divides (crudely a richer coast and a poorer southern and western interior), formalized in the *hukou* system of residence permits. But while the surge of rural migrant labour to Chinese factories has underpinned the country's economic miracle, their systematic insecurity and weakness as a working class (given co-opted, government-controlled trade unions) is what secures the wage repression that keeps the low-cost, export-driven economy ticking. Combined with financial repression that makes access to credit hard for private enterprise (Brandt and Li 2003; Li and Xia 2008; Huang 2008) and individuals, and it is no wonder that many observers of China's economy see a permanent and systemic imbalance away from consumption (Boyer 2016; Chu 2013).

So the bigger picture of Chinese innovation and political economy undoubtedly displays a system that is very much 'not what the CCP wants' – *as well as* what it does. But, crucially, beyond even these macro assessments there is also some significant innovation success in China. It is just that it is not where the fetishism of high-technology and new-to-the-world innovation of both Chinese government policy and neoliberal common-sense chooses to look. In this sense too, therefore, the system is again 'not what the CCP wants'. What is this success?

Note

1 The Four Modernizations was one of the high-level slogans of the day – a key power/ knowledge device of Chinese government – that sought to rally popular initiative to four key elements of society that needed modernization in the post-Mao era: agriculture, industry, national defence and science and technology (Hughes 2007).

References

Abrami, R., W. Kirby and F.W. McFarlan (2014) *Can China Lead?: Reaching the Limits of Power and Growth*. Cambridge, MA: Harvard Business Review Press.

Bathelt, H., A. Malmberg and P. Maskell (2004) 'Clusters and knowledge: local buzz, global pipelines and the process of knowledge creation', *Progress in Human Geography* 28(1): 31–56.

Boldrin, M. and D. Levine (2010) *Against Intellectual Monopoly*, Cambridge: Cambridge University Press.

Bound, K., T. Saunders, J. Wilsdon and J. Adams (2013) *China's Absorptive State*, London: NESTA.

Boyer, R. (2016) 'How the specificity of Chinese capitalism explains its position in the world economy', *Working Paper*, available at: http://robertboyer.org/download/How%20the%20 specificity%20of%20Chinese%20capitalism%20explains%20its%20position.pdf.

Brandt, L. and H. Li (2003) 'Bank discrimination in transition economies: ideology, information, or incentives?', *Journal of comparative economics* 31(3): 387–413.

Breznitz, D. (2007) *Innovation and the State*, New Haven, CT: Yale University Press.

Breznitz, D. and M. Murphree (2011) *Run of the Red Queen*, New Haven, CT: Yale University Press.

Brown, K. (2016) *CEO, China: The Rise of Xi Jinping*, London: I.B. Tauris & Co.

Butollo, F. (2014) *The End of Cheap Labour? Industrial Transformation and 'Social Upgrading' in China*, Frankfurt and New York: Campus.

Butollo, F. and B. Lüthje (2016) ' "Made in China 2025": Intelligent manufacturing and work': 42–61.

Cao, C. (2015) 'China', *UNESCO Science Report, Towards 2030*. http://unesdoc.unesco.org/images/0023/002354/235406e.pdf.

Cao, C., R.P. Suttmeier and D.F. Simon (2006) 'China's 15-year science and technology plan', *Physics Today*, 59(12): 38.

Cao, X. (2016) 'The impact of government ownership on venture performance: Evidence from China', presentation at the *International Symposium on Innovation-Driven Development*, Sun Yat Sen University, Guangzhou, 13–15 June.

Chen, A., D. Patton and M. Kenney (2015) 'Chinese university technology transfer: A literature review and taxonomy', *Working Paper*. University of California, Davis.

Chen, L. (2008) 'Bureaucratic system and negotiation network: A theoretical framework for China's industrial policy', *Working Paper*, School of Public Policy and Management, Tsinghua University, Beijing.

Cheng, M.L. and C. Huang (2016) 'Transforming China's IP system to stimulate innovation', in *China's Innovation Challenge*, A.Y. Lewin, M. Kenney and J.P. Murmann (eds) Cambridge: Cambridge University Press: 152–188.

Chu, B. (2013) *Chinese Whispers*, London: Weidenfeld & Nicholson.

Development Research Center of the State Council (DRC) and the World Bank (2013) *China 2030: Building a Modern, Harmonious, and Creative Society*. Washington DC: World Bank.

The Economist (2015a) 'This article is guilty of spreading panic and disorder', 5 December.

The Economist (2015b) 'That's entertainment', 5 December.

The Economist (2016a) 'The coming debt bust', 7 May.

The Economist (2016b) 'Taking flight', 23 January.

The Economist (2016c) 'Suppress and support', 13 August.

The Economist (2016d) 'Reagan's Chinese echo', 2 January.

Ernst, D. (2006) 'Innovation offshoring. Asia's emerging role in global innovation networks', *East-West Center, East-West Center Special Report, 10*.

Freeman, R.B. and W. Huang (2015) 'China's "Great Leap Forward" in science and engineering', *NBER Working Paper 21081*, Cambridge, MA: NBER.

Fu, X.L. (2016) *China's Path to Innovation*, Cambridge: Cambridge University Press.

Fuller, D.B. (2009) 'How law, politics and transnational networks affect technology entrepreneurship: Explaining divergent venture capital investing strategies in China', *Asia Pacific Journal of Management* 27: 445–459.

Fuller, D.B. (2016) *Paper Tigers, Hidden Dragons: Firms and the Political Economy of China's Technological Development*, Oxford: Oxford University Press.

Fuller, D.B., V. Shih and R. Tao (2015) 'Market governance and firm performance under China's state capitalism', *Management and Organization Review* 11(4): 711–713.

Geall, S. (2016) 'Green innovation in China: China's wind power industry and the global transition to a low-carbon economy', *Pacific Affairs* 89(2): 417–419.

Herrigel, G., V. Wittke and U. Voskamp (2013) 'The process of Chinese manufacturing upgrading: Transitioning from unilateral to recursive mutual learning relations', *Global Strategy Journal*, 3(1): 109–125.

Huang, Y. (2008) *Capitalism with Chinese characteristics: Entrepreneurship and the state*, Cambridge: Cambridge University Press.

Hughes, C. (2007) *Chinese Nationalism in the Global Era*, London: Routledge.

International Monetary Fund (IMF) (2014) 'World economic outlook: legacies, clouds, uncertainties', www.imf.org/external/pubs/ft/weo/2014/02/.

Jacques, M. (2009) *When China Rules the World*, London: Allen Lane.

Jaffe, A. and J. Lerner (2007) *Innovation and its Discontents*, Princeton, NJ: Princeton University Press.

Jakobson, L. (ed.) (2007) *Innovation with Chinese Characteristics*, Basingstoke: Palgrave Macmillan.

Kahn, M.E. and S. Zheng (2016) *Blue Skies over Beijing*, Oxford and Princeton, NJ: Princeton University Press.

Lazonick, W., Y. Zhou and Y.F. Sun (2016) *China as an Innovation Nation*, Oxford: Oxford University Press.

Lewin, A.Y., M. Kenney and J.P. Murmann (2016) 'Introduction', in *China's Innovation Challenge*, A.Y. Lewin, M. Kenney and J.P. Murmann (eds) Cambridge: Cambridge University Press: 1–31.

Lewis, J.I. (2013) *Green innovation in China: China's wind power industry and the global transition to a low-carbon economy*, New York, NY: Columbia University Press.

Li, D.M. and L.J. Xia (2008) 'Ownership type, the institutional environment and R&D intensity of Chinese listed Firms', *Journal of Finance and Economics*, 4: 010.

Li, H.J. (2015) *China's New Energy Revolution*, New York: McGraw-Hill.

Li, K.Q. (2015) Symposium on science and technology strategy. Xinhua News Service Beijing, China, 27 July. http://news.xinhuanet.com/english/2015-07/28/c_134455919.htm.

Liang, Z. and L. Xue (2010) 'The evolution of China's IPR system and its impact on the patenting behaviours and strategies of multinationals in China', *International Journal of Technology Management* 51(2–4): 469–496.

Lieberthal, K. (1992) 'Introduction: The "fragmented authoritarianism" model and its limitations', in K. Lieberthal and A. Thurston (eds) *Bureaucracy, Politics and Decision Making in Post-Mao China*, Berkeley, CA: University of California Press: 1–30.

Lin, J.Y. (2016) 'New structural economics: the future of the Chinese economy', in *China's Innovation Challenge*, A.Y. Lewin, M. Kenney and J.P. Murmann (eds), Cambridge: Cambridge University Press: 32–55.

Lin, W.L. and C. Wong (2012) 'Are Beijing's equalization policies reaching the poor?', *China Journal* 67: 23–45.

Liu, X.L. (2006) 'Path-following or leapfrogging in catching-up: The case of Chinese telecommunication equipment industry', Paper presented at the CIRCLE seminar Series, Lund, Sweden.

Mazzucato, M. (2011) *The Entrepreneurial State*, London: Demos.

Mertha, A. (2009) '"Fragmented authoritarianism 2.0": political pluralization in the Chinese policy process', *The China Quarterly*, 200: 995–1012.

Midler, P. (2011) *Poorly Made in China*, Hoboken, NJ: John Wiley & Sons, Inc.

Mitchell, T. (2016) 'Xi's China', Special report in 3 parts, *Financial Times*, July.

Mowery, D., R. Nelson, B. Sampat and A. Ziedonis (2004) *Ivory Tower and Industrial Innovation*, Stanford: Stanford University Press.

National Natural Science Foundation of China (NSFC) (2016) 'NSFC fulfils allocation of more than 18.3 billion Yuan of its annual budget in 2015' www.nsfc.gov.cn/publish/portal1/tab158/info50047.htm.

Naughton, B. (2006) *The Chinese Economy: Transitions and Growth*, Cambridge (MA): MIT Press.

Naughton, B. (2011) China's economic policy today: The new state activism. *Eurasian Geography and Economics*, 52(3): 313–329.

Organization for Economic Cooperation and Development (OECD) (2008) *OECD Reviews of Innovation Policy*. Paris: OECD.

OECD (2014) *OECD science, technology and industry outlook 2014*. DOI: 10.1787/sti_outlook-2014-en.

Oi, J.C. (1992) 'Fiscal reform and the economic foundations of local state corporatism in China', *World Politics* 45(01): 99–126.

Rabinovitch, S. (2016) 'Big but brittle', Special Report: Finance in China, *The Economist* 7 May.

Schwaag-Serger, S. and M. Breidne (2007) 'China's Fifteen-Year Plan for science and technology: An assessment', *Asia Policy* 4: 135–164.

Segal, A. (2003) *Digital Dragon*, Ithaca, NY: Cornell University Press.

Segal, A. (2012) *Advantage: How American Innovation Can Overcome the Asian Challenge*, New York: WW Norton & Company.

Solinger, D.J. (1996) 'Despite decentralization: disadvantages, dependence and ongoing central power in the inland-the case of Wuhan', *The China Quarterly* 145: 1–34.

Steinfeld, E.S. (2004) 'China's shallow integration: networked production and the new challenges for late industrialization', *World Development* 32(11): 1971–1987.

Stirling, A. (2009) *Direction, Distribution and Diversity! Pluralising Progress in Innovation, Sustainability and Development*, STEPS Working Paper 32, Brighton: STEPS Centre.

Suttmeier, R.P. (2017) 'The transformation of Chinese science', in D. Tyfield, R. Lave, S. Randalls and C. Thorpe (eds), *The Routledge Handbook of the Political Economy of Science*, London: Routledge: 360–377.

Thun, E. (2006) *Changing Lanes in China: Foreign Direct Investment, Local Governments, and Auto Sector Development*, Cambridge: Cambridge University Press.

Tse, E. (2016) *China's Disruptors*, London: Portfolio Penguin.

Tyfield, D., J. Jin and T. Rooker (2010) *Game-Changing China: Lessons from China about Disruptive Low-Carbon Innovation*, London: NESTA.

Urban, F., S. Geall, and Y. Wang (2016) 'Solar PV and solar water heaters in China: Different pathways to low carbon energy', *Renewable and Sustainable Energy Reviews*, 64: 531–542.

Van Noorden, R. (2016) 'China by the Numbers', *Nature*, 534: 7608. www.nature.com/news/china-by-the-numbers-1.20122.

World Bank (2015) 'GDP growth (annual percent)', http://data.worldbank.org/indicator/NY.GDP.MKTP.KD.ZG/.

Xi, J.P. (2015) Speech before the Central Party Mass Organizations Work Conference, 6–7 July, http://news.ifeng.com/a/20151103/46096419_0.shtml.

Xi, J.P. (2016) *Zai sheng bu ji zhuyao lingdao ganbu xuexi guanche dang de shiba jie wu zhongquanhui jingshen zhuanti yantao banshang de jianghua* (Speech at leading provincial cadres studies meeting to promote the spirit of the 5th central committee meeting of the 18th party congress). Beijing: Xinhua.

Xie, Y. and X. Zhou (2014) 'Income inequality in today's China', *Proceedings of the National Academy of Sciences*, 111(19): 6928–6933.

Xing, Y.Q. (2011) *China's High-Tech Exports: Myth and Reality*, GRIPS Discussion Paper 11–05, Tokyo: GRIPS.

Xu, H.G. and T. Ye (2017) 'Changes in Chinese higher education in the era of globalization', in D. Tyfield, R. Lave, S. Randalls and C. Thorpe (eds), *The Routledge Handbook of the Political Economy of Science*, London: Routledge: 156–168.

Yu, E. (2016) 'Huawei tops global list of patent applications', *zdnet.com* www.zdnet.com/article/huawei-tops-global-list-of-patent-applications/.

Zeng, M. and P. Williamson, P. (2007) *Dragons at Your Door: How Chinese Cost Innovation is Disrupting Global Competition*, Cambridge, MA: Harvard Business School Press.

Zhang, S. and Y. He (2013) 'Analysis on the development and policy of solar PV power in China', *Renewable and Sustainble Energy Reviews* 21: 393–401.

Zhao, Y. (2010) 'China's pursuits of indigenous innovations in information technology developments: hopes, follies and uncertainties', *Chinese Journal of Communication* 3(3): 266–289.

6 The unexpected innovation hegemon

Enter the Chinese disrupters

At its most obvious, the success of the Chinese innovation system (see Chapter 5) consists of the small but growing list of Chinese companies that are not only successful technology companies, often clustered in ICT, but giants of global stature and increasingly household brand names, even in the global North. As Tse (2016) describes, for instance, in the likes of Huawei, Alibaba, Tencent, Baidu, Lenovo, Haier and Midea, China is slowly but surely developing a group of companies that are both globally competitive in hi-tech sectors, such as smart-phones and internet platforms, clustered in specific regions (especially Shenzhen and the PRD (Butollo and Lüthje 2016)), and led by management teams that are highly – even fundamentally – innovative and ambitious. And this group is growing, with recent start-ups already storming the market and in sectors expanding beyond ICTs, such as Xiaomi (a maker of smartphones) or Yihaodian (an online supermarket), into cleantech (e.g. Broad air conditioners) or personal finance (Noah) (Tse 2016; see also Rein 2015; Yip and McKern 2016).

It is hard not to be impressed by many of these companies. For instance, Baidu, Alibaba and Tencent (together often abbreviated as the "BATs") together command considerable dominance of the Chinese internet, with the largest online community of any one country at over 700 million users in 2014, easily dwarfing the online population of the US and EU combined. Their success is also much more than the inevitable domination of home-based internet companies protected by the 'Great Firewall of China' making knock-off copies of original ideas emerging from Silicon Valley; even as for each of Google, Amazon, eBay, Facebook, Twitter, WhatsApp etc.… there are indeed Chinese equivalents owned by these companies. Rather these are companies that are increasingly global leaders, and in both size and innovation (*The Economist* 2016a).

Alibaba's New York stock exchange listing was the biggest ever IPO, at $25 billion, in turn raising its market capitalization to fourth in the world amongst technology companies. It is also highly profitable, deploying a business model that is not dependent on advertizing revenue, as are those of Facebook and Google, with profits over twice those of Facebook in 2013 (Tse 2016: 10); Tencent's revenues and profits were even bigger. Together, Alibaba and Tencent's WeChat app have also transformed e-commerce and cashless purchasing, bringing it into mainstream life in China in ways that continue to elude their Western competitors. Alibaba was host to over $350 billion of e-commerce in 2014, or more than eBay and Amazon combined, while WeChat is now supporting cashless payments and offering an increasingly sophisticated and user-friendly range of features and services between a growing network of over 500 million users (Tse 2016: 10).

If new heights of internet, 'Tech'-based innovation are what is needed to convince the China naysayers that innovation in China is indeed a serious global competitor, then, there can no longer be any doubt that this is the case. But more important than these headlines (important and consequential though such headline narratives are) is the much bigger ecosystem of Chinese innovation of which these companies are simply the tip of the iceberg. What all of this much bigger mass of Chinese companies has in common is its business model, whence they get the name 'disrupters', following in the foot-steps of business strategy insights initially summarized by Clayton Christensen (1997) as 'disruptive innovation'. This language has subsequently been adopted, and twisted, by the new generation of Tech innovation, epitomized by Uber, AirBnB and Spotify, in which existing assets and modes of service delivery are 'disrupted' by a new start-up with a few algorithms and an internet-based platform for massive data collection and processing (Straw and Baxter 2014). Yet its original meaning concerns the provision of services and goods in cheaper, easier-to-use and novel combinations that break away from the existing 'common-sense' improvement of technologies along existing product development trajectories and thereby transform – or 'disrupt' – the very meaning of those goods and services (Christensen 1997). Archetypically, this would include the digital camera, which initially offered photos of much *poorer* quality than film, but entirely transformed the very practice of taking photos and their sharing and consumption.

In similar vein, a vast number of Chinese companies have emerged and established themselves as based upon essentially defensive approaches to innovation that instantiate many of the key elements of disruptive innovation. This innovation is thus demonstrably not – even deliberately and necessarily not – new-to-the-world innovation at (what is currently defined as) the hi-tech cutting edge. Instead, it is 'good enough' innovation, targeting the mid-range market through forms of unrelenting but incremental process innovation in manufacturing and service delivery that provides low-cost, personally customized and/or easier-to-use and/or – repair offerings of ever-improving quality (Zeng and Williamson 2007). This kind of innovation thus focuses efforts into developing engineering capabilities for quick translation to mass manufacturing, where their competitive advantage lies in servicing customers with 'tempo, volume and cost' (Nahm and Steinfeld 2014); and continually improving quality.

This is generally not innovation that it is in the interests or, possibly, the capacity of established hi-tech companies to provide. For it both sacrifices the unique advantages of the super-rents of proprietary innovation from sales to elite and cutting-edge consumers, and demands cost-cutting and operational flexibility and *lower* quality standards that would harm their established position (Brandt and Thun 2016; Christensen 1997). But for up-and-coming companies which are fast-followers and nimble learners keen to amass any business going (see e.g. Tse 2016 on Huawei's strategy of growth), there are multiple such opportunities for growth and corporate success. This has included businesses across a wide range of sectors, including, importantly cleantech and renewable energy. For instance, China's wind energy sector is the world's largest experiment in renewable energy and the main challenger to China's entrenched coal-based energy production, which is the single greatest GHG emitter in the world (Nahm and Steinfeld 2014; Kirkegaard 2016; ClimateGroup 2014, 2015; see also

Tyfield *et al.* 2010). Indeed, their very success reminds us that the incumbent dominant definition of innovation as hi-tech and new-to-the-world is indeed partial at best, technologically fetishist and simply misleading – conflating innovation with invention – at worst.

Moreover, it is clear that there are systemic reasons why Chinese innovation should prove particularly characterized by and adept at this kind of innovation; reasons precisely captured in the *conjunction* of the optimist and pessimist cases and the broader systemic problematic of the unstoppable force and the immoveable object to which they (respectively and then together) point us. It is the unique combination of forces at macro- and meso-level that produces both the context of persistently 'uninstitutionalized' (Lieberthal 2004) 'structured uncertainty' (Breznitz and Murphree 2011) underlying much innovative weakness vis-à-vis prevailing conventional metrics, together with openings *and* drivers to *other* kinds of innovation as the basis of its strengths. Indeed, the gravitation towards this model of innovation as that which succeeds in China is arguably just the incremental accumulation over time of successful experiments amongst Chinese entrepreneurs in 'rational reaction … to the incentives and restraints they face, both internally and externally' (Breznitz and Murphree 2011: 12).

This would include all of the conditions discussed above. For instance, with a massive domestic market largely in search of low-cost and 'good-enough' quality – if to start with, but increasingly no longer, 'knock-off' or *shanzhai* – goods, Chinese companies could responsively service this market and thereby grow to a size not possible in almost any other country (Brandt and Thun 2010, 2016). Moreover, along the way, with relatively good profits not dependent on – foundationally independent of – government subsidy or favour, many of these companies could begin to invest in their own innovation departments in ways not possible amongst the cut-throat competitive export sector (Zeng and Williamson 2007); thereby establishing relatively self-sustaining dynamics of deepening innovation competitiveness and catch-up.

Similarly, the complexity and illegibility of the political economy and its fragmentation into local developmental states has both compelled adoption of a strategy that is experimentalist and opportunistic (as characterizes contemporary Chinese government more broadly (Heilmann 2008)) and provided the gaps and interstices into which successful disruptive businesses have been able to insert themselves. These businesses have thus even been able to adapt to and grow off what otherwise seems to be the insuperable obstacle of the division into two economies, state-favoured and not. This includes effective and mutually beneficial connections with both state-owned enterprises (e.g. through patronage of a state-owned 'mother-in-law' (Segal 2003; Brandt and Thun 2010: 1570)) and/or foreign trans-national corporations, patching the very problems of their overseas partner that the 'two economies' predicament constructs for these corporations. In this way, the disrupter companies have been able to exploit the systematic division, turning it even from a strategic disadvantage (for both them and the Chinese economy more generally) into a unique advantage. In particular, these companies can thereby draw on funding, not least through overseas ethnic Chinese networks, that is both unavailable at home and exposes them to the spur of hard budget constraints (Tse 2016; Fuller 2016), while also taking advantage of access to the domestic market and even the state-favoured sector that is largely inaccessible to completely foreign enterprise.

Furthermore, inverting the straightforward weakness into a strength, even the fragmented authoritarianism itself can then become a unique advantage available to these firms as specifically Chinese. The unique scale and power of the national state in fora of international regulatory architecture can secure privileges and leeway not available to firms from other, much smaller developing countries. Conversely, keen to support and claim credit for local success stories, local Governments will support growing private enterprise in their territories even if this means unusual concessions to their foreign partners (Oh 2013).

What is crucial, however, is that none of these alchemical transformations from disadvantage to advantage, at firm or national/regional innovation system level, is possible without a foundational orientation to a restless search for incremental advantage and innovation. In short, these companies are not and can never be at rest, secure in their establishment of a competitive advantage; even to the extent of the relatively unassailable, if undeniably dynamic, heights of transnational corporations in hi-tech sectors through the 1990s and 2000s; and even in the case of the internet giants (Tse 2016) given the hubbub of entrepreneurial activity in this space today. This is thus an *essentially dynamic* innovation model that instantiates and feeds, in positive feedback loops, both the existing 'strengths' *and* 'weaknesses', as conventionally defined, of the Chinese innovation system and their increasingly intense interaction that is the dynamism of the system's evolution itself.

But recall that this is irreducibly *innovation-as-politics*. In CP/KS terms, we are here looking at a seminal instance of the quintessentially liberal logic of innovation and counter-innovation (Chapter 3), with innovation upgrade simply the ultimate defensive competitive strategy and innovation per se the key competitive resource (Butollo 2013: 145). Here, we find (to date) a context for Chinese business of largely *foreign* global-leading hi-tech innovation and new-to-the-world product/service development alongside the permanent pressure in China to be the cheapest, fastest and most convenient follower, learner, imitator and improver of such innovation. With innovation specifically as *politics* – as it is in China as nowhere else in terms of techno-nationalist policy – this incumbent field of global competition then also generates continually strengthening efforts of (central) Party-state-led innovation policy and national upgrade directed to the state-favoured sector, and the deepening of that regime more generally. Yet, in turn, this must and does elicit an equal-and-opposite effort of counter-innovation – in strategy and actual product or service innovation – amongst the much larger and more dynamic set of private and hybrid foreign/domestic disrupters.

A perfect example of this process, for instance, is how Alibaba innovated a cashless payment system specifically tailored to a massive but low-trust society (itself a legacy specifically of the Cultural Revolution, but more generally of the systematic discouragement of trust in strangers from the Party-state-induced absence of the rule of law) with strong and trusted state banks (to keep out global neoliberal finance). Alibaba – not Visa or Silicon Valley or the state-favoured sector itself – thereby created a platform for what is now the world's largest e-commerce market. Here, in other words, we see how this group of companies has not only learnt successfully to negotiate both the neoliberal global competition stacked against it and an often misfiring Party-state innovation policy driven through scarce *guanxi*, but has even been incubated by precisely this *prima facie* hostile environment.

Described thus, it is clear we are talking about innovation in the crucible of the immoveable object and the unstoppable force, where explicitly noting this metalogic further illuminates the process and trajectory of Chinese innovation upgrade. First, because it then becomes clear that just as China's disrupters are fed by this tension, so too *it* is fed in turn by their continuing growth and success. The immoveable object of the CCP Party-state and its systemically hamstrung innovation policy (as *per* the pessimists' case) is both kept afloat and further entrenched by the entirely inadvertent success of Chinese disrupters that these policies nonetheless produce, albeit largely despite, in the interstices of and under negative pressure from, their actual direct effects and express intentions. The continued growth of not just the Chinese economy, sponsored by this movement, but also concentrations of economic *power* that can be repeatedly co-opted by a successfully adapting Party-state (Shambaugh 2009; cf. Shambaugh 2015; Pei 2006) – as for instance in the doctrine of the "3 Represents", incorporating capitalists and entrepreneurs into the CCP as one of its three pillars – also feeds the IO *and* its deepening immoveability in the entrenched power of the cadre-capitalist elite (e.g. Dickson 2008; Tsai 2007; So 2003).

Similarly, the unstoppable force of global neoliberalism and its innovation model has also been supported and kept alive by these disrupters for some time: generating low-cost innovations that mitigate stagnating wages in the global North, while, through fast-following innovation, also accelerating the expansion of the Chinese market and *its* growing appetite for foreign, cutting-edge technology, thereby preparing the 'next billion customers' (e.g. McGregor 2007) of global TNC wet dreams. This, in turn, feeds the FDI treadmill with just such promises of a fast-developing consumer and business market in China. By inserting themselves successfully into the fragmented, disassembled value chains of global manufacturing too, they have staved off a total collapse in profitability of this sector, thereby extending the shelf-life of this offshoring model. And the disrupters themselves, in turn, have benefitted from the cheap labour, which the combination of global neoliberalism and CCP provides in the super-exploitation of Chinese workers, in terms of their specifically low-cost strategies.

Finally, then, the dynamic itself of their intensifying confrontation is also sustained: by feeding both IO and UF in the current incarnations and their productive-cum-destructive symbiosis in those existing forms, without demanding fundamental qualitative change and/or accommodation of the other.

But, second, returning to this meta-dynamic of IO vs UF also illuminates the peculiar and distinctive process and trajectory of this disruptive innovation and China's innovation upgrade more generally, as veering from one 'industry quake' to another (Kirkegaard 2016); a dynamic especially noticeable in sectors of emerging technologies, such as renewable energy (wind, solar PV) (Kirkegaard 2016; Nahm and Steinfeld 2014; Urban *et al.* 2016) or LEDs (Butollo 2013). This 'non-linear' (Kirkegaard 2016) process is characterized by surges of over- and mis-allocated investment (private and state (ten Brink and Butollo 2016)), followed by brutal market shakedowns rocking these sectors to their very foundations and, finally, as the dust settles, the emergence of a sector that has nonetheless been produced … until the next wave of another boom-bust cycle. These booms are driven by overenthusiastic and misaligned government incentives and policy-drives, 'burning money' (ten Brink and Butollo 2016), incentivizing quantitative growth (and perhaps of the wrong things) at all costs, as in the drive to expand wind power installations regardless of connectivity to the grid or quality and durability of the resulting wind farms (Kirkegaard 2016). As such, again we encounter here the fundamentally turbulent but productive dynamic of innovation(-as-politics) and counter-innovation.

But it is precisely the conjunction of the unstoppable force as permanent foil and aspiration for Chinese business, setting the global innovation policy agenda, and its subsequent implementation in the context of the immoveable object and the deliberate, contrived but partial *exclusion* of global capitalist pressures that produces the contortions generative of this oblique, stuttering rhythm of innovation advance. For it is only this *combination* in particular that produced both the imperatives *and* perverse incentives regarding innovation upgrade unique to China. Recursively, then, we see further how it is in this context that China specifically has had to, and has succeeded in (in itself no surprise, given its size, the obvious ingenuity and pragmatism of Chinese people and the socio-economic pressures involved), cultivating precisely the kind of entrepreneur that is responsive and resilient to *this* essentially unpredictable, high stakes, complex, and fast-shifting business context.

Beyond disrupting just innovation

Acknowledging both this non-linear process and its fundamental basis specifically in the (co-production of the) meta-dynamic of the IO vs UF, however, is crucial. Absent this logic it is tempting to conclude that disruptive, middle-range innovation for the massive domestic market is a sustainable advantage for Chinese innovation over the medium-to-long-term (Breznitz and Murphree 2011); perhaps even to see it as systematically complementary with existing neoliberal global TNC innovation – as it surely has been – but in a stable equilibrium rather than a dynamic and accelerating power/knowledge system productivity. But once we set the development of this kind of innovation in the broader system context of the IO vs UF, it emerges clearly that there are immanent limits to this trajectory. These are both from without, in terms of ever-increasing tensions between IO and UF and their manifestation in deepening systemic problems of China's and the global political economy's *régulation* regime respectively; and from within, regarding the ceilings of innovation upgrade against which the very success of such disrupters will necessarily press. In both cases, moreover, there is ample and growing evidence that this 'tipping point' is already being reached.

Immanently, the continuation of this dynamic presupposes that Chinese firms will remain, and will be happy to remain, limited to the second-tier role of global fast-followers. For the system-functional role that Chinese disruptive innovation plays, fixing – i.e. both feeding and alleviating – the intensity of the IO vs UF tension, depends upon the continual upgrade of their innovation capacities ... but only up to a point; namely up to the frontier of global innovation cutting-edge *and no further*. To push beyond this boundary is not only profoundly to unsettle the unspeakable truth, for both CCP Party-state and global neoliberalism, regarding these firms as unacknowledged pillars of these respective complex power/knowledge systems: i.e. as private and/or hybrid firms at the commanding heights of the Chinese economy, *not* state-owned or even necessarily state-favoured firms; and as *Chinese* firms with opaque and complex, but indubitable, connections to the Communist party-state, not transnational corporations (just) subject to the disciplines of neoliberal global finance capital. But it also demands completely different relations of support and feedback with both of these, such as in terms of a genuinely entrepreneurial state that is anathema to both.

Precisely in the emergence of such undoubted disrupter success stories as can be detailed by a business commentator like Tse (2016) or Rein (2015), though, it is clear that Chinese companies are now pressing at these limits and determined, as a sheer matter of their continued growth and increasing competitiveness (in what, recall, is not just a 'run …' but '*accelerate* or die' business climate), to breach these boundaries into unquestioned global innovation leadership. Concretely, for instance, as the mobile internet is increasingly challenging the computer-based internet, it is blurring sector definitions. This is shaking up even the only-recently-established division of the spoils (of innovation) amongst the BATs Chinese internet giants, now increasingly in direct competition with each other, as well as with fast-rising start-ups (Tse 2016: 83). Nor is this just limited to a few companies; it is rather a broader, more structural shift, as marked by the fall in recent years of exports of low-value-added goods, down now to under one-third of total exports; itself an extraordinary measure of the success of Chinese disrupter firms.

Moreover, this includes multiple pivotal emerging industries of twenty-first century technology, including renewable energy, where a significant shift is also clearly evident towards a search for deeper integration into global innovation networks and 'quality' (Kirkegaard 2016; Nahm 2016; ten Brink and Butollo 2016). For instance, there is a new surge of outward FDI from many of these disrupter firms, purchasing and partnering with R&D expertise (Tse 2016) in Silicon Valley, New England, Germany, Japan etc…. Similarly, there is organic growth of capacity and demand for deepening participation in the global innovation networks (GINs) that are necessary to do 'cutting edge' innovation. This thus involves drawing on and building access to the global markets for high-skilled knowledge workers or 'talent', despite the continuing unattractiveness of life in megacity China especially for that in-demand, globe-trotting group.

Limits to this model are also being reached from external changes and pressures, though these may be traced in part (given the systemic feedback loops) also to spill-over from these trends within the hi-tech innovation sectors. Here we mean deepening systemic imbalances, both within the Chinese political economy and, more broadly, as China – and hence China's economic and innovation engine of these disruptive innovators – has inserted itself ever-more centrally into global capitalism, in that global system as a whole. Compounding the immanent demand for innovation upgrade is the equally strident demand from the Chinese Government itself, treating innovation as a supposed panacea for the host of profound systemic political economic problems and imbalances now facing China.

Together, these problems are often framed and/or summarized in terms of the 'middle income trap' (DRC/World Bank 2013): the danger, evident in economic history, that fast-growing developing economies reach a certain middling level of national income, socio-economic development and innovation capacity but then become trapped, never advancing beyond this level into the serried ranks of the developed economies. Given that those that have bucked this trend are actually so few – limited to a few economies in East Asia and Europe in the Cold War period – the trap appears to be the more usual outcome.[1] Yet this coincides with a deepening acknowledgement by the Chinese government of the sheer unsustainability – economic, environmental and social – of its current economic model and *régulation* regime (Boyer 2016), presenting it with a clearly acknowledged and loudly proclaimed crossroads (McNally *et al.* 2013; Fenby 2014; Butollo and Lüthje 2016), towards a 'new normal' and 'ecological civilization' as opposed to sheer GDP growth.

This is manifest in multiple concrete problems at macro-level now demanding increasingly urgent redress, and also in new *ways* since existing policy and institutional levers, as in the post-2008 stimulus, are merely exacerbating the system dysfunction.[2] For instance, it is this massive government stimulus and its very success in keeping investment and a construction and real-estate boom afloat that has catalyzed not just the re-emergence of China's problem of corporate and local government debt, and hence non-performing loans, but its growth to systemically threatening levels above even those of the US at the peak of the sub-prime credit boom (*The Economist* 2016b); as well as driving the persistent imbalance from consumption to investment in heavy industry and construction, conditioned by the central policies of wage (including social security) and financial repression (Pettis 2009) on the one hand, and free-flowing credit to state-favoured industries in precisely those sectors of overcapacity on the other, and with financial security-seeking private investment in real estate bridging the two in the form of urban property bubbles.

This also then feeds the *global* systemic imbalances that keep alive the neoliberal accumulation regime even as it disintegrates from within. In terms even of the ultimate guarantor of this system of Wall Street supremacy, dollar seigniorage (Hung 2016; Pettis 2013), the Chinese government's feeding of the economy's habit of dependence on cheap exports keeps the yuan cheap against the dollar (McNally 2012: 758; Hung 2016), supporting a systemic one-way bet in favour of holding dollars; while continuing Chinese appetite for US debt is essential to the sustainability of American programmes of quantitative easing that are both staving off stagnation and generating further system imbalances.

What is crucial is that 'innovation' is treated here as the solution to all these problems by the Chinese government, in the redirection of the economy towards an 'innovation-oriented economy'. It is undoubtedly recognized in such policy redirection that Chinese innovation capacity needs significant and novel forms of support to take the next step and realize these policy goals. What is systematically overlooked – and necessarily so, given the complex power/knowledge system from which such pronouncements are emitted and towards the preservation of which they are primarily aimed – however, is how the implementation of these policies will be as fundamentally conditioned by and co-produced with the system dynamics of the Chinese CP/KS, *and its encounter with global capitalism* – i.e. by the dynamic of immoveable object and unstoppable force – as has any prior policy drive to date. In other words, any Governmental attempt to address the 'new upgrading problem' (ten Brink and Butollo 2016; see also Rein 2015; Yip and McKern 2016) will necessarily be performed by, mediated through and co-productive of precisely the *same* context of structured uncertainty and non-linear, essentially-contested boom-and-bust. Whatever such innovation does produce, it will *not produce what the CCP wants* in terms of world-leading innovation 'solving' the middle-income trap while entrenching the Party-state regime.

And this holds not just regarding the characterization of the resulting innovation capacity per se. In other words, intensification – and Government-sponsored, no less – of the focus on China's innovation upgrade and renewed system-driven determination to reinvent again China's disrupters will tend to lead not to the hoped-for mitigation of the underlying socio-political-economic tensions but, from the very logic of systemic co-production, *their* intensification as well (see Figure 6.1). Moreover, this dynamic may now be at the point of climax and qualitative emergence. This would then be the historical convergence of all the system forces at play – unstoppable force and immoveable object, Chinese political economy, Chinese innovation capacity and upgrade, global innovation model and the dynamic evolution of both

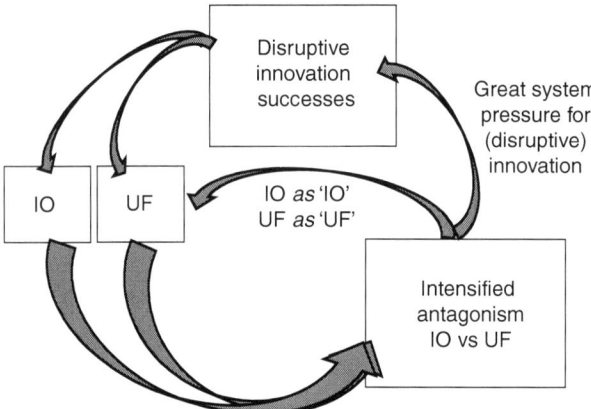

Figure 6.1 Cycles of disruptive innovation amidst the Immoveable Object vs Unstoppable Force.

global and Chinese CP/KS alongside the Four Challenges – all condensing on the issue of China's disruptive innovation.

What next? Impetus, capacity, agency, process and conditions

From this perspective, however, it is clear that there is a now constant stream of emergent evidence of just such an imminent climax in the encounter of immoveable object and unstoppable force, China and capitalism. And evidence not just of the impetus towards crisis/emergence but also, and equally crucially, of the capacity, agency, process and conditions for it to take place; and where all of these are specific to China.

First, regarding the *impetus*, starting from without, we have considered the peaking of the Four Challenges and the crisis of global neoliberalism, and their systemic pull – power/knowledge systems abhorring a vacuum – for system transition and a new model of innovation. In Chapter 4 we opened this discussion of contemporary China by highlighting the specific intensity of the Four Challenges there, and their particular manifestation as *political* problems that are existentially challenging to the very constitution of that polity, rather than contested issues affecting the mere fortunes of specific parties and personalities in Government. These problems, however, also manifest directly in the intensity of the UF/IO encounter, and not just because of the growing centrality of China in global capitalism and the unique importance and difficulties of addressing the Four Challenges for a new restablilized capitalist regime in China.

Immanent in the move towards twenty-first century innovation at the global cutting-edge of harnessed complexity, there are also particular challenges to the CCP Party-state regime's oversight of innovation policy. As discussed in Chapter 1, innovation that is within and co-productive of contemporary cosmopolitized knowledge-capitalist risk-societies presupposes governments that can effectively manage the twin challenges of complexity and its government: on the one hand, of being powerful enough to harness and govern through *consumption* of that complexity; and, on the other, being light-touch enough to minimize compounding system complexity with its 'unnecessary' proliferation by the action and structure of the state itself.

Such 'Goldilocks' (i.e. 'just right') balancing between too little and too much complexity, however, is something that the party-state regime systematically drives too far in both directions: as in heavy-handed top-down misallocation and systemic institutional bias and disincentive on the one hand, and illegible governmental complexity with high stakes on the other. Even from a CP/KS perspective that can explain the resilience to date of the 'immoveable object' – a persistence that we have seen is fundamentally baffling to a less pragmatic, tacitly Euro-American-centric analysis assuming established categories of social and political economic theory (Chapter 4) – it is hard to see how complex innovation capitalism can be achieved in China without significant transformation of its power/knowledge relations and practices of government; which is not, of course, to say without the overthrow of the CCP and the institution instead of familiar Western models of liberal democracy (see Chapter 10).

Crucially, though, all these pressures to profound change and climax are experienced and treated by the Party-state itself, and most other agencies besides, as problems of innovation governance, far removed from direct political challenge to the incumbent Chinese CP/KS, as are issues of human rights or worker or explicitly anti-regime agitation. Indeed, to the contrary, in this case the Government is wholeheartedly *behind*, not *against*, propagation of this system momentum.

Second, regarding *capacity* for system change, this, of course, concerns the significant innovation capacity and competitiveness that China has now unquestionably developed. There is thus deepening system and agential capacity pushing to new frontiers of innovation both immanently and with full Government support, and with leading firms concentrated in many of precisely the areas and sectors needed for twenty-first century industrial leadership. This includes not only 'Tech', but also green energy and robotics, and – no less importantly given blurring industry boundaries – with existing giants such as Alibaba or Baidu themselves expressly committed to expanding into these emerging sectors. From the perspective of a reshaped global innovation model, the size and global strength of many of these firms also stamps a definitive Chinese role in their systemic emergence, not least because low-carbon transition remains just as embryonic or stalled elsewhere in the world given the hostile climate of zombie, exterminist neoliberalism.

In terms of the prospective emergence and convergence of Chinese innovation strengths in the medium-term, there are clearly specific opportunities for Chinese firms in the key fields of the emerging and much-hyped (and perhaps 'green') Internet of Things (IoT) and new ('third' or even 'fourth') 'industrial revolution' (e.g. Straw and Baxter 2014; Rifkin 2011; Schwab 2016); an industrial system transition that is clearly and expressly addressed to the Four Challenges, albeit at present usually in piecemeal fashion, not grappling with all four together. Industrial automation seems a particular opening here for China, not least because Chinese leadership and exports would face reduced risk of popular rejection overseas given the remoteness of this sector from the consumer front-line.

On the one hand, Chinese Government and firms display perhaps the world's most ambitious plans for industrial automation through the impressive national 'China 2025' project, aiming to take China from a 'made in China' world to 'made by China' with the world's most advanced manufacturing sector. Current trends already show China's industrial robot market has grown at 30 per cent p.a. to become the world's largest, and one-fifth of the global total (Tse 2016: 109), with a projected further rise from $193 billion in 2015 to $361 billion in 2020, and even possibly $736 billion by 2030 (*The Economist* 2016c). And this is being supported, in familiar local state corporatist fashion, by provincial plans. For instance, Guangdong is competing with Zhejiang, where $82 billion has been pledged over five years for advanced manufacturing and an industrial IoT (Tse 2016: 109). Nor is it just Chinese companies involved, with GE opening a 'digital foundry' in Shanghai, and other foreign giants such as Siemens, HP, Honeywell and Cisco all investing heavily in China (*The Economist* 2016c). On the other hand, Chinese business and individual customers also have an exceptional appetite and openness to automation and AI, displaying considerably less concern about, or interest in, an apocalyptic transhuman transcendence than Westerners (in what seems a particularly American and Russian framing and obsession).

A sober analysis of the current state of industrial automation in China certainly reveals a picture still far removed from the sci-fi imaginaries of this policy drive's most ardent proponents. Moreover, and of supreme importance, for all the policy sophistication and learning evident in the *China 2025* plan – formulated through systematic consultation with leading industry parties, both Chinese and foreign – the equally systematic absence from this policy is any consideration of the position of labour in all this technological upgrading and deliberate replacement of 'man' with 'machine' (*ren huan jiqi*) (Butollo and Lüthje 2016). Not only does this thus leave the crucial consideration of 'social upgrading' a peripheral concern, but it also exposes an Achilles Heel to the policy: namely the lack of inclusion, both as a matter of cultivated highly-skilled labour essential to the incremental process of industrial upgrade and politically regarding worker unrest, of what is the world's largest industrial working class in a country characterized by the governmental problem of how to manage a surplus of labour, and a working class with revolutionary history besides. While a green industrial IoT would unquestionably provide multiple direct responses to the Four Challenges, therefore, and to China's current political economic problems, this is not to propose the seamless slide into a glorious new harmonious socio-economic equilibrium. To the contrary, much more likely is precisely a replay of the same non-linear, essentially contested, surging and crashing dynamic of China's innovation upgrade to date; only perhaps now with higher and more explicitly political stakes.

This leads us to the *agency* of this historical climax, which is, of course, precisely the growing group of Chinese disruptive entrepreneurs and the broader group of high-skilled knowledge workers identifying the bright opportunities of their future with these firms and the intoxicating excitement of China's growing innovation capacity. This group of highly cosmopolitized – perhaps with foreign degrees, or work experience in foreign companies and/or overseas, or with essential business partnerships and financial connections abroad – innovative and (self-)entrepreneurial Chinese is both the engine and product of China's disruptive innovation success and the nation's continued prosperity, and the personal incarnations of the bridges between China and global capitalism, UF and IO, that have been so instrumental in building that success.

This is thus an increasingly large (in absolute, if not necessarily relative, numbers), empowered and publicly lionized group and culture or emerging social identity within China (see also next chapter) that is already in place. That success to date has also been built precisely on their deftness in managing the scale, pace of change and complexity of contemporary China; i.e. precisely the skills of management and strategy needed for successful business enterprise in the age of the Four Challenges (cf. Tse 2016: 108). As Tse notes (2016: 97), this group instantiates the very acme of contemporary business strategy of 'exploiting temporary competitive advantage' (McGrath 2013) with 'an across the board need to cope with multiple challenges simultaneously.' Indeed, we could call this the very Confucian ethic of twenty-first century knowledge capitalism (cf. Weber 2002), of responsive pragmatism to a system accepted as unintelligibly complex but which will thrive, and you with it, only through constant, disciplined and judicious strategic intervention (Tse 2016: 61). In these circumstances, therefore, it is easy to conceive also how this group in particular is well-placed – not just in China, but globally – to lead, and frame itself as leaders in, the incremental but unstoppable construction of a system transition to green IoT capitalism.

Mediated and co-produced with the meta-dynamic of intensifying UF vs IO, the process of emergence of this group to date – perhaps now on the cusp of breaking out of their shell into the light of 'universal' public adulation – reveals clearly the core challenge to CCP Party-state and US-centric neoliberalism alike, but specifically the former. With every cycle of innovation upgrade (see Figure 6.1 above), what is produced is the deepening *asymmetric* inter-dependence between the CCP regime and this driver of national prosperity, where such prosperity is in turn the CCP's self-declared primary goal and the de facto pillar of its popular legitimacy.

To be sure, this *is* an *inter*-dependence, with this group essentially politically conservative, dependent on social peace and fearful of disturbance that could jeopardize their gains (Chen and Lu 2011); and, indeed, many may actually be CCP members.[3] They are also specifically dependent on *growing*, not weakening, state support as necessary conditions for the kinds of systemic, new industry innovation to which they are pressing from within and to which the objective Challenges are drawing them from without. But the profound intensity and novelty of the present impetus impresses upon us – and perhaps increasingly also them themselves – the insistent if surprising question of '*which* social order are they committed to upholding?' Indeed, the state support they crave may be precisely what the *Party*-state cannot deliver and directly frustrates, in terms of the 'Goldilocks' complexity mitigation discussed above. For instance, consider their specifically trenchant demands for equality before the law, secure property rights, access to dependable public information and predictable law-making and a (more) level-playing field of arm's-length relations to the Government (cf. Tse 2016; Breznitz and Murphree 2011: 11).

Fourth, in terms of *process* too, precisely this political tension and the 'dislogic' of non-linear progress (Kirkegaard 2016) characteristic of Chinese innovation upgrade also become strengths of sorts regarding low-carbon transition. In the West it is interpreted as how to guide a necessarily politically disruptive low-carbon system transition in the context of profound political economic turbulence. This is generating only slow progress, at best. Meanwhile, and ironically, the determination of the CCP to chart a similar course *without* political disruption conditions a process that is in fact highly system productive (i.e. in terms of power/knowledge momentum) albeit experimental, wasteful, uncontrollable and destructive – and is understood, accepted and worked with as such by the disruptive innovators. In China specifically, therefore, there is a process of socio-technical system transition that is singularly productive precisely *both due to* its authoritarian top-down focus on (and popular, bottom-up distaste for) social disorder, which forces and/or affords acceptance of what is otherwise wrenching and socially intolerable socio-economic change; and *in spite of it*, in terms of its unique structured uncertainty and complex power/knowledge system-productiveness.

Finally, regarding the necessary *conditions*, there is the systemic product of this whole process over the past few decades of innovation upgrade; namely the deepening systemic centrality, both within global capitalism and the Chinese polity and thence of each within the other, of this growing constituency of firms and personalities, and specifically in the new emerging sectors at the apex of an emerging structure of green, 2.0, twenty-first century capitalism.

Conclusion

Only by considering the full system complexity of the 'supply-side' of Chinese innovation, as systems of power/knowledge relations, amidst the deepening historical encounter of China and global capitalism can we come to something approximating a systemic understanding. And where this is an understanding not just of where we are now, in terms of how strong or weak Chinese innovation capacity 'really is'; but also where this is tending to lead – i.e. albeit with no guarantees regarding actual futures (see Chapter 11). In this way, we have seen how this whole process of innovation-as-politics has led to date and is leading from here to a situation in which nothing is just as it seems. Most importantly, strengths are not just strengths, nor weaknesses weaknesses, but in a dizzying dis-logic founded on and productive of an essential and globally-unrivalled system dynamism, strengths are weaknesses are strengths are weaknesses (or yang begets yin begets yang ... (Brincat and Ling 2014)) *ad infinitum* (see Table 6.1 and Figure 6.1 above).

Most importantly though, insofar as this essentially complex systemic process may be summarized, it is transforming weaknesses of the Chinese innovation system into

Table 6.1 Quadrant of Chinese innovation – supply-side

	Direct effects (at agent level)	*Indirect effects (at system level)*
Intended (immoveable object)	Increasing globally significant state investment and support for innovation at unrivalled scale and pace	Constraints and opportunities feed private/hybrid-overseas disruptive innovators creating resilient, highly dynamic and competitive firms of increasing systemic importance and innovation capacity
Responded (unstoppable force)	Misallocation and hamstrung central planning, plus unwarranted focus on hi-tech supply/push, generating deepening political economic imbalances, in China and globally	Increasing capacity bridging '2 economies' of domestic and global demand, and responding to immanent demand to 'move up the value chain', not least into new emerging industry sectors; together with deepening systemic dysfunction, pushing beyond the incumbent CP/KSs of both China and global capitalism

potential strategic strengths that are of pivotal significance for the renewal and resettlement of global capitalism. On the one hand, the absolute weaknesses of the Chinese innovation system *really are* weaknesses, both agential and structural, and perhaps irreparably so when set against the impossible goal of beating the US at its own neoliberal innovation game. But set alongside the implacable, existential determination of the CCP to do just that, this becomes an explosive, uniquely dynamic system context, of permanently shiftless and unsettled renewal, i.e. towards the innovation of innovation itself.

On the other hand, in this way, within the rapidly shifting systemic context that this process constructs, the emerging strengths that are *relative* weaknesses vis-à-vis global *neoliberal* innovation also morph and flip into relative strengths and strategic advantages. Chinese disruptive innovation emerges as uniquely pivotal, promising and adept with the new strategic challenges of complexity. It is thus Chinese innovation specifically, emergent from and feeding into the climaxing of the IO vs UF, that is driving innovation-as-politics and power/knowledge system emergence with the intensity, scale and depth needed for and capable of overcoming all the profound system lock-ins (Unruh 2000), both within China and in global financialized, high-carbon neoliberalism, towards a green knowledge-industrial capitalism and system transition.

Notes

1 Moreover, and especially challenging to China, is that there has been no case to date in which a country has broken this ceiling without essentially liberal democratic political reform (Lewin *et al.* 2016) – a point to which we return below.
2 This is in perfect parallel with the situation of normalized and increasingly ineffective 'extraordinary' monetary policy in the US, Eurozone, UK and Japan (Wolf 2016, Turner 2015), with negative side effects that are overwhelming any temporary respite and life-support they offer.
3 Though this does not mean they are not politically critical – and may be even more so (Huang and Chen 2015).

References

Boyer, R. (2016) 'How the specificity of Chinese capitalism explains its position in the world economy', *Working Paper*, available at: http://robertboyer.org/download/How%20the%20 specificity%20of%20Chinese%20capitalism%20explains%20its%20position.pdf.

Brandt, L. and E. Thun (2010) 'The fight for the middle: upgrading, competition, and industrial development in China', *World Development* 38(11): 1555–1574.

Brandt, L. and E. Thun (2016) 'Constructing a ladder for growth: Policy, markets, and industrial upgrading in China', *World Development* 80: 78–95.

Breznitz, D. and M. Murphree (2011) *Run of the Red Queen*, New Haven, CT: Yale University Press.

Brincat, S. and L.H.M. Ling (2014) 'Dialectics for IR: Hegel and the Dao', *Globalizations* 11(5): 661–687.

Butollo, F. (2013) 'Moving beyond cheap labour? Industrial and social upgrading in the garment and LED industries of the Pearl River Delta', *Journal of Current Chinese Affairs*, 42(4): 139–170.

Butollo, F. and B. Lüthje (2016) '"Made in China 2025": Intelligent manufacturing and work': 42–61.

Chen, J. and C. Lu (2011) 'Democratization and the middle class in China', *Political Research Quarterly* 64(3): 705–719.

Christensen, C. (1997) *The Innovator's Dilemma: When New Technologies Cause Great Firms to Fail*, Cambridge, MA: Harvard Business Review Press.

ClimateGroup (2014) 'Eco-civilization: China's blueprint for a new era', Beijing: Climate-Group.

ClimateGroup (2015) 'Inside China', Beijing: ClimateGroup.

Development Research Center of the State Council (DRC) and the World Bank (2013) *China 2030: Building a Modern, Harmonious, and Creative Society*, Washington DC: World Bank.

Dickson, B. (2008) *Wealth into Power*, Cambridge: Cambridge University Press.

The Economist (2016a) 'China's tech trailblazers', 6 August.

The Economist (2016b) 'China's debt binge: Putting off the inevitable', 9 March.

The Economist (2016c) 'The great convergence', 23 July.

Fenby, J. (2014) *Will China Dominate the 21st Century*, Hoboken, NJ: John Wiley & Sons.

Fuller, D.B. (2016) *Paper Tigers, Hidden Dragons: Firms and the Political Economy of China's Technological Development*, Oxford: Oxford University Press.

Heilmann, S. (2008) 'Policy experimentation in China's economic rise', *Studies in Comparative International Development* 43(1): 1–26.

Huang, D.Y. and C.M. Chen (2015) 'Revolving out of the Party-State: the *Xiahai* entrepreneurs and circumscribing government power in China', *Journal of Contemporary China* 25(97): 41–58.

Hung, H.F. (2016) *The China Boom*, New York: Columbia University Press.

Kirkegaard, J.K. (2016) 'Rapid upgrading through experiment (self-)disruptive impasse: The case of China's wind turbine industry', paper presented to the *International Symposium on Innovation-Driven Development*, Sun Yat Sen University, Guangzhou, 13–15 June.

Lewin, A.Y., M. Kenney and J.P. Murmann (2016) 'Introduction', in *China's Innovation Challenge*, A.Y. Lewin, M. Kenney and J.P. Murmann (eds) Cambridge: Cambridge University Press: 1–31.

Lieberthal, K. (2004) *Governing China* (2nd edition), New York: Norton.

McGrath, R.G. (2013) 'Transient advantage', *Harvard Business Review* 91(6): 62–70.

McGregor, J. (2007) *One billion customers: Lessons from the front lines of doing business in China*, London: Simon and Schuster.

McNally, C.A. (2012) 'Sino-Capitalism: China's re-emergence and the international political economy', *World Politics* 64(4): 741–76.

McNally, C.A., B. Lüthje and T. Ten Brink. (2013) 'Rebalancing China's emergent capitalism: State power, economic liberalization and social upgrading', *Journal of Current Chinese Affairs* 42(4): 3–15.

Nahm, J. (2016) 'Exploiting the implementation gap: Policy divergence and industrial upgrading in China's wind and solar energy sectors', paper presented to the *International Symposium on Innovation-Driven Development*, Sun Yat Sen University, Guangzhou, 13–15 June.

Nahm, J. and E.S. Steinfeld (2014) 'Scale-up nation: China's specialization in innovative manufacturing', *World Development* 54: 288–300.

Oh, S.Y. (2013) 'Fragmented liberalization in the Chinese automotive industry: The political logic behind Beijing Hyundai's success in the Chinese market', *The China Quarterly* 216: 920–945.

Pei, M.X. (2006) *China's Trapped Transition: The limits of developmental autocracy*, Cambridge: Cambridge University Press.

Pettis, M. (2009) *The Great Rebalancing*, Princeton, NJ: Princeton University Press.

Rein, S. (2015) *The End of Copycat China*, Hoboken, NJ: John Wiley & Sons, Inc.

Rifkin, J. (2011) *The Third Industrial Revolution: How Lateral Power is Transforming Energy, the Economy, and the World*, Basingstoke: Macmillan.

Schwab, K. (2016) *The Fourth Industrial Revolution*, Geneva: WEF.

Segal, A. (2003) *Digital Dragon*, Ithaca, NY: Cornell University Press.

Shambaugh, D. (2009) *China's Communist Party: Atrophy and Adaptation*, Berkeley, CA: University of California Press.

Shambaugh, D. (2015) 'The coming Chinese Crack-up', *Wall Street Journal*, 7 March.

So, A. (2003) 'The changing pattern of class and class conflict in China', *Journal of Contemporary Asia* 33.3: 363–376.

Straw, J. and M. Baxter (2014) *iDisrupted*, New York: New Generation Publishing.

ten Brink, T. and F. Butollo (2016) 'A new upgrading paradigm? Innovation policy and domestic demand in strategic emerging industries', paper presented to the *International Symposium on Innovation-Driven Development*, Sun Yat Sen University, Guangzhou, 13–15 June.

Tsai, K. (2007) *Capitalism without Democracy: The Private Sector in Contemporary China*, Ithaca: Cornell University Press.

Tse, E. (2016) *China's Disruptors*, London: Portfolio Penguin.

Turner, A. (2015) *Between Debt and the Devil*, Princeton, NJ: Princeton University Press.

Tyfield, D., J. Jin and T. Rooker (2010) *Game-Changing China: Lessons from China about Disruptive Low-Carbon Innovation*, London: NESTA.

Unruh, G. (2000) 'Understanding carbon lock-in', *Energy Policy* 28: 817–830.

Urban, F., S. Geall and Y. Wang (2016) 'Solar PV and solar water heaters in China: Different pathways to low carbon energy', *Renewable and Sustainable Energy Reviews* 64: 531–542.

Weber, M. (2002) *The Protestant Ethic and the Spirit of Capitalism: and other writings*, London: Penguin.

Wolf, M. (2016) 'How low can rates go?', *BBC Radio 4*, 21 July, www.bbc.co.uk/programmes/b07krycg.

Yip, G.S. and B. McKern (2016) *China's Next Strategic Advantage: From Imitation to Innovation*, Cambridge, MA: MIT Press.

Zeng, M. and P. Williamson (2007) *Dragons at Your Door: How Chinese Cost Innovation is Disrupting Global Competition*, Cambridge, MA: Harvard Business School Press.

7 The demand side

The emergence of risk/ innovation-class in China

The importance of the demand-side of innovation

The discussion above regarding the supply- or firm-side of innovation is an undoubtedly crucial aspect of any assessment of China's innovation upgrade. But there is a second aspect that needs just as much attention, even though it is generally overlooked in discussions of innovation, especially those conventionally focused on hi-tech, new-to-the-world innovation. This is the demand side, regarding the consumption and active purchase and adoption of innovations. Once we acknowledge that innovation really is a lot more than invention, the question of demand becomes in many ways the most important issue regarding success or failure of innovation, perhaps in terms of 'venturesome consumption' (Bhidé 2009). Moreover, opening up beyond production or supply of innovation in this way is to open the floodgates of understanding of innovation towards a much broader, systemic and socio-historical or qualitative analysis of this process, all perspectives fundamental to a faithfully representative, or even just usefully strategic, understanding of innovation.

Importantly, in terms of the bigger picture that is our concern here – of complexity system transition – this includes an appreciation of how innovation that has systemic effect is necessarily socio-technical and, indeed, power/knowledge relational. And, reciprocally, from such a perspective it is clear that demand is the crucial missing piece in practice, policy and understanding of low-carbon transition (Shove and Walker 2015; Tyfield *et al.* 2015), whether in terms of the still-stuttering initiatives of smart energy grids or green urban e-mobility (see next chapter). Moreover, from a CP/KS perspective, it is clear that just as a focus on demand pushes the analyst beyond a purely techno-economistic analysis of innovation, so too, and *vice versa*, a broader conception of innovation-as-politics reframes the concept of demand. No longer, in other words, is it adequate in this discussion to conceive of demand (for innovation) as a purely utilitarian matter of rational maximization (or bounded rational satisficing) of preferences, or perhaps investigations into how consumer tastes could be 'nudged' to be more green than they currently are. Rather 'demand' must be understood itself in CP/KS terms, as connoting the qualitative change of the systemic emergence and substantive shaping of the very constituencies that will *constitute* that market demand for these innovations and, in turn, be enabled by them, in the positive feedback loops of power/knowledge of growing power momentum (see Chapter 1).

From this perspective, then, the central question of demand for innovation upgrade and/or low-carbon system transition alike is 'what is the changing *socio-political* profile of demand for innovation, and how is it emerging in co-production with such innovation (upgrade)?' In the context of China, the answer to this question appears to be obvious. For here we find a social group, characterized by its high demand for innovation, that is emerging at the singularly Chinese scale and pace that we have now encountered repeatedly above and is transforming China's innovation demand profile in the process. This is the Chinese 'middle class' (Li 2010; Guo 2008), whose formation is surely the most important story of global socio-economic change of the early twenty-first century (Therborn 2012): for innovation per se, for low-carbon system transition, for China and for global capitalism.

Obvious, perhaps. Yet, to understand both the process of the emergence of the Chinese middle class, regarding 'where are we now?' and where this is tendentially leading, in a genealogy of the emerging present, in fact requires some quite profound rethinking, in a process facilitated by meso-level CP/KS analysis. The very *meaning* of 'innovation' has had to be reconceived in our consideration of the supply-side of Chinese upgrading efforts, with China's innovation strengths very far from where we (or the CCP) would usually be looking. So too on the demand side, with the issue of China's middle class. Regarding this issue we find even fundamental disagreement about the most basic of issues, namely '*is* there a Chinese middle-class or not?' in perfect parallel with the essential contestation of whether or not 'China('s innovation) will rule the world'. And, of course, this debate reflects similarly profound hopes and fears from all sides, as we shall see. Similarly, with evidence to suit all tastes (and deeply sedimented political hopes and fears) what is needed is not merely a presentation of the evidence but its theoretical organization into a form that offers a credible appreciation of the complex totality. This is what we aim to provide in this discussion.

Our argument consists of 4 steps, over this chapter and the next. First, we argue that the Chinese middle class is not a class at all, in the familiar sense of the unequal distribution of assets and diverse forms of 'capital' (Bourdieu 2001), but rather an entirely new social phenomenon: a *risk*-class, or more properly still, a risk-innovation-class. But, second, just as it is not a familiar 'class', neither is it straightforwardly 'middle', in any substantive sense recognizable to Euro-American scholars and laity used to their dominant sense of this term. Rather, this middle class is the specific beneficiary and agent of this new system of risk-innovation-class, positioning itself as the new stalwarts of complexity-attentive responsibility and meritocratic respectability as against a feckless, decadent elite above and an uncivilized mass beneath.

Third, on studying the dynamics of this powerful socio-political dynamic, however, we find again, as with the continued improvement of China's disruptive innovation along existing trajectories, that it is now pushing at a ceiling that is calling for and calling forth more profound and disruptive qualitative *socio-political* change. But, fourth, here too we find that the dynamic itself has also thereby furnished the impetus, capacity, agency, process and conditions for that deep systemic change to be realized, with innovation-as-politics crucial in all cases.

Beyond 'class' in the West and China

Starting with the issue of 'class', this is generally understood in lay parlance, and then illuminated in the sociological literature, as the division of (generally a national) society into a clear, if more-or-less complicated, hierarchy of strata, each of which is distinguished by the access to, and self-reinforcing privileges thereby enacted regarding, diverse forms of social good and/or asset. This is not the place to offer a comprehensive summary of the complicated, and important, debates about class thus understood (Savage 2000; Wright 2005). But whether based in a Marxian analysis of capitalist society as divided into holders of (financial) capital vs those forced to sell their labour on a daily basis, or a more sociological, Weberian account that gives equal or more weight to issues of semiotic status and other non-economics forms of 'capital', such as the cultural capital of socially-esteemed high-brow knowledge or social capital in powerful personal connections (Bourdieu 2001, 1984), several features unite most of these approaches.

First, class is a form of stratification specifically appropriate to the individualized societies of capitalist modernity, as a powerful mechanism of reproduction and differentiation of such societies, binding into relatively stable and sedimented collective identities the otherwise unruly anarchy of individualized persons that capitalism also produces. Second, 'class' here refers specifically to the feedback loops of systemic *goods* and their unequal distribution, such as money, property, socially-sanctioned forms of knowledge and/or culture etc.... (Curran 2013); systemic goods, moreover, that may be produced with all the historically exceptional productivity of capitalism, making the inequality of their distribution all the more offensive, as in grinding Victorian poverty amidst unprecedented plenty. Similarly, living class identities meaningful to those thus labelled and probably intentionally adopted by them – as must be the case for 'class' as a system to subsist – are adopted and fashioned around legitimation – to oneself and others, of like and different class – of the specific access (or not) to various capitals that one's class position affords.

Most important for our purposes, however, is that, crudely defined though it remains here, this familiar understanding – or power/knowledge technology – of 'class' is profoundly problematic in the early twenty-first century, both in the now-neoliberalized core of the global North from which it hails, and even more so in China. Indeed, one is today hard-pressed to find a sociologist or a class theorist for whom contemporary debate does not start by admitting the challenges to relevance of familiar class theory, as sedimented in mid-twentieth century sociology, to contemporary Western societies (let alone elsewhere) (e.g. Savage 2000).

Under the pressures of globalization, neoliberalism, deindustrialization, social fragmentation, individualization and reflexive modernity there can be no doubt that established class hierarchies (e.g. of 'upper', 'middle professional', 'middle clerical' and 'working' classes) and self-adopted identities offer more confusion than insight, and more heat than light, regarding contemporary Western societies and their dysfunctions and profound inequalities (Beck 2013). Within neoliberalism, class – and 'working class' in particular – no longer offers a meaningful rallying point or prediction of voting intention.[1] And even as class politics have attained a new piquancy in the post-Crash world, evident in core countries such as the US or UK (e.g. Freeman 2014; Jones 2012; Tyler 2013), and arguably the new populist nationalism (Sasson 2016), but also elsewhere (e.g. Brazil (*The Economist* 2014a), this is more a matter of a resurgent culture war, amplified through the essentially polarizing and outraged platforms of networked social media, than any meaningful return to relatively sedimented and stable class hierarchy. Indeed, precisely to the contrary, it serves to mask the continuation and deepening of *class*'s socially confusing, politically disorientating absence, despite yawning and evident socio-economic *inequality*.

But if things are puzzling in the global North today regarding class, in China they are even more complicated and confounding. At the root of this bafflement is, again, the historical conjunction of the Chinese CCP Party-state and its deepening relation to (specifically neoliberal) global capitalism. While the latter has exposed China to the full raw force of capitalism's insatiable exploitation, and in ways unmediated by a functioning welfare state and independent trade unions, the former means that the discourse of class is essentially the property of that governmental status quo overseeing this turbo-capitalist experiment (Blecher 2002). Even as Dengist reform China has moved far beyond Maoism and revolutionary fervour – even now officially repudiating the Cultural Revolution as an unqualified historical 'mistake' (Hornby 2016) – the language of class (*jieji*) remains indissolubly connected with such connotations of class struggle (e.g. Kraus 1977; Mitter 2005). And insofar as the one-party-state remains officially and explicitly a *Communist* one, with Marxism-Leninism-Maoism as its founding basis written into the Constitution, this is a connection and meaning that cannot be explicitly gainsaid; while, conversely, it is played upon strategically by workers, particularly at state-owned enterprises, in campaigns about their grievances or (Government-sponsored) mass redundancy (Lee 2007).

The discourse of 'class', in other words, is essentially system conservative and so quite useless – and so generally rejected – as a power/knowledge technology of public societal critique (or even approval!) in the way that class and class analysis have become conventionally understood to function in Euro-American societies. Instead, official and academic Chinese discussion has deliberately latched on to the much less loaded term of 'strata' (*jieceng*) (Liang 1997, in Anagnost 2008). Yet, for all this, in recent years, class has tenaciously reappeared, albeit in the guise of a newly significant discussion about the *middle*, not the revolutionary working, class.

This has been driven (in conditions of the IO vs UF) by a pidgin sociological effort by Party social scientists to appropriate essentially conservative Western (and American in particular) sociological theories. These theories supposedly establish how the (capitalist) middle class is, on the one hand, the socio-economic constituency that underpins social and political stability of market economic societies – precisely what the CCP most avidly seeks for China – and hence, on the other, the group that needs to be cultivated to achieve that harmonious order. This is then expressed in terms of the CCP's wish for an 'olive-shaped' *xiaokang* society, with a broad-based, middling prosperity topped and tailed by comparatively small groups.

But it is not, or even any longer primarily, in this official discourse that the Chinese middle 'class' is most important. Class is increasingly a vibrant and self-ascribed term of lay parlance, even as there is systematic confusion and bewilderment about what it actually means (Yu 2014). This includes lack of clarity about the necessary corollary of the meaning of 'class' per se, namely its clear delineation into well-understood, identifiable, mutually inter-defined and richly substantiated 'class*es*' and their social meanings and associations (to which we return below). Notwithstanding this unclear meaning, however, the concept of class per se has undoubtedly achieved a stunning renaissance in contemporary China. What are we to make of this?

The suggestion of this discussion is that this resurrection of the social category of class today in China is actually the harbinger and embryonic emergence of an entirely new social category of the stratification of capitalist societies. At present this is still only empirically observable as a class-*in*-itself, but as it becomes a class-*for*-itself, under pressure of the unique social dynamism of China-in-global-risk-society, it will play out with global significance over the next few decades. This new social category is risk-innovation class, or simply risk-class (Curran 2013, 2016).

Not 'class' but 'risk-class'

Discussion of risk-class has emerged from an attempt to understand the sustained reproduction *and deepening* of systemic inequalities in individualized, capitalist societies – as per 'class' – but amidst the new social context of cosmopolitized, mobile risk-societies that have been produced by and in parallel with the dominance of neoliberal financialized globalization – hence 'risk'. It thus builds upon explorations of the profound social transformation described and explained by seminal sociological work on the latter set of issues (e.g. Beck 2009; Urry 2002; Sassen 2001; Brenner 2014), but takes seriously the possibility for these to be refracted through a reformed hierarchy of 'class' given the historical and theoretical necessity of 'class' for functioning capitalist systems. In doing so, it also reframes the theorization of class and the problems that the former theorists persuasively identify in existing theories (e.g. Beck 2013) regarding an intrinsically reproductionist bent that renders these familiar approaches (e.g. Bourdieu 1984; Goldthorpe 2002; Atkinson 2007) unable and/or uninterested to explore the profound novelty – not least of forms of inequality – of global risk-society.

As Curran (2013) in particular has insightfully shown, the key to this conundrum of a concept that is undoubtedly one of 'class' and yet dynamic, system-productive (not just *re*-productive) and attentive to the new predicaments of global system complexity (Urry 2002) is to see how the familiar concept of class-as-differential-distribution-of-system-goods can be profoundly reframed simply by also incorporating attention to the differential distribution of system *bads*. For starters, attention to system bads is directly to attend to the key aspect of this new complex globalizing world: the systematic proliferation of (often socio-technical) risks and outright existential dangers that are the very source and product of neoliberal power/knowledge dynamism and its disaster capitalism (Klein 2007; Pellizzoni 2011). Essential to global risk-society is the dynamic immanent in the very *success* of the mid-twentieth century technocratic capitalist application of innovations, namely that it generates proliferating risks, new uncertainties and systemic gaps in understanding and knowledge – ignorance and non-knowledge (Gross 2010) – that, in turn, may be more-or-less successfully managed with further innovations, or at least exploited for entrepreneurial profit under neoliberalism, in a self-propelling spiral of growing system differentiation and complexity (see Chapter 1). Risk-society, therefore, generates proliferating and increasingly individualized exposure to risk as perhaps the key characteristic of this new social condition.

Moreover, small differences in 'initial' exposure to risk, e.g. at birth, tend to compound and cluster. For instance, poverty goes together with poor housing in higher-risk areas, whether for crime, environmental hazard, poor standards of education and health, dependence on loan-shark finance, exposure to work insecurity and/or displacement by other poor immigrant workers willing to work for even less... all in positive feedback loops. This generates systematic differentiation in life-chances through positive feedback loops of relative exposure to or escape from not just these risks themselves, but *also* inclusion or exclusion from the system *goods* that accrue to some from the existence and growth of these system bads, especially to those sponsoring their propagation in the first place. Quintessentially, this would include (global) environmental and financial risk (Curran 2016), for instance, where both of these are produced overwhelmingly by groups that are *not* the groups then most exposed to the full force of their potentially catastrophic outcome, as climate change

(Roberts and Park 2007) and the Great Financial Crash (Stephens 2010) exemplify. Risk-class, thus, is a short-hand for these complex – but not simply chaotic or unpredictable – incremental iterative accretions of relative systemic advantage and disadvantage of individualized cosmopolitized selves into system logics of widening inequality.

In fact, implicit in this seemingly minor redefinition, then, is a profound reshaping of the whole logic of class (see Table 7.1), where risk-class must be understood as emergent on top of, and so resituating and not simply displacing, residual class-as-goods system logics. In particular, as risk-class, the class system is imparted with an intrinsic dynamism, or impetus to constant acceleration, absent even in the class-as-goods societies emergent with industrial capitalism in the late eighteenth, nineteenth and twentieth centuries. For while the individualized-cum-collectivized contestation of class position in the latter case focuses on and is driven by competition for distribution of system goods (e.g. return on capital vs the fruits of one's labour as wages), in the case of risk-class there is the added element of contestation concerning the distribution of system bads. The former thus is a structured and structuring socio-political competition about *how much* of what everyone obviously wants is distributed to whom. As such, it is possible, if not by any means guaranteed, to be content with one's lot and class-defined allocation. Class compromise and an associated moral economy of 'fair shares', as in the post-war Keynesian welfare-warfare national state regime (Jessop 2002), is thus possible, generating relatively stable, if always passing, capitalist social order.

But the very premise of risk-class is that one now lives in a global, complex system that produces multiple systemic risks and straightforward system bads – potentially existential security threats – as the inescapable flipside of (some of) the novel system *goods* also emergent from that complexity. *Risk*-class thus necessarily incorporates the socially contested and uneven distribution of what *nobody* wants personally, but which *everybody* indirectly wills through their aspiration for the system goods that are their inescapable corollary. And this in turn complicates, unsettles and so dynamizes the moral economic calculations of fair shares regarding those system goods, in terms of the extent to which benefiting from systemic risks is – or today in late neo-liberalism is very clearly *not* – matched with exposure to the downside.

In other words, in risk-class we have a system that is attendant not just to the comparatively manageable (if hotly contested) social calculus of 'fair' balancing of (what the system as a whole judges as) what one 'puts in' and 'gets out'; but the inordinately more complex, unstable and difficult-to-measure one of one's contribution to the production of system goods *and bads* as against one's consumption of goods and exposure to bads, which everyone wants to minimize as a matter of sheer existential security. Moreover, the latter calculus depends upon essentially contested forms of (possibly quite recondite) expert knowledge – e.g. regarding (global) environmental risks – in ways that go beyond the lay knowledge needed for contestation of class-as-goods, opening up a whole new dimension of political jockeying and shiftless positioning regarding these knowledges themselves. And this complexity is compounded yet further by the added calculation of one's responsibility for production of system goods *mitigating* system bads, as in the production and consumption of, say, low-carbon innovations.

Finally, as a whole, while distribution of goods alone could sponsor the powerful Enlightenment hope in progress towards a future of definitively fair shares and abundance – whether as liberal/conservative or radical/socialist political visions –

Table 7.1 Risk-class vs class-as-goods

	(Global-)Risk-Class	Class
Definition/Formally understood in terms of:	Differential individualized distribution of (production and/vs. consumption of) goods *and* bads	Differential individualized distribution of (production and/vs. consumption of) goods
Dominant, system-productive political epistemologies:	Systemic (technoscience-mediated) bads as clear and present, alongside goods, and in need of (scientific) understanding → importance of post-Enlightenment knowledge-politics	Goods, as socio-material and largely self-evident, can be enjoyed without specialized understanding, while collective goods of historically unprecedented economic productivity demonstrate *collective rationality* of this political economy ↑ priority of Enlightenment secular materialism
Key questions of distribution for (system-productive) acceptance of this specific form of power/knowledge system:	Bads will not be accepted per se but only insofar as (i) their production per se is seen as 'necessary' or inescapable systemically for the specific political economic model and its associated benefits; and (ii) their 'fair' distribution, regarding benefits and costs, is seen to approximate with responsibility for their production and/or efforts at their mitigation	System goods are accepted and desired per se, hence struggle is only over their 'fair' distribution vis-à-vis (contested understandings of) 'who really produced them', and with a shared premise in such struggle of the possibility of a perfectly fair and collectively rational matching of effort put 'in' and gains taken 'out', in a society with ever-diminishing costs, limits, scarcities or downsides. Bads are systematically present but seen as eliminable backwardness not ineliminable systemic products
A moral economy of (contested judgements about):	Complex, dynamic, uncertain and imperfectible systems and their goods *and bads* at both collective and individual levels	Societies (as political economies) of potentially increasingly rational distribution of production and consumption of goods
A politics regarding:	Contending and thus intrinsically dynamic situated perspectives (at present still emerging) of optimal government of complex cosmopolitized-social systems and global-risks	Rival ideological understandings of 'fair shares' and 'rationally optimal' organization of production (and consumption); hence between preservation, incremental ameliorative reform or radical subversion of the class system

This new 'class' form is middle-class-driven in initial formation as the self-assembling group who are:

1 Individualized and seeking security/success in individualized fashion that is fundamentally supportive of capitalist social relations

2 Below (and self-consciously differentiated from) the elite, so systematically less enabled and more concerned regarding proliferating global risks that are, in turn, largely produced by and overwhelmingly benefitting the elite

3 Hence systematically motivated to construct new power/knowledge technologies capable of both illuminating the personal impacts of and/or responding to these systemic risks

4 But also sufficiently enabled and systematically important to contemporary global knowledge-economy growth to be able to both *demand and take* action regarding global risks *and* action that fundamentally promises to *preserve and improve* their existing standards of living on given definitions – and thus to preserve incumbent systemic power relations.

1 Actively individualizing, as historical vanguard of this new system-productive moral economy of liberal capitalism

2 Below an *Ancien* elite while, via private property, finance and mastery of new rational power/knowledges, fast emerging as de facto the most powerful and system-productive social group

3 Hence systematically motivated to construct new power/knowledge technologies capable of 'rationalizing' socio-economic life to the specific advantage of these 'rational' individuals

4 Standalone torchbearers of the construction of the 'modern' 'rational' 'society', in which they also happen to be primary beneficiaries – and thus as primary drivers of the construction of new modern, liberal power relations.

risk-class conditions a system that is intrinsically more sombre and disillusioned, fundamentally changing the lived understanding of one's class identity as well. For the system bads are here to stay, the inescapable flipside of the emergence of complex system goods. In short, no longer can either the bourgeoisie or the working class formulate class theories and positions with Grand Narratives that pamper their self-importance as the Historical agents of the realization of a glorious final universal human emancipation.

Liberty-security driving risk-class in China – security

This emerging system logic is particularly striking in contemporary China, with the unique intensity of (global) system bads there (see Chapter 4) making lived concerns and anxieties about the uneven distribution of these security threats, and the fairness or otherwise of that distribution, amongst the most characteristic aspects of the phenomenology of everyday life. This is seeding a highly dynamic and power/knowledge-system-productive logic, driven specifically by the increasingly individualized twinned pursuit of one's personal (and familial) 'liberty' – the deepening of one's autonomy and personal opportunity – and 'security' – in terms of optimized but never definitively secured shelter from these existentially threatening system bads. It is the combination and restless recursive interplay of these twin forces that drives the emergence of risk-class as a system.

Consider first the security threat aspect. We have considered above the particular intensity of the Four Challenges in China, in co-production with the intensifying clash of the immoveable object (IO) of the CCP Party-state complex power/knowledge system and the unstoppable force (UF) of neoliberal globalization, also conceptualized as CP/KS. And there can be little doubt that contemporary China is indeed a place of intense global risk exposure. Indeed, China is undergoing a unique 'compressed modernity' (Chang 2010) unfamiliar in the West, in which the challenges of both industrial 'first modernity' and the 'second' or 'reflexive' modernity, emerging from the former's success, are encountered at the same time, deepening and complicating both (Han and Shim 2010).

For instance, while the West could 'pollute first, clean-up later' – or industrialize first, then de-industrialize and resolve the profound pollution problems of the former stage in ways that supported grappling with the novel problems of de-industrialization, yielding the contested but dominant discourse of the 'environmental Kuznets curve' (Stern 2004) – China has had, and will get, no such leeway. The result is not merely to slow down and complicate the 'cleaning up' but actually to exacerbate the initial polluting. Hence an exceptionally breakneck industrialization together with environmental pollution of a cost, in economic terms alone, that on some (even official) measures almost entirely negates even the record-breaking economic growth it has notched up (Economy 2007). In short, China's problems with (global) risks and complexity-system bads are new and 'we' in the West have *not* been there before (cf. Kahn and Zheng 2016: 3), however much both Western and Chinese decision-makers wish to believe it.

Regarding environmental quality, across almost any metric or issue at which one might choose to look, the challenges in China are immense (e.g. Shapiro 2016, Economy 2011, Watts 2010, Kahn and Zheng 2016). Statistics illustrating these problems are now familiar, even in the West. Notoriously, 12 of the 20 most polluted cities in the world are in China (World Bank 2007). In air pollution, only 1 per cent of China's urban population live in cities with air-quality that would meet EU standards (Kahn and Zheng 2016: 3). Concentrations of noxious gases and carcinogenic micro particulate matter (e.g. $PM_{2.5}$) in the air of China's cities exceed World Health Organization recommended maxima as a matter of course, with not-infrequent spikes that are positively hazardous. Beijing, with a population of 21 million, had $PM_{2.5}$ levels of over four times the threshold of a public health emergency in January 2013, and in October 2011 had levels so high that they were judged 'beyond index' by the US Environmental Protection Agency standards (Kahn and Zheng 2016: 3, 11), with similar peaks – of 'airpocalypse' – more recently. And even as air quality improves as one moves south along the Eastern coast down to Hong Kong, at its best, air pollution from the mainland is still causing 1,200 premature deaths there (Kahn and Zheng 2016: 3; citing Edgilis 2009).

In terms of greenhouse gas emissions, too, China is now in absolute terms the biggest emitter of GHGs, overtaking the US around 2006 (Hornby and Shepherd 2015). To be sure, this must be set against both the (near) order-of-magnitude greater size of the Chinese population vs even other industrialized countries, and its status as offshored workshop of the world (e.g. Wang and Watson 2007). Historical accumulation of GHG emissions also matters, given that CO_2 at least remains in the atmosphere for approximately 100 years, and China is a late-starter. Yet total annual emissions now exceed those of the US and EU combined, and even *per capita* emissions overtook those of the EU in 2014 (Clark 2014). Bridging the issues of air pollution and GHG emissions in particular is the massive dependence on coal combustion, for electricity, heat and industrial processes – arguably the greatest single source of GHGs in the world, by 2012 reaching levels of consumption equivalent to nearly 50 per cent of global totals (Kahn and Zheng 2016: 25) – and, to a lesser extent, the booming demand for internal combustion engine private cars. Such locked-in problems spell significant challenges in the medium-term for any prospect of Chinese low-carbon transition, even as this is increasingly a global (and national) emergency.

Similarly hazardous levels of pollution exist across multiple other issues (Lu, Jenkins *et al.* 2015). Regarding water quality, for instance, in 2012 the Chinese Ministry of Environmental Protection (promoted to ministry level in 2008 to reflect the growing seriousness with which the central government takes these issues) rated 57 per cent of groundwater in nearly 200 cities as 'bad' or 'extremely bad', and over 30 per cent of the major rivers as 'polluted' or 'seriously polluted' (Kahn and Zheng 2016: 3; Li 2010; Khan *et al.* 2009). Soil quality (Hornby 2015) and pollution from overuse of fossil-fuel-based nitrogen fertilizers (Shen *et al.* 2013) and pesticides (Li *et al.* 2014) or high concentrations of industrial heavy metals (He *et al.* 2013) are also massive problems. Soil quality also faces intense challenges from desertification, salinization or simple quantitative loss from urban sprawl and local-government-driven development. These issues spill over into a key everyday concern of food safety (Yan 2012; Lu, Song, *et al.* 2015) including staples such as rice, maize/corn and pork. For example, Guangzhou is China's third city, with many sophisticated and demanding consumers in amongst the richest parts of the country and with a particularly strong cultural love of food and culinary pride, even for China. Yet 44 per cent of its rice samples were found to contain poisonous levels of cadmium (Kahn and Zheng 2016: 43).

Finally, awareness of exposure to system risks has taken on a new level in recent years, specifically facilitated by the now ubiquitous smart mobile phone. Whether through commentary and expressions of outrage regarding pollution issues on social media, that are largely tolerated by the heavily-monitoring authorities, or in new smart phone apps, that convey real-time date about air quality, and from multiple sources deemed reliable (e.g. the US Embassy), citizens of China's megacities are now bombarded with daily reminders of just how bad the pollution is. This has also broken through in terms of an embryonic public sphere (Geall 2013, Calhoun and Yang 2007) and civil society in China, of public commentary and over 500,000 non-governmental organizations (*The Economist* 2014b; Lu 2009), that is specifically focused on environmental issues, issues often deemed tolerable by or even indirectly supportive of central Government.[2]

It is not just pollution and its side-effects that expose the average Chinese citizen to intense risk. For in finance, education, health, housing ... even marriage (Kahn and Zheng 2016: 57), the 'structured uncertainty' (Breznitz and Murphree 2011) discussed above regarding innovation similarly shapes everyday life and competition for access to these assets for oneself and one's (probably one-child) family. Alto-gether, this conditions an intense focus on one's own lot that is fed specifically by the distinctive form of individualization that has emerged in CCP China's encounter with global risk-society as an 'individualization without individualism' (Yan 2010). On the one hand, individuals have become increasingly exposed to the double-edged sword of growing socio-economic autonomy, as opposed to having their life mapped out any longer by their position in the Party-state, and thus 'individualized'. Yet, on the other, the immoveable presence of the latter persists in systematically discourag-ing expressions of individual*ism* in terms of the moral and political priority of the individual – and, therefore, also of new collective associations and identities of indi-viduals in civil society. This, in turn, engenders a general mood of individualized hyper-competition, stress and anxiety, further feeding lived concern about these pro-found security risks to oneself and family. And while there has been significant pro-gress in some of these issues, that progress is nowhere near as fast as the growth of awareness of the problems.[3] In short, therefore, Chinese life – of the urban, 'middle-classes' especially (see below) – is security aware with an unrivalled intensity.

But this is all the more so given that exposure to these objective system risks is also highly differentiated and unequal, reposing on and interacting with the deep inequalities of capitalist China. This uneven, and recursively entrenching, distribu-tion of system bads, however, is precisely what feeds the translation of these security hazards into dynamics of risk-class formation, and in multiple, complex ways.

It is specifically *class* that is thereby cultivated, in that it is the sharp-elbowed competition for comparative status of individuals exposed to a capitalist political economy. But it is *risk*-class in terms of the essential mediation and focus of this process on issues of complex power/knowledge technologies, relations and out-comes. For instance, the very risks and hazards from which one seeks to protect oneself are the dynamic, emergent and not-easily-readable products of global risk society – in China, compounded by their compressed overlay on the problems of first modernity. But both understanding these risks – such that one can then act to minimize exposure – and access to levers with which actually to achieve some self-protection are intrinsically issues of differential power/knowledge and one's (dynamic) position in such systems.

For instance, as Kahn and Zheng (2016: 104, 139, 141; see also Kahn 2002) note, educational attainment is a key capacity differentiating those who take action, including consumption choices and even political action, to secure themselves from environmental risks, including in China. Set inextricably within complex systems of dynamic power/knowledge relations and technologies, in other words, agency to protect oneself tends to beget and accumulate capacity to do so, but where that agency in the first place is always and necessarily conditioned by *existing and unequal* capacity, depending on one's system position.

Objectively severe system risks coupled with an increasingly concentrated subjective attention, where *both* of these are power/knowledge mediated, thus tend to a systemic dynamism of polarization into clear social strata of those most enabled and least exposed and *vice versa*. But, this is not due to growing epistemic mastery on the hand and deepening self-harming ignorance on the other; we are here discussing power/knowledge not simply objective knowledge. To the contrary, the complex world of locally manifest global risks remains systematically uncertain and too complex for anyone to understand comprehensively, once and for all, even as some become more *capable* and enabled to keep pace with the restless tempo of change that threatens to leave those who don't run ever *further* behind. Indeed, that the world confronting the individualized selves, who are now charged with governing for themselves, is so complex is a further powerful driver towards finding useful power/knowledge technologies and rules-of-thumb that can mitigate that complexity somewhat: including, of course, the relative refuge of a *class* identity for oneself and others.

Liberty-security driving risk-class in China – liberty

On the security side, therefore, we see powerful, self-propelling dynamics specifically towards the emergence and active personal adoption of the category of risk-class. But these are compounded and given further impetus by the other side to this whole process, namely the multiple system goods, including specifically those produced by complex risk-innovation societies, and the similarly uneven distribution of these. This is thus individualized pursuit of one's growing autonomy, or liberty. While scholarly and journalistic attention to China from the West seems especially (if understandably) focused on the many problems confronting this fast-changing society and the anxieties it elicits within Chinese hearts, this alone is a partial and misleadingly bleak half-truth. Quite to the contrary, we find in global surveys about happiness and especially optimism regarding the future that contemporary China consistently scores well, and far better than the gloomy, stagnating West (Dahlgreen 2016; Tse 2016: 208).

There are obvious reasons for this too, again both objectively and subjectively. Objectively there can be no doubt that life and living standards in China have improved immeasurably in an extremely short period of time. The Mao years must be credited with a dramatic reversal and improvement in life expectancy, which had fallen catastrophically in the early twentieth century amidst the collapse of the imperial state, but it has risen still further, from 66 years to 73.3 years in the past 20 years (Kahn and Zheng 2016: 135). This has also gone hand-in-hand with a considerable *reduction* in many forms of risk, as would be associated with pre-industrialized societies, such as 'improvements in medical care, better diets, declining smoking rates' and falls in infant mortality (ibid.).

The poverty rate has fallen significantly and with global effect, with China's improvement in incomes single-handedly reversing otherwise global trends under neoliberalism towards the increase of inter-national inequality (between average incomes in rich and poor countries) and global inequality (regarding the global distribution of incomes at the individual level) between 1980 and 2000 (Milanovic 2013). Average incomes similarly have risen, and especially in the now globally-interconnected megacities along the coast. This includes China's rocketing up the rank of countries by numbers of dollar billionaires, with multiple fortunes being made. But, more modestly, those on incomes of comparable prosperity to Western average incomes or above have also grown fast. And all of this in the context of a strategic landscape of profound and rapid change wherever one looks and in qualitative ways that even the spectacular figure of three decades of 8 per cent GDP growth utterly fails to capture.

In this context, then, the individualized pursuit of security is matched by a zeitgeist of opportunity, growing autonomy and the expectation of its continued growth – of 'I want my share' (e.g. Osnos 2015) – that is also alien to contemporary Westerners. This sense of almost Wild West individual possibility, moreover, is especially sensed – as almost a common-sense – amongst the young, post-Mao, one-child generation who have never known a China different to this and have been the focus of their entire family's undivided pressure and indulgence. This therefore marks a particularly vivid contrast with their contemporaries in the US and Europe, convinced their life prospects are much worse than their parents' were. It is this generation in China too that has highly developed cosmopolitan tastes and experiences and who are now digital natives, enabled by 2.0 internet platforms in their day-to-day lives; in both respects placing them squarely and constitutively within global complex systems and identified with and dependent on the system goods emergent from that new socio-technical order.

This 'liberty' dynamic thus specifically drives adoption of risk-class categorization, both in itself and in the crucial interplay between it and the 'security' dynamic. For, in itself, the enormous dangling carrot of growing 'liberty' provides precisely the individualized incentive for the active embrace, and not merely heavy-hearted acceptance, of global risk-society in terms of the unprecedented opportunities it offers to you precisely as an individual embracing innovation. This thus personalizes the necessary agency of complex risk-society in willed identification with it as a whole, driving the constitution and reproduction of this system in its totality. As risk-*innovation*-class, therefore, with risk and innovation treated simply as flipsides of each other (see Chapters 1 and 3), the pursuit of *this* individualized liberty conditions and emerges as a self-ascribed and proudly displayed collective identity as being amongst the risk-taking winners of this new society.

But, security anxieties are the inevitable flipside of this risky, unprotected chasing of liberty. Indeed, the opportunity itself becomes a new security threat in terms of the danger of missing out, e.g. not having ridden a property boom and so being now excluded from that market opportunity and decent downtown housing (e.g. Kahn and Zheng 2016: 56). So too in more complicated but iterative interactions, the polarization of accumulating exposure to security threats is understood not only as an added 'stick', impelling one to do everything one can to lock oneself instead into the opposite positive feedback loop of accumulating opportunity. But the two also feed directly into each other. For instance, the opportunity of high-quality, elite education (which is also the *threat* that *others* will get this but not you) simultaneously builds the cultural and social capital that in turn opens employment and investment opportunities, but also secures oneself, both directly and indirectly, from falling into cycles of deepening exposure to, and conditioning by, system bads. It is, thus, specifically the combination of intense pursuit of a never-definitively-achieved security and an uncertain-but-endless horizon of personal opportunity that conditions the emergence of a common-sense 'risk-class' categorization in China especially.

In fact, we have seen this dynamic already in considering the supply side of Chinese innovation. There unquestionably remains strong demand (from individuals and their parents alike) for the security and self-advancement of a good state-sector job. But today in China we also find a large (in absolute numbers) and growing culture of entrepreneurship, especially amongst the young, urban and globally educated. These entrepreneurs are specifically in search of a personal autonomy and prospects of financial opportunity understood by this generation to be uniquely available in contemporary risky China *and* unavailable to the same extent elsewhere or in other employment (Blau 2016).

But to this employment consideration, we can add a set of characteristics that are idiosyncratically Chinese that tend towards risk-class regarding the issues of demand for innovation, our concern here. This concerns the unfamiliar nature of individualization in China, under the shadow of the IO vs UF tension. That Chinese individualization has emerged 'without individualism' (as above) tends towards its primary expression *not* in civic or political self-assertion (as in the West, from the 1960s especially), but constrained to issues of specifically economic and consumer autonomy. As such, issues of consumption, taste and consumerist display take on a heightened personal and social significance, doing even more affective and self-shaping work – in terms of the 'conduct of conduct' of the self, in Foucauldian terms. In other words, consumption is a key aspect of the dynamics of liberty/security.

On the one hand, 'liberty' manifests particularly in the fast-changing profile of consumer demand and a heightened fascination with brands (and *global* brands, in particular) as badges of high-quality. This has passed in relatively quick succession from the flood of counterfeit (and *shanzhai*) goods, to global-leading demand for the 'real thing' of elite brands, and recently to an even more exclusive preference for deliberately less ostentatious brands (Yu 2014: 22, 102; see below). On the other, the headlong dash of compressed modernity of huge industrial and manufacturing growth feeding a massive domestic market hungry for low-cost goods together with poorly developed institutions of consumer protection has conditioned in China, as nowhere else, a constant stream of scandals concerning the safety and quality of Chinese consumer goods, including poisonous food, baby milk, toys, household equipment etc.... The very objective proliferation of multiple security threats from such consumer choices thus simply feeds directly back into the prevailing fetishism with brands as marks of quality and safety – of liberty and security respectively.

To this can be added a further iteration in terms of how China's particular process of individualization both feeds the need, and profoundly shapes a huge appetite, for connection, affirmational comparison and collective identity and status, again primarily through consumption. Thrown before the pitiless isolation of an emerging market society without a meaningful system of social welfare but where the essentially exclusive personal ties of *guanxi* to the Party-state (or teachers or doctors ...) still matter enormously, relative refuge is sought in more legible markers of commonality as displayed in one's consumption choices and gifts thereof – one's clothes, drink, housing and car (see Chapter 9). This, of course, is precisely the logic of social stratification, actively performed and policed by anxious, autonomy-seeking selves, of risk-class.

And this dynamic has been further compounded by the advent of first the internet and social media and then the mobile internet (all provided on Chinese platforms, Chapter 5). As Yu describes (2014: 57–59), this facilitates the cycling of an essential characteristic of this consumer-display-based individualization, in that it is essentially other-directed, in terms of oriented primarily to seeking the approval, and escaping the ridicule, of others with whom one wants to be identified. This thus explains both the huge amount of time spent online sharing selfie photos and commenting on others', and the specifically commercial form of the 2.0 internet in China; and also the practice of shopping itself in China, which amongst the urban young especially is increasingly a seamless weaving of 'flesh' and 'online' worlds, with shoppers constantly seeking real-time approval and adulation for their purchases (Yu 2014).

This networked-individuality (Yu 2014: 23, 45, 57–59; Rainie and Wellman 2014; cf. Papacharissi 2010) thus yields a specific model of not just conspicuous consumption but 'conspicuous *achievement*' through one's consumption choices. Importantly, these displays are also understood as badges of one's personal (even moral) merit, not simply one's risk-taking luck and/or supreme individual entitlement as in cultures of neoliberalism. Moreover, this essentially networked individuality both feeds the specific form of *risk*-class, and can be contrasted with Enlightenment sovereign individualism in ways parallel to risk-class vs class-as-goods. In the former case, networked individuality is both a form of capitalist individualization that specifically feeds the social stratification that is 'class', and *also* essentially relational and conditional on others, and hence dependent upon the essentially risky and un-securable approval of other networked-individuals and the complex, capricious, emergent (2.0) network as whole. In contrast to the latter, then, it is no longer the rational, natural sovereign individual and their objective, self-directed accumulation of goods that is the ideological basis and supposed achievement of this system but a form of individuality that is *constitutively* conditioned and relational, and *understood* to be such.

Conclusion

In short, as regards demand for innovation and the emergence and shaping of a new socio-political constituency co-produced with system-transition innovations, this specifically Chinese dynamic of hyper-competition regarding consumption – and especially of intensely power/knowledge-mediated (i.e. hi-tech, branded etc....) goods and services – plays out in a dynamic that directly feeds the necessarily twin dynamics of risk-class formation: i.e. at the level of the *individual's* wilful appropriation of risk-'class' labels, and at the level of the emergence of a *system* of risk-class.

Regarding the former, with consumption of commodities via the market as the dominant mechanism, the goods and services most ardently pursued are both likely relatively expensive and marks of scarce societally-, network-approved success. This thus affords specifically liberal, capitalist dynamics of social differentiation, where the ability to pay not only affords the direct benefit of the high-quality commodity itself, but also the status that comes with it, setting up the positive feedback loops of accumulating liberty and reducing risk exposure. Moreover, this applies not just to obvious consumer items, like clothes or gadgets, but also and more importantly to assets that are also individually appropriated capacities for risk-taking, in iterative cycles of deepening privilege or disadvantage. And indeed, this is precisely the dynamic we find in China regarding housing (Tomba 2009, 2010; Zhang 2012; Kahn and Zheng 2016: 11,12), health (Kahn and Zheng 2016), schooling and higher education (Crabb 2014; Tsang 2013; Xu and Ye 2017; Kahn and Zheng 2016; *The Economist* 2016) at home and abroad (Blau 2016), and experience of the world, work and travel (Xu and Wu 2016; Liu-Farrer 2016).

Regarding the latter, meanwhile, each of these systems in China has not only become a structure propagating cycles of privilege rather than socio-economic mobility, and widely understood as such, thereby catapulting yet further hyper-competition to ensure one's family is on the side of the system winners; as for instance in the national *gaokao* exams in which approximately 10 million high-school students compete to get into university, with only two elite universities, both in Beijing and with admission stacked towards Beijing residents. Through this dynamic, a new system is also taking shape of socially-stratified differentiation that is precisely focused on hungry, self-propelling demand for goods that will further one's liberty and reduce one's exposure to objectively severe system bads or, preferably, do both. Risk-class, in other words, is what underpins the emergence of a power/knowledge constituency constitutively identified with the pursuit of the personal mitigation of system risks of sufficient power momentum to have systemic effect.

Notes

1 For instance, in the UK elections of 2010 and 2015, class was a far worse predictor of voting preference than whether one lived in the South vs the North and/or in major cities (voting Conservative and Labour respectively) (*The Economist* 2013).
2 In one particularly notable case, in March 2015 a former news anchor, Chai Jing, released a documentary, called 'Under the Dome', that had carefully solicited official approval but was still powerfully critical about air pollution after her unborn child was discovered to have a tumour most likely caused by Beijing smog. But indicative of the continuing political sensitivity of these issues and the fine but unclear line between official toleration and rejection, when nearly 300 million people watched it on China's Youku free-to-access video-streaming site in its opening weekend, it was deemed too popular and promptly censored.
3 Government efforts can also feed this dynamic of impatience, as when the skies of Beijing were cleared of pollution for the Olympics and especially for the 2014 APEC summit. Citizens can see that the government can make the pollution go away for foreign grandees like the US President, painting the skies the new colour of 'APEC blue' (Gan 2014), but not for them.

References

Anagnost, A. (2008) 'From "class" to "social strata": Grasping the social totality in reform-era China', *Third World Quarterly* 29: 497–519.

Atkinson, W. (2007) 'Beck, individualization and the death of class: A critique', *British Journal of Sociology* 58(3): 349–366.

Beck, U. (2009) *World at risk*, Cambridge: Polity.

Beck, U. (2013) 'Why "class" is too soft a category to capture the explosiveness of social inequality at the beginning of the twenty-first century', *British Journal of Sociology* 64(1): 63–74.

Bhidé, A. (2009) *The Venturesome Economy*, Princeton, NJ: Princeton University Press.

Blau, R. (2016) 'The new class war', Special Report: 'Chinese Society', *The Economist*, 9 July.

Blecher, M. (2002) 'Hegemony & workers' politics in China', *China Quarterly* 170: 283–303.

Bourdieu, P. (1984) *Distinction: a Social Critique of the Judgement of Taste*, Cambridge, MA: Harvard University Press.

Bourdieu, P. (2001) 'The forms of capital', in M. Granovetter and R. Swedberg (eds), *The Sociology of Economic Life* (2nd Edition), Boulder, CO: Westview Press [first published in 1983]: 96–111.

Brenner, N. (2014) *Implosions/Explosions: Towards a Study of Planetary Urbanization*, Berlin: JOVIS Verlag.

Breznitz, D. and M. Murphree (2011) *Run of the Red Queen*, New Haven, CT: Yale University Press.

Calhoun, C. and G. Yang (2007) 'Media, civil society, and the rise of a green public sphere in China', *China Information* 2: 211–236.

Chang, K.S. (2010) 'The second modern vondition? Compressed modernity as internalized reflexive cosmopolitization', *British Journal of Sociology* 61(3): 444–464.

Clark, P. (2014) 'China's emissions outstrip those of EU and US', *Financial Times*, 22 September.

Crabb, M.W. (2014) 'Governing the middle-class family in urban China: Educational Reform and questions of choice', *Economy & Society* 39(3): 385–402.

Curran, D. (2013) 'Risk Society and the Distribution of Bads: Theorizing Class in the Risk Society', *British Journal of Sociology* 64(1): 44–62.

Curran, D. (2016) *Risk, Power and Inequality in the 21st Century*, Basingstoke: Palgrave Macmillan.

Dahlgren, W. (2016) 'China is nearly twice as likely as any other country to say the world is getting better – while Britain is the fifth least optimistic', *YouGov.Uk* <https://yougov.co.uk/news/2016/01/05/chinese-people-are-most-optimistic-world/>.

The Economist (2013) 'Divided Kingdom', 18 September.

The Economist (2014a) 'The Kids are All Right', 25 January.

The Economist (2014b) 'Beneath the Glacier', 12 April.

The Economist (2016) 'The Glass Ceiling', 4 June.

Economy, E.C. (2007) 'The Great Leap Backward? The Costs of China's environmental Crisis', *Foreign Affairs*: 38–59.

Economy, E.C. (2011) *The River Runs Black: the Environmental Challenge to China's Future*, Ithaca NY: Cornell University Press.

Edgilis (2009) *Outdoor Air Pollution in Asian Cities*, Singapore: Edgilis.

Freeman, H. (2014) 'Come on, Britain – it's the 21st century. Stop this obsession with social Class', *Guardian*, 26 November.

Gan, N. (2014) 'Smoggy skies replace "APEC blue" in Beijing after pollution curbs end', *South China Morning Post*, 16 November.

Geall, S. (ed.), (2013) *China and the Environment*, London: Zed Books.

Goldthorpe, J.H. (2002) 'Globalization and social class', *West European Politics* 25(3): 1–28.

Gross, M. (2010) *Ignorance and Surprise*, Cambridge, MA: MIT Press.

Guo, Y. (2008) 'Farewell to class, except the middle class: The politics of class analysis in contemporary China', *Asia-Pacific Journal* 26(2): 1–19.

Han, S.J. and Y.H. Shim (2010) 'Redefining second modernity for East Asia: a critical assessment', *British Journal of Sociology* 61(3): 465–488.

He, B., Z. Yun, J. Shi, and G. Jiang (2013) 'Research progress of heavy metal pollution in China: Sources, analytical methods, status, and toxicity', *Chinese Science Bulletin* 58(2): 134–140.

Hornby, L. (2015) 'Chinese Environment: Ground Operation', *Financial Times*, 1 September, www.ft.com/cms/s/2d096f594-4be0-11e5b558-8a9722977189.html.

Hornby, L. (2016) 'China declares Cultural Revolution a "total mistake"', *Financial Times* 17 May.

Hornby, L. and C. Shepherd (2015) 'China learns lessons of past failures ahead of Paris climate talks', *Financial Times* 29 November, www.ft.com/content/480e813a-8f81-11e5-a549-b89a1dfede9b.

Jessop, B. (2002) *The Future of the Capitalist State*, Cambridge: Polity.

Jones, O. (2012) *Chavs: The Demonization of the Working Class*, London: Verso.

Kahn, M.E. (2002) 'Demographic change and the demand for environmental regulation', *Journal of Policy Analysis and Management* 21(1): 45–62.

Kahn, M.E. and S. Zheng (2016) *Blue Skies over Beijing*, Oxford and Princeton, NJ: Princeton University Press.

Khan, S., M.A. Hanjra and J. Mu (2009) 'Water management and crop production for food security in China: A Review', *Agricultural Water Management* 96: 349–360.

Klein, N. (2007) *The Shock Doctrine*, London: Penguin.

Kraus, R.C. (1977) 'Class conflict and the vocabulary of social analysis in China', *The China Quarterly*, 69: 54–74.

Lee, C.K. (2007) *Against the Law: Labor Protests in China's Rustbelt and Sunbelt*, Berkeley, CA: University of California Press.

Li, C. (ed.) (2010) *China's Emerging Middle Class*, New York: Brookings.

Li H., E. Zeng and J. You (2014) 'Mitigating pesticide pollution in China requires law enforcement, farmer training, and technological innovation', *Environmental Toxicology and Chemistry* 33(5): 963–971.

Li, J. (2010) 'Food security water shortages loom as Northern China's aquifers are sucked dry', *Science* 328: 1462–1463.

Liang, A. (1997) *A Comprehensive Analysis of Social Stratification in China* [Zhongguo shehui ge jieceng fenxi], Beijing: Jingji ribao chubanshe.

Liu-Farrer, G. (2016) 'Migration as class-based consumption: The emigration of the rich in contemporary China', *The China Quarterly* 226: 499–518.

Lu, Y. (2009) *Non-Governmental Organizations in China*, London: Routledge.

Lu, Y.L., A. Jenkins, *et al.* (2015) 'Addressing China's grand Challenge of achieving food security while ensuring environmental sustainability', *Science Advances* 1(1). Available at: http://advances.sciencemag.org/content/1/1/e1400039.

Lu, Y.L., S. Song, *et al.* (2015) 'Impacts of soil and water pollution on food safety and health risks in China', *Environment International* 77: 5–15.

Milanovic, B. (2013) 'Global income inequality in numbers: in history and now', *Global Policy* 4(2): 198–208.

Mitter, R. (2005) *A Bitter Revolution*, Oxford: Oxford University Press.

Osnos, E. (2015) *Age of Ambition: Chasing Fortune, Truth and Faith in the New China*, New York: Vintage.

Pellizzoni, L. (2011) 'Governing through disorder: Neoliberal environmental governance and social theory', *Global Environmental Change* 21: 795–803.

Papacharissi, Z.A. (2010) *A Private Sphere*, Cambridge: Polity.

Rainie, L. and B. Wellman (2014) *Networked*, Cambridge, MA: MIT Press.

Roberts, J.T. and B.C. Parks (2007) *A Climate of Injustice: Global Inequality, North-South Politics and Climate Policy*, Cambridge, MA & London: MIT Press.

Sassen, S. (2001) *The Global City*, Princeton, NJ: Princeton University Press.

Sasson, E. (2016) 'Blame Trump's victory on college-educated whites, not the working class', *New Republic*, 15 November.

Savage, M. (2000) *Class Analysis and Social Transformation*, Buckingham: Open University Press.

Shapiro, J. (2016) *China's Environmental Challenges*, Hoboken, NJ: John Wiley & Sons.

Shen, J., Z.L. Cui, Y.X. Miao, G.H. Mi, H.Y. Zhang, M.S. Fan, C.C. Zhang, R.F. Jiang, W.F. Zhang, H.G. Li, X.P. Chen, X.L. Li and F.S. Zhang (2013) 'Transforming agriculture in China: From solely high yield to both high yield and high resource use efficiency', *Global Food Security* 2: 1–8.

Shove, E., and G. Walker (2014) 'What is energy for? Social practice and energy demand', *Theory, Culture & Society*, 31(5): 41–58.

Stephens, P. (2010) 'Three years on, the markets are masters again', *Financial Times*, 29 July.

Stern, D.I. (2004) 'The rise and fall of the environmental Kuznets curve', *World Development* 32(8): 1419–1439.

Therborn, G. (2012) 'Class in the 21st century', *New Left Review* 78: 5–29.

Tomba, L. (2009) 'Of quality, harmony, and community: Civilization and the middle class in urban China', *Positions* 17(3): 591–616.

Tomba, L. (2010) 'Gating urban spaces in China: Inclusion, exclusion and government', in S. Bagaeen, and O. Uduku (eds) *Gated Communities*, Abingdon: Earthscan: 27–38.

Tsang, E.Y.H. (2013) 'The quest for higher education by the Chinese middle class: Retrenching social mobility?', *Higher Education* 66(6): 653–668.

Tse, E. (2016) *China's Disruptors*, London: Portfolio Penguin.

Tyfield, D., A. Ely and S. Geall (2015) 'Low carbon innovation in China: From overlooked opportunities and challenges to transitions in power relations and practices', *Sustainable Development* 23(4): 206–216.

Tyler, I. (2013) *Revolting Subjects: Social Abjection and Resistance in Neoliberal Britain*. London: Zed Books.

Urry, J. (2002) *Global Complexity*, Cambridge: Polity.

Wang, T. and J. Watson (2007) 'Who owns China's carbon emissions', *Tyndall Briefing Note* 23.

Watts, J.S. (2010) *When a Billion Chinese Jump: How China will Save Mankind – or Destroy It*, London: Simon and Schuster.

World Bank (2007) *World Development Indicators 2007*, Washington DC: World Bank.

Wright, E.O. (ed.), (2005) *Approaches to Class Analysis*, Cambridge: Cambridge University Press.

Xu, H., and Y. Wu (2016) 'Lifestyle mobility in China: Context, perspective and prospects', *Mobilities*, 11(4): 509–520.

Xu, H.G. and T. Ye (2017) 'Changes in Chinese higher education in the era of globalization', in D. Tyfield, R. Lave, S. Randalls and C. Thorpe (eds), *The Routledge Handbook of the Political Economy of Science*, London: Routledge: 156–168.

Yan, Y. (2010) 'The Chinese Path to Individualization', *British Journal of Sociology* 61(3): 489–512.

Yan, Y. (2012) 'Food safety and social risk in contemporary China', *The Journal of Asian Studies* 71(03): 705–729.

Yu, L. (2014) *Consumption in China*, Cambridge: Polity.

Zhang, L. (2012) *In Search of Paradise: Middle-Class Living in a Chinese Metropolis*, Ithaca NY: Cornell University Press.

8 The emerging historic bloc

China's middle risk/innovation-class

Not 'middle class' …

We have explored the essentially divisive, polarizing and restless dynamic of risk-class and its emergence as a social category (of a class-*in*-itself) that could constitute 'demand' for innovation leading to socio-technical system transition in China. But the very dynamic of power/knowledge-mediated, network-individualized and moralistic hyper-competition of market purchases points to a second key feature of this unfolding process in China; namely that it is focused, primarily benefits and is driven by the '*middle* class', likewise constituting and shaping this category and its substantive hierarchy in the process.

On the one hand, the 'middle' (risk-)class are simultaneously both exposed to the risks, and hence ceaselessly seeking shelter from them in ways that the elite, with multiple escape routes and unassailable positions of power/knowledge empowerment, need not concern themselves. On the other, they are also sufficiently enabled to mobilize against (anything they perceive to be) the 'unfair' distribution (and/or production in the first place) of system bads and (to hope, more-or-less realistically) to attach themselves to the positive feedback dynamics of *rising* opportunity and liberty. Given their number and systemic positioning, it is the middle risk-class who are then most systemically enabled to drive the profound power/knowledge changes needed for a socio-technical system transition *simply through pursuing their own, qualitatively evolving interests*, rather than through 'rational' or 'virtuous' action for the good of others and of the system as a whole – though in the process it also importantly affords them dominant ownership and definition of these self-promoting labels too, as powerful power/knowledge technologies.

Just as we are here not talking about a familiar concept of 'class', though, so too this 'middle class' is unfamiliar and perplexing, demanding another round of conceptual redefinition. There is a growing literature, both scholarly and journalistic, concerning the Chinese 'middle class', often discussed as part of a broader 'global' middle class emerging across much of the global South (e.g. Ravallion 2010; Barton *et al.* 2013; Kharas 2010; Therborn 2012). Indeed, the emergence of this group is, like the rise of China more generally, again an issue of significant interest, hope and fear, both in China (Guo 2008; Ren 2013) and overseas, for multiple parties.

In China, as discussed above, there is the literature that explores the socio-economic development and population profile of the country and, using various metrics, aims to furnish evidence that a middle class, who will supposedly be bulwarks of social stability and embodiments of the Party-state's promise of generalized prosperity, is indeed emerging (Li 2010). Alongside this dominant approach, however, there is an unrepentant 'conservative' (i.e. hard left) literature which despairs at the evidence of betrayal of the working class, and reads the same evidence (or lack thereof) through diametrically opposite normative spectacles. In both cases, however, whatever conventional definition of 'middle class' (being middle class-as-goods, of course) is adopted, unequivocal evidence for its emergence is lacking and always subject to fundamental objection given that there is no consensus amongst the snark-hunting sociologists of what this supposedly familiar, but actually elusive, beast consists of and how to prove its existence, with which measurements (Guo 2008).

Overseas, meanwhile, a similar hunt also wades into the same difficulties and objections. For instance, in what sense is it meaningful to describe a population earning, say, more than $4 a day (hence more than twice a widely-used but contested global poverty line) 'middle class'? And what does that designation have to do with broader social connotations and significance of this term that are derived from countries in the global North where such a salary, or even 20 times that income, would probably not qualify them as 'middle class'? In which case, where should we draw the line? And on what basis, such that this new income level in turn is robust? And does it leave more than a tiny fraction of people now included? And do even these people enjoy the other aspects of 'middle class' life, such as educational credentials, property holdings, travel and holidays, professional experience and accreditation etc....?

As Ravallion (2010) points out, for instance, those who may have achieved a break beyond the poverty line will probably still be vulnerable to falling back below it in a way that is relatively unthinkable for a middle-class Westerner. And adding further to the complication, this is all now being discussed in the context of a parallel debate about the *demise and decline* of the middle class in the global North in the aftermath of the 2008 crash (Porter 2011; Fukuyama 2012), further eviscerating the concept of 'middle class' of meaningful substance. The challenge, in other words, is how to capture what may be a significant global shift in levels of intra-national and global development (Therborn 2012, 2014) without using familiar sociological terminology that tends to confuse and occlude more than it illuminates.

Moreover, discussion outside China about the emerging 'middle class' can be just as baldly partisan as it is within the country. Whether eager to find the next billion customers (McGregor 2007) that will save your business and (what is presumed to be Western-dominated) capitalist global order, or the agents of the frustratingly delayed but supposedly inevitable revolution against the CCP regime towards liberal democracy – understood as the inextricable partner of China's unquestionably capitalist economy – the Chinese middle class are vested with hopes from these quarters no less strong than by those from the CCP, albeit often exactly opposite ones.

This leaves only a more rigorous and credible analysis that simply looks at the empirical evidence and finds a very confusing landscape that is in no way usefully summarized simply as the 'middle class', perhaps even not finding one at all (Goodman 2008). Instead, it sees a very diverse group (even just in terms of income distribution) of a growing stratum of Chinese society that is no longer absolutely poor but also certainly not members of a national and/or global elite (Goodman 2015; Therborn 2012; Milanovic 2013). This group, depending on the definitions employed and the income strata applied, can be massive in absolute and relative terms, but then nowhere near 'middle class' and inclusive of a massive range of living standards to the point of analytical uselessness; or still big in absolute but small in relative terms, and then simply 'middling' in Chinese terms; or small in both absolute and relative terms but approaching a fuller equivalence to 'middle class' status in the West. And yet 'middle class' remains a term widely self-ascribed by citizens in China.

What emerges from this confusing panorama, however, is all these facts together: a growing prosperity, taking most but not all of the (massive) Chinese population beyond absolute definitions of poverty, that is generating a wide spectrum of those between global super-wealth and this absolute poverty line and hence 'in the middle of' a steeply unequal and qualitatively as-yet-undefined-and-unsettled socio-economic hierarchy. Moreover, while Tocqueville's dictum – that thwarted expectations of rising prosperity are when a population are at their most volatile – is firmly etched into CCP minds, this loosely defined but unequivocal middling stratum has to date proven the CCP correct and the 'democracy'-exporting neoliberals wrong. For the preponderance of scholarly evidence – and the simple absence of anything resembling a movement of fundamental regime challenge since 1989 – weighs heavily on the side of their essential political conservatism, supporting, if only passively, the political status quo (Chen and Lu 2011; Whyte 2010a) rather than the makings of a system-critical, unified class force.

… but prolepsis for a new Chinese complexity bourgeoisie

Of course, given the argument above, part of the reason that we would argue this middle class (or middle classes) has been so difficult to find is precisely because it is not a 'class' at all but, at most, a middle *risk*-class. We would hypothesize, in other words, that systematic analysis regarding objective differences in and subjective anxieties regarding exposure to systemic risks, and fairness of that exposure, could significantly illuminate the emerging 'class' (and not just stratum) stratification of Chinese society today.

Even more importantly, though, just as the risk-class system is in-formation and emergent, so too the vast majority of discussion about China's middle class misses the essential point about it: that the 'middle class' is bearer of the ongoing construction of the new order of social stratification and is thus a process (Liechty 2003) and a 'prolepsis', i.e. a social category that is currently deployed but 'has not yet come into view as if it existed in fact' (Ren 2013). In other words, the Chinese middle class currently both exists and does not exist: it exists as a socially-efficacious discourse of interpersonal differentiation and socio-political order, but not yet as a materially and institutionally sedimented and empirically-observable social hierarchy.

On this analysis, the 'middle class' category is a power/knowledge technology (or *dispositif* (Ren 2013)) that is performing systemically crucial work of 'anticipatory staging' (Anagnost 2004: 200) in the ongoing, dynamic *construction* and *unification* of a national class of 'responsible', self-governing citizens from what remains a highly complex, unruly, systematically under-institutionalized and uniquely self-advancing and global risk-exposed collection of individualizing agents. It is no wonder, therefore, that the 'middle class' cannot be found because it neither exists for empirical analysis nor even are its characteristics settled. Rather, both the 'middle class' per se and what it substantively means are in the process of their parallel co-production and co-emergence.

This process, however, is precisely the highly dynamic and essentially contested one of liberty/security and network-individualized hyper-competition of the emergence of risk-class described above. It is the '*middle*' (risk-)class that are particularly well-placed to *benefit* from the emergence of that novel social category, while they are likewise especially *enabled* to drive and construct the systemic logic of risk-class. Hence risk-class is most clearly emergent empirically regarding specifically 'middle-class' concerns, as in issues of parking of cars, which remain financially inaccessible to the majority, or home ownership and property associations (Wang and Yuan 2013; Cai 2005; Heberer 2009; Tomba 2009). It is precisely this combination – of inescapable personal insecurities regarding complex system bads and a *relative* personal enablement regarding access to system goods that is understood must also be defended, as itself insecure – that makes a *middle-class* individual a powerful proponent of and contributor to a system that rationalizes and re-orders the existing profound inequalities of distribution of both goods and bads. And that does so in ways that *build on*, rather than collectivistically and solidaristically directly challenge, the highly individualized spirit of the age and the deeply sedimented bases of that unequal individualized allocation in the first place.

The emergence of risk-class thus forges a system that works with the grain of existing conditions: still producing winners and losers, but changing their identity and definition such that there is now the widely perceived, if not necessarily objective, possibility of individually being amongst the former. A popular and widely adopted evolving discourse-cum-practice of 'middle' (tacitly risk-)class thus sets free the productive dynamics of (network-) individualized pursuit of being one of the winners, and desperate flight from being one of the losers, that bootstraps the very system of risk-class into existence, with the *new* winners *and losers* (Hui 2016; Chan *et al.* 2013; Solinger 2012; Sun 2009) also thereby constructed.

Moreover, this is not just a middle class individual but a *Chinese* one most graphically or archetypically given the especially intense conditions driving such dynamics in China. These are intense, again, because of the unique meta-dynamics and conditions of China as the site of the historically pivotal encounter of the immoveable object and the unstoppable force, together with the unique conditions of China's size and pragmatic culture on the one hand, and the neoliberal-sponsored emergence of the Four Challenges and knowledge capitalism on the other.

For starters, this has generated the existing 'class' structure of China that is so crucial in conditioning the dynamics of middle risk-class emergence so powerfully. This existing stratification divides China systematically and formally, not just de facto, into two across two key dimensions. On the one hand, the household registration (*hukou*) system (Goodman 2014; Whyte 2010b) allots to every mainlander an official classification as rural or urban resident. This system was originally introduced in the late 1950s to stem the flood of migrants into the cities in search of better perks given the essential urban bias of the Maoist project of concentration of rural surplus into urban industrial capital (Hung 2016). Greater dispensations to urban workers were also necessary to mollify the highly radicalized and tightly clustered communities of urban workers upon which the early PRC itself depended, preventing overspill into destabilizing political mobilization.

The household registration system was thus introduced to lock down rural folk to their existing locations (and some sent down to the country from the cities as well). In the reform era, however, the system has been loosened enough so as not completely to prevent rural migration to the burgeoning factories that have been the engines of the country's economic miracle. Yet it remains in place, despite repeated (local and national) policy drives to reform or scrap it, because it has proven so important in disciplining this massive, restless and potentially restive 'floating population' of rural-to-urban migrant workers (Nyiri 2010), and forcing them to return to their remote, countryside 'homes' when out of work.

Urban *hukou* holders thus remain systematically advantaged in terms of access to better public and social services and security of dwelling and employment in the fast-growing cities that are the primary loci of the massive development opportunities (*fazhan jihui*) – the 'liberty' – characteristic of contemporary China as seemingly nowhere else on Earth. The *hukou* also assigns a specific place to its holder; hence one may have, for instance, a rural or urban Beijing *hukou* or a rural or urban *hukou* from elsewhere. Given the uneven economic geography of contemporary China, conditioned by its fragmented authoritarianism and local state corporatism, this means that even urban status in a less developed province is clearly less attractive than in a big city on the coast or the Yangtze. The result is a systematically 'graduated citizenship' (Ong 2006) of a quasi-apartheid division into two countries (Whyte 2010b).

On the other hand, though, in the shadow of the Party-state, there is also a continuing division into those within and outwith the 'system' (*tizhinei* vs *tizhiwai* (e.g. Tomba 2004)), with again more job security and privileged access to powerful decision-makers for the former. To be sure, the growth to economic dominance of the 'private' sector has significantly diminished this division over the past decades. Yet its persistence is clear in the continuing preference of the young and their parents for secure and respectable *tizhinei* jobs that promise steady and successful careers, notwithstanding the growing appetite for entrepreneurial autonomy (above); or in the continuing advantages (if not necessity) of business partnerships with those who do have such privileged within-the-system connections. Like the *hukou* system, this division is also graded rather than binary, not least due to the hierarchies of the Party-state across its entire apparatus and the fuzziness of the distinctions between 'state' and 'private' sectors.

Together, then, these two specifically Chinese divisions have conditioned what may be crudely summarized as a three-fold stratification, specifically observable within mega-city urban China. At the top are a 'cadre-capitalist elite' (So 2003; Dickson 2008; Tsai 2007; Ho 2013), consisting of the Party grandees and their families who have amassed great fortunes, legally and illegally, over the past 30 years as well as those who have first become rich and then joined the Party (e.g. under the '3 Represents' policy) or at least secured for themselves the necessary Party-state connections to maintain and safeguard their wealth. Meanwhile, at the bottom, are the rural migrant workers, systematically insecure and subjected to all the wage repression and lack of representation – let alone lack of powerful personal contacts – that the twin systems above have made possible. This massive group of over 200 million, in turn, is the connection between China's increasingly developed coastal mega-cities and the massive hinterland, where still nearly 50 per cent of China's population live with rural *hukou* status or in smaller towns (including 'left behind' grandparents and children).

While the top and bottom are clear, and clearly defined by the existing CP/KS, however, the third consists of everyone else 'in the middle' – those with neither great wealth and Party-state power nor those with low, insecure livelihoods and scant levers of political influence. This is thus, of course, a highly diverse group, even just over the two systems just described. It incorporates many who are *tizhinei* but also many who are not, and many who are urban outside the biggest cities or who are still designated rural within them or the most developed provinces. This is thus a group that is prima facie defined and unified by what they are *not*. And, with this fundamentally *via negativa* definition connoting not just a bland, descriptive differentiation but also increasingly a highly normative one.

This middle is neither the systematically denigrated urban 'problem' of the 'uncivilized' rural migrant worker, with their 'coarse' habits and demeanour, thick accents, 'farmer's hands', lack of education and 'undisciplined' practices (e.g. Ngai 2005; Chang 2010); all power/knowledge discourses officially propagated as disciplinary technologies. But nor is the 'middle' part of what is increasingly viewed as a corrupt and self-serving elite – hence the urgency of the current anti-corruption campaign driven by President Xi Jinping, as a matter of existential security for the CCP Party-state and its continued popular legitimacy. It is not just the super-rich and/or Princeling elite (i.e. second or third generations of families with strong and publicly heralded connections to the CCP's revolutionary past and the founding of the PRC) that indulge in and benefit from this corruption and 'rule by man' (or even 'rule *by* law' vs rule *of* law).[1] Rather, through the *tizhinei/wai* distinction, petty corruption and expectation of bribes or personal gifts to curry personal favour is commonplace and a familiar feature of life for many of this 'middle', e.g. in terms of queues for doctors or schools, or marks from exams etc....

Moreover, these dynamics of corruption are mediated by the system of *guanxi*. On the one hand, it must be noted how, from a CP/KS perspective, the crucial Chinese system of personal *guanxi* connections may be seen to be highly system-functional under the stabilized conditions that have often prevailed over the prior centuries and millennia (Yang 2002); namely, a moral economy of pragmatic system government of strictly hierarchical rule-by-man, dependent on both formal *and informal* personal connections of *guanxi*, with the latter necessary to fill in the inevitable gaps of the persistently under-institutionalized former, and moralized through a Confucian ethical culture that tended to disperse money and power through disdain for the merchant and the individual pursuit of wealth (see Chapter 4). *Guanxi* in turn was a crucial resource of survival during the Cultural Revolution atop radically politicized, collectivized everyday life, giving this power/knowledge system a newly modern twist and lease of life (Yang 2002; Gold *et al.* 2002).

But, on the other hand, while still filling the gaps of persistent under-institutionalization (Zhan 2012) in the Reform period, under conditions in which power/knowledge is *concentrated* through an increasingly capitalist market economy into positive feedback loops of money and power, *guanxi* becomes instead an engine of deepening power/knowledge inequality and naked self-interest. Indeed, some contemporary Chinese social theorists and critics even see here a generalized 'moral crisis' (Ci 2009; Luo 2008; Ford 2011). This, in turn, creates a situation in which frustrations and system unfairness and irrationality may arise simply as a matter of information asymmetry between the system agent and the external petitioner, without any ill intent (Blau 2016), but with such asymmetry understood as a dynamic of *power*/knowledge not just instrumental objective reason.

The middle, in short, is thus that group which is most subject to and frustrated by such petty graft and uninstitutionalized-cum-bureaucratic frustration, being simultaneously those able and eager to participate in competition for these scarce and bureaucratically-controlled goods and liberties but not so well-connected as to be able to systematically win from that system. It is not just the middle themselves, however, who are increasingly convinced of their unique virtue regarding the

Chinese CP/KS. To the Party-state too, as we have seen, they are both the group on whom is placed the hope and burden of CP/KS stability and socio-political conservative order, and the group – precisely as *not* a cadre-capitalist elite, though this can hardly be stated explicitly – on whom and for whom the great new task of the CCP itself 'cleaning up' the system disease of petty corruption is pinned.

Moreover, as the bedrock of the private sector and the educated professions – both of which are central to the 'new normal' of an 'innovation-driven' economy to which the Chinese political economy is oriented, under pressure from its multiple systemic dysfunctions (discussed above) – this is also the group that is increasingly systematically central for that Party-state regime. While, on the other hand, for the 'bottom' of the rural migrant service, factory or construction worker, the dream of achievement focuses precisely on such apparently attainable 'middling' liberty and security, not of course the rarified and inconceivable heights of the cadre-capitalist elite.

For effectively all existing strata across the steeply unequal socio-economic hierarchy of contemporary Chinese society, therefore, it is the middle – and the wealthier, more educated upper ranks – specifically that emerges as the 'respectable' (cf. Hanley 2016) class identity. And this identity and discourse *in its very breadth and lack of concrete positive substantive definition* today affords its widespread self-appropriation by hundreds of millions who may simply be aspiring to such status but are very far from attaining it on any objective criteria, let alone all the dimensions conventionally connected with 'middle classes' in the Euro-American sociological literature (e.g. profession, education, property ownership etc....). For instance, as much as two-thirds of the population self-define as 'middle class' against a best estimate of at most (a steady, not rising) 12 per cent who actually fulfil socio-economic criteria with which such achieved status could reasonably be associated (Goodman 2015). In these circumstances the impossibility of understanding the Chinese 'middle class' in conventional terms of class-as-goods and settled social fact rather than as an ongoing socio-political strategy is thus palpable.

Three dynamics of middle risk-class emergence in China

Atop this existing Chinese socio-economic stratification, then, we can see how the dynamics of the 'middle' risk-class emergence are particularly intense. Three ways, in particular, stand out in this regard. First, consider the dynamics of liberty/security regarding compressed and intensified global risk-society, in complex, multi-factorial and iteratively interacting and compounding positive feedback loops. Here we find that the uniquely severe exposure to 'liberty' and 'security' of global risk-society in contemporary (urban) China, when overlain on the highly unequal yet dynamic and still-forming socio-economic stratification of that society, is particularly productive of a middle risk-class.

To recap, by 'liberty' we mean the intoxicating promise of growing personal autonomy that is both an extraordinary, yet now utterly normalized, development in modern Chinese history – in turn normalizing in people's lives a titanic rate of social change – and a unique experience in the world today, vs both a stagnating global North and political economies across the global South, even in the other 'BRICS', that are now deeply unsettled. And by 'security' we mean, the manifestation of (global) risks, of pollution, financial system turbulence, climate and weather, food, water and energy/resource security and 2.0 emergence of the post-human 'internet of things'. And regarding both together, the potentially existential 'onto-political' (Tyfield 2017) opportunities and threats of being personally included or excluded from the cycles of deepening accumulation of the assets – and specifically the power/ knowledge assets, such as education, social networks, experience – that underpin being able to surf upon and prosper from, not sink beneath, the waves of global complex society. Finally, the very intensity of 'liberty' and 'security' are themselves, of course, profoundly conditioned by and feed back into the unique conjuncture of IO and UF specific to early twenty-first century China.

In this situation, therefore, both carrots and sticks for risk-class emergence are uniquely big, and it is the *middle* risk-class, as potential but not guaranteed winner of this conjunction of liberty and security, that is most enabled and incentivized to pursue them. This is all the more so given the reflexive dynamics of global risk liberty/security through issues of social stratification themselves, and again in ways that are exceptionally Chinese. Not only do these circumstances of liberty and security objectively condition positive feedback loops of pursuit of (networked-)indi-vidual advantage, compounded by flight from rising tides of global risk (proverbial, and possibly literal in a sinking Shanghai or the entire Pearl River Delta, two of the most densely populated and flood-risked regions on the planet). But this is also necessarily *in competition with others* for intrinsically scarce assets and advantages: whether now and specific to China, as in good, powerful *guanxi* connections; or stretching in the near-term future and more general, regarding what will necessarily be more-or-less limited places at elite education institutions or residence in prime-located, clean and/or liveable environments. In a society as uniquely fast-developing, populous and socio-economically unequal as China, this means hyper-competition with hundreds of millions of others who are all and always seeking to 'move up' the social hierarchy in search of greater liberty and security – and who you *know* to be nipping at your heels no matter how high or fast you climb.

As such, we see that (risk-)class is not only a process *objectively*, but is experienced *subjectively*, as a (perhaps rapidly) depreciating asset that is simply one's strategic advantage in the *next* round of competition. One's middle risk-class status itself is thus never a secure achievement not just because of the objective complexity, uncer-tainty and unpredictability of global risk-society itself but also because of the hyper-competitive individualized strategizing of others. Moreover, this very dynamism for the seemingly attainable relative security of middle-class status renders the substance and practices of 'middle classness' itself highly dynamic; and in a country where such status has never been sedimented but is itself new, fluid and, for the time being, incorporating of a diverse multitude. A new but omnipresent risk regarding risk-class thus also emerges in terms of the danger not just of falling behind objectively, but simply of failing to stay abreast of what *counts* as middle class now, and which (power/knowledge) assets one should be cultivating or displaying. Given the uniquely relational, vigilant social network-regarding individualization that is charac-teristic of China today, therefore, this all feeds directly the dynamic of the collective and systemic emergence of the new risk-class system centred on its winners, the middle class (Figure 8.1).

This leads to the second issue and manifestation of liberty/security, namely the essential translation of these dynamics into an emergent socio-political discourse that attaches one's success, achieved and/or felt-to-be-deserved, to the specifically indi-vidual basis of one's moral and cultural quality, or *suzhi*. Again, this discourse bears the marks of the immoveable object of the Party-state, as a key term in the broader

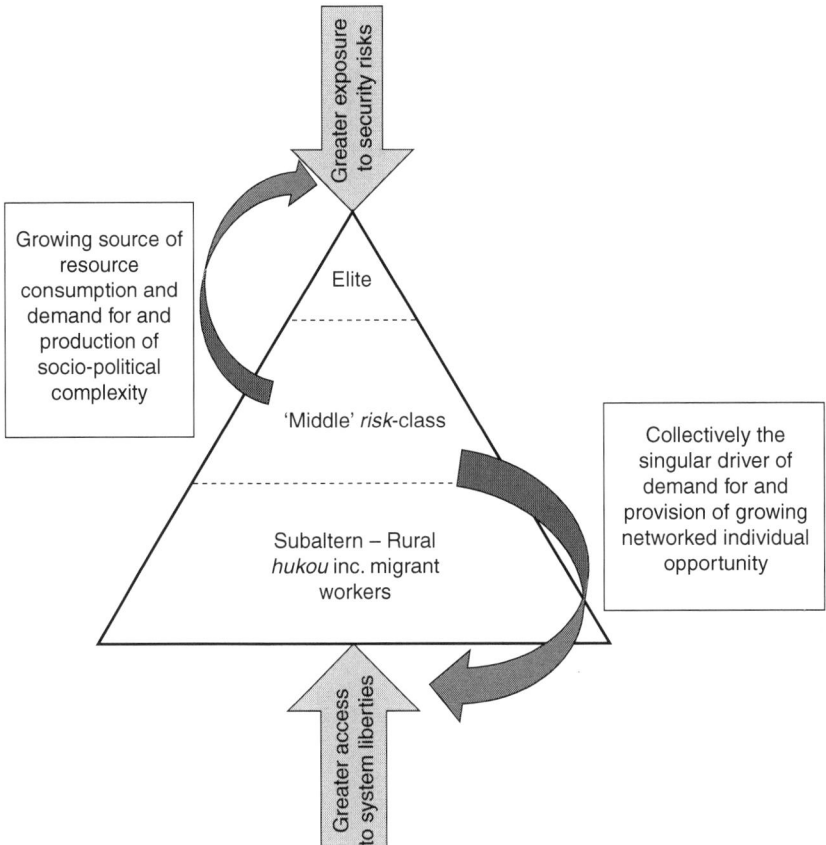

Figure 8.1 Middle *risk*-class.

governmental project, central to the ongoing renewal of its relation with its citizenry, of producing 'civilized' citizens capable of being 'governed at a distance' (Ren 2013; Tomba 2009) and entrusted to be 'responsible'. For this too comes with the systematic distinction of those who *cannot* be thus trusted and instead are constructed as the legitimate objects of paternalistic government, of whom the epitome, given China's system of graduated citizenship, is the rural migrant worker. *Suzhi* thus denotes a power/knowledge technology of personal character judgement that has sought to settle a supposedly meritocratic basis for collective stratification and stability of urban China's existing yawning inequalities (Anagnost 2008; Kipnis 2006; Jacka 2009; Crabb 2010).

Given the importance of economic autonomy and networked-individualized (possibly 2.0-mediated) consumer display and 'conspicuous achievement' (Yu 2014) described above, this is then also especially performed and displayed in one's taste and consumption choices. Compounding, therefore, the social and political significance and complexity of consumption in China, this adds a particular dynamism to

Chinese consumer tastes. An accelerating treadmill emerges from individual pursuit of the twin goals of being able to display an unusual level of achievement manifest in the latest, fashionable displays of success but in ways that are sufficiently well-networked to be recognized as such. Brands, of course, are a key power/knowledge technology in this process, but in this technology *which* brands count as 'middle class' is a fast-changing field. As recent developments in taste show (*The Economist* 2016a), however, away from the more ostentatious trappings of Western luxury brands (e.g. the duty-free airport fare of Louis Vuitton bags, Rolex watches, French make-up and perfume or Italian tailoring or sports cars etc....) to more deliberately understated and possibly East Asian brands (such as Korean make-up), it is the modest, hard-working middle class that are framed as the highest 'quality' persons here not the elite. *Suzhi* thus becomes a key motivating factor shaping consumption decisions – or demand – just as, *vice versa*, conspicuous achievement is a crucial factor in the governmental work of *suzhi*, shaping and constituting the hyper-competitive subjectivity that is constantly searching for the middle-class success (Figure 8.2).

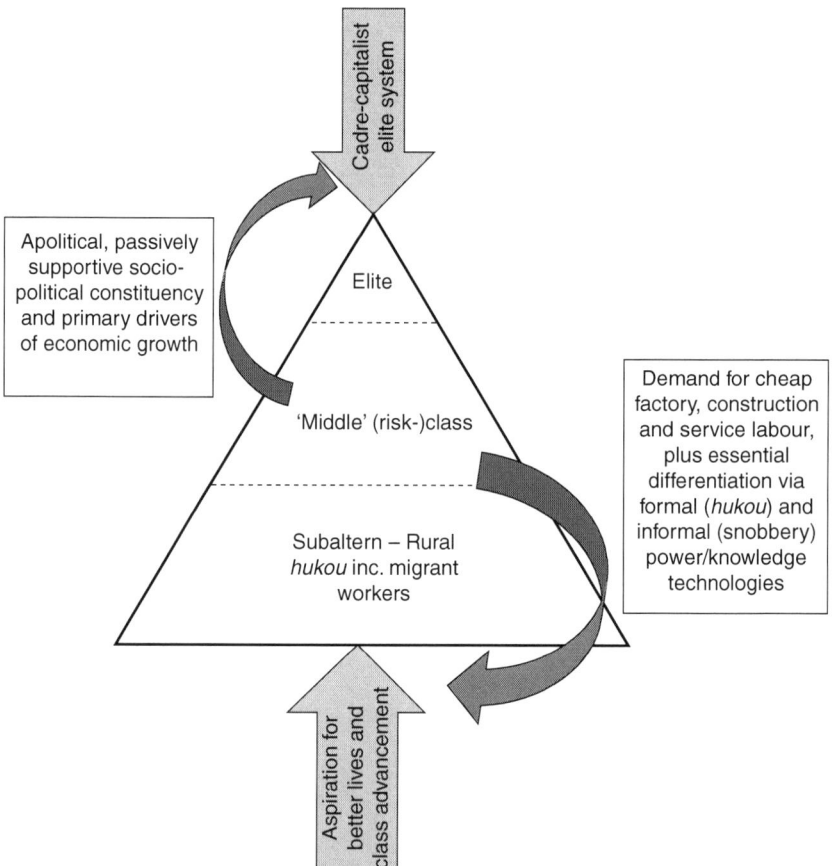

Figure 8.2 Middle class and consumption.

The importance of consumption in the broader government of contemporary China, shaping and constituting anew both the apparatus of a capital-G Government of the state and the everyday orderly conduct of increasingly individualized Chinese subjectivities, points to the final issue. This concerns how 'class' (and thus what is implicitly *risk*-class) is particularly pivotal as an arena for power/knowledge action in China, as discussed above. On the one hand, the deepening incorporation into global capitalism has unquestionably subjected Chinese society to its characteristic progressive individualization and uneven development. Indeed, as we have seen, amidst a compressed modernity that encounters global risk individualization without first having constructed a comprehensive welfare state, and with social interpersonal trust already ragged after the ravages of the Cultural Revolution, this is a process that is particularly pitiless and brutal. It has inevitably, thus, catalyzed a process of growing individualized demands in active pursuit of their growing autonomy and individual self-definition and in defence of what they have achieved thus far.

This process (again, precisely of liberty and security) thus underpins a constant escalation of individualized but potentially collective grievance from the personal level of sub-politics to higher, more organized, explicit levels of government. In the West, this has historically led to the growing explicit contestation of an increasingly self-conscious bourgeoisie, both at the level of a newly constituted 'public sphere' and 'civil society' and at the heights of Government, in the form of political parties representing newly collective interest (e.g. Weber 1978). But on the other hand, the high politics of Government in China today is clearly excluded as a zone of legitimate political action for the majority, limited instead to the concealed and illegible wranglings of various Party factions. And, indeed, the greater the pressure from 'bottom-up', the greater the impetus 'top-down' to contain and exclude such explicitly (capital 'P') Political expression and to renew the Party-state regime itself; a key dynamic of the IO vs UF again, as in the contemporary drive of authoritarian illiberalism from the Xi administration.

The result, therefore, is the systematic containment of these bubbling political forces – forces constitutively essential to the continued *vitality* of the CCP Party-state, as much as essentially in tension with it – at the level of new forms of collective identification regarding permissible forms of individual expression, notably spending. These new forms of collective identity and socio-institutional relations emerge as making sense to, and of, the lives the newly-autonomous live and affording new forms of power-relational enablement without demanding explicit political organization or mobilization. The primary form of this, thus, in the individualized and competitive context in play is precisely that of 'class'. Moreover, at the tier above 'class', such embryonic public sphere as has emerged in recent years, facilitated by the (specifically Chinese) social media accessible on the ubiquitous smart phones of the urban 'middle' classes, especially pertains to issues of (global) risk, such as the environment, as an area of circumscribed but comparative openness of debate (Geall 2013, Calhoun and Yang 2007). And, of course, it is specifically the 'middle' class – young, educated, urban, cosmopolitized and digitally native – that dominate those discussions (Figure 8.3).

Together then, these three dynamics, all set within the relatively concrete context of the highly unequal Chinese socio-economic stratification and the more abstract (but no less real, as diversely manifest) tension of the IO vs UF, condition in China as nowhere else a dynamism towards the emergence and self-assertion of a new 'middle class' that, as a middle *risk*-class, is also the vehicle and driver of the emergence of that new social category and power/knowledge system logic per se.

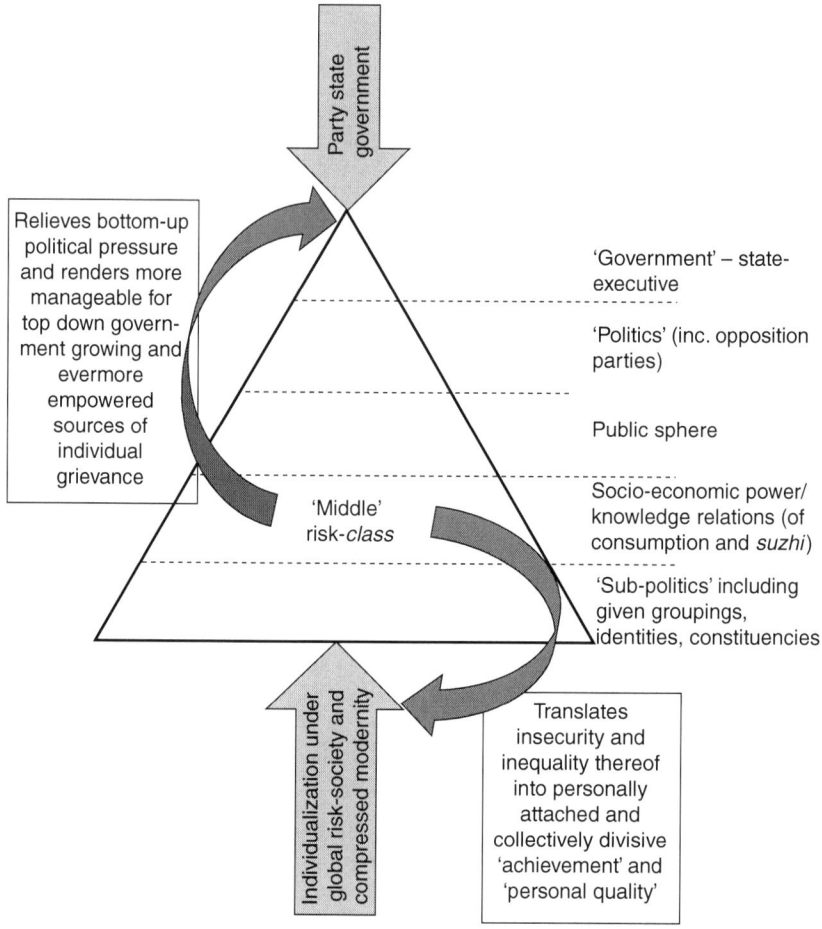

Relieves bottom-up political pressure and renders more manageable for top down govern-ment growing and evermore empowered sources of individual grievance

'Government' – state-executive

'Politics' (inc. opposition parties)

Public sphere

Socio-economic power/ knowledge relations (of consumption and *suzhi*)

'Sub-politics' including given groupings, identities, constituencies

'Middle' risk-*class*

Party state government

Individualization under global risk-society and compressed modernity

Translates insecurity and inequality thereof into personally attached and collectively divisive 'achievement' and 'personal quality'

Figure 8.3 Middle risk-*class*.

The imminent and immanent emergence of the middle risk-class-for-itself

Before we can tie the threads of this discussion back together, though, and return to our central issue of the shape and emergence of demand in China for low-carbon and complexity-attentive innovation towards system transition, we must attend to one final issue; namely the imminent historical climax of the trends described above, with the emergence into self-consciousness of what is already forming as a middle risk-class-*in*-itself as a class-*for*-itself.

In fact, the trends and tendencies towards this historical tipping point are mul-tiple and clear and increasingly apparent empirically. We have already considered, for instance, how it is the middle-class of highly-skilled knowledge workers that are the explicit bearers of the latest incarnation of the CCP's national project – tying together with our conclusions on the supply-side of China's innovation competitiveness, especially regarding what Tse (2016: 51) calls the '4th wave of Chinese entrepreneurs'. This is true for almost every goal set out by the Xi/Li administration over the past few years and formulated and formalized in the most recent 13th Five-Year Plan and its associated flood of legislation, policy announcements and funding drives. Hence, whether regarding the 'China Dream' of generalized, 'moderately well-off' (*xiaokang*) prosperity ... or the 'new normal' of slower but more stable and service-oriented growth ... or the escape from the 'middle income trap' via an 'innovation-oriented society' rebalanced to high-quality, high-value-added goods and their consumption ... or the building of a globally-leading 'ecological civilization' ... in every case, it is explicitly the middle class that are imagined as the goal and agents of these slogans.

Indeed, perhaps the most vivid example of the centrality of the 'middle class' to the contemporary CCP regime is just how bourgeois and strikingly *un*-socialist are Xi Jinping's key propaganda drive of '12 core socialist virtues' – inescapably pasted on every possible wall, billboard and digital dot screen today in public space, public transport and even private shopfronts. For instance, 'prosperity', 'patriotism', 'democracy', 'rule of law' and 'harmony' feature prominently – even as the actual meaning of many of these is unclear at best – while 'revolution', 'internationalism', 'dictatorship of the proletariat', 'rule of the Party' and 'class struggle' get nary a mention. If the former list read like motherhood and apple pie it is precisely because they are ... to a liberal, capitalist global common-sense.

But perhaps the best way to see the imminent climax of the dynamics of the middle risk-class is to explore the immanent convergence of each of the three trends outlined above (see Figure 8.4), of:

a   intense exposure to global risk society;
b   *suzhi* as government through consumption-based ascription of personal character;
c   the concentration of the politics unleashed by individualization onto issues of class as systematically *de*-politicized collective social identity.

As regards (a) and (c), first, we find the increasing tension between qualitatively changing, increasingly self-assertive and increasingly system-central middle class aspirations and what the incumbent conditions of both the IO within China and the UF without, and their heightening tension in turn, is able to deliver. This is so even as both the CCP policy and global neoliberal 'common-sense', as formulated in G20 communiques or IMF, World Bank or WTO pronouncements, are increasingly explicit in their support of Chinese 'middle class' aspirations *for their own* self-preservation. Yet the condition of hyper-competition besetting the Chinese middle class is eliciting ever-greater orientation to system risks and to the solutions for them.

This especially includes fairer, better and simply more access to the power/knowledge assets that equip one as a potential system winner in global risk-society. In such key issues as world-class education, health, environment, media and circulation of objective information and comment, and opportunities for profitable but not high-risk investment, saving or credit, however, these are all assets that the CCP will quintessentially struggle to provide within China without constitutively weakening its grip on state power; just as, conversely, the neoliberal system equally is designed

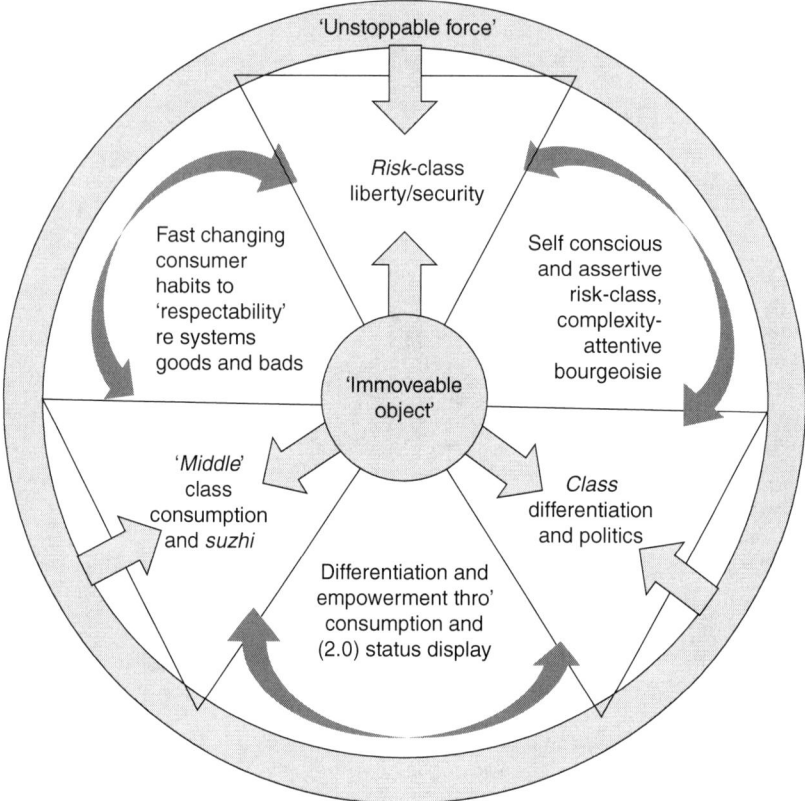

Figure 8.4 Middle risk-class and IO vs UF.

to provide them globally only to a global elite. And the intrinsically fundamental challenge posed to both complex power/knowledge systems by these conditions is simply compounded by the climaxing tension of both, feeding each other given their existing systemic interdependence. In these circumstances, therefore, while the growth of the middle risk-class continues to support both IO and UF and their mounting clash, the attempts by both to rectify the Four Challenges just tend to exacerbate the system dysfunctions – as in the cack-handed attempts to manage the Chinese stock market volatility and crashes of autumn 2015 – while they continue both to fail to support actively and actively to hinder system transition innovation.

Whether in terms of the growing demands amongst the middle-class, and on essentially *a*political, pragmatic grounds of frustrated self-advancement, for greater legal certainty and transparency, security of property, reliable information, predictable government and voice over 'legitimate' defence of liberties won to date; or in terms of enabling, and not holding back, ambitions, fed by official government policies and initiatives, of higher quality urban environments and innovation-oriented knowledge-intensive work … In all these ways, the progressive development of (a) and (c) above is converging in the qualitative emergence of a self-conscious and assertive risk-class bourgeoisie, characterized by and specifically enabled by their distinctively complexity-attentive and liberal-capitalist concerns. At its limit, with system dysfunction being exacerbated, not mitigated, by immanent tensions, there is a particularly strong possibility of a system crash (of globally significant proportions) – e.g. of real estate bubbles and/or the ballooning corporate and local government debt – in China from which the middle class have most to lose. Yet they are also becoming ever more central to the plans of how to avoid such catastrophic system collapse and would, in any case, likely be further empowered by its realization.

Second, as regards *suzhi* and class as the primary locus of the politics of individualization, these already combine in ways eliminating what is, in principle at least, an alternative route to rebalancing the Chinese political economy towards consumption, high-quality services and high-skilled knowledge employment; namely a broader-based socio-economic emancipation of the industrial and service workers and those still living in rural poverty. For if the reluctant but unrelenting empowerment of the essentially system-conservative urban 'middle' classes is a problem for the incumbent regime, a broader emancipation of those below them would be a much more explosive and uncontrollable route to take, while also going directly against the main thrust of both Government and socio-cultural sub-politics already in place, as in the instituted division of society and the snobbery and fear of the 'floating population' this has entrenched.

As such, even if (elements of the) Party-state would prefer this route over empowering just the middle class – perhaps based on some residual ideological attachment to socialism – it would likely be strongly *resisted* by the urban middle-class, who depend on that class exclusion in their everyday lives (e.g. in the form of domestic service and childcare) and for their economic prosperity (e.g. in terms of the systematic repression of wages that enables in turn solid profits for them or their domestic or foreign employers to cream off). By contrast, rebalancing towards greater economic empowerment of the middle classes may be rebalancing *enough* from the perspective of re-regularizing capital accumulation and power/knowledge relations, even as it will certainly fall far short of the broader 'social upgrading' (Butollo and Lüthje 2016) that many see as needed, on normative grounds if nothing else.[2]

So, on the one hand, *suzhi* and concentration on the politics of class feed the division of society in the ongoing constitution of the system *losers*, and the deepening of this division through establishing the *legitimacy* of that division. For those with low *suzhi* can thereby be dismissed as failures in life and society because of their own personal failings and character shortfalls. But the flipside of this is equally in train, namely the growing sedimentation of the common sense that the middle class are system winners because of their *high* personal quality and hence deserve their success and its further enablement. At present, then, the middle class have a growing sense of self-righteousness regarding their uniquely commendable contributions to the national political economy and socio-political order, again with this self-consciousness reaffirmed, not gainsaid, by official rhetoric and programmes of public education and civilization (*wenming*).

Not only does this cultivate a deepening asymmetric interdependence between Party-state and middle classes (as we saw also on the supply-side), but insofar as the added complexity and frustration to middle class urban life introduced by the former for the latter persists, it also cultivates a growing pragmatic, experimental experience of pushing beyond simply network-individualized consumption choices and display into other forms of pseudo-collective self-assertion. This is particularly noticeable in areas and ways, discussed above, that either concern issues that have tacit acceptance from the Party-state, such as environmental issues or home ownership, or are not fundamentally and directly antagonistic to that regime, including especially via online 2.0 social media comment in an embryonic but still highly policed 2.0 public sphere (MacKinnon 2013; Xiao 2011; *The Economist* 2016b), or both.

But in each case, it is incrementally accumulating a growing emergent self-aware collective (and nationalist) identity and imagined community (Anderson 1981) amongst middle-class strangers, as well as their empowerment *within* the state apparatus, via forms of 2.0 consultative Leninism (Tsang 2010) or the Party-state's monitoring of social media to gauge (a specifically middle-class) public opinion (Mertha 2009; King *et al.* 2013). Together then, these manifest most importantly in a growing self-righteous pride and sense of universalistic legitimate entitlement that is broadly shared also by many outside this group, both above and below, and hence unifies the fissiparous, individualizing social totality as a whole in ways that are self-confirming. Here, in other words, are the burgeoning seeds of a self-conscious 2.0-networked Chinese middle class as Gramscian historic bloc (Gramsci 1971; Arrighi 1994; Rupert 1993), who are also increasingly self-aware of their collective identity and systemic power as such – and both within China and, via China's pivotal place in global capitalism, the world.

Moreover, as with the convergence of (a) and (c) above, this must be set against the short-to-medium term prospects for *worsening* frustrations to those growing demands, which leads to the final conjunction of (a) and (b), intense global risk exposure and *suzhi*. In this final instance, we find the tendential synthesis of the fast-evolving consumer tastes with deepening exposure to global risk society towards the growing prioritization specifically amongst the Chinese middle classes of post-material concerns for 'liveable', attractive, clean urban environments (that they have perhaps already developed a taste for overseas or even set one foot in elsewhere, with a second home in Vancouver, Singapore or London) and their mobile or touristic consumption. This includes also a shift already evident, going beyond or reshaping 'conspicuous achievement' in the realm of consumption into forms of 'conspicuous conservation' (cf. Griskevicius *et al.* 2010), or the (perhaps deliberately-reserved-yet-still-ostentatious, or inconspicuously conspicuous) display of one's green credentials and environmental consciousness, *as well as* one's success and style.

Of course, the significance of that display is precisely to signal a meta-level shift in the nature of the concerns of this middle class. This is the move away from flashy brands flaunting sheer monetary gain to a more explicitly normative concern for the 'good of society' and of 'the planet'. The latter lends itself in turn to a further qualitative elevation in the quasi-universal *moral* standing of the middle class vis-à-vis the collective interests of Chinese society as whole; a crucial element of any possible hegemony. Moreover, this new moral legitimacy is specifically regarding the new collective, systemic issues of global risk-society. Hence the middle risk-class emerge as a *new* constituency that stands out in being *constitutively* oriented to the *novel* predicament of proliferating system bads in ways that other, existing groups (in China and globally) are not. For insofar as many low-carbon innovations or environmental products and services remain relatively expensive commodities and assets purchasable on the market – not provided as public goods – they will necessarily be accessible to, and attractive to, those who have a reasonable disposable income already and are thus also oriented to such post-materialistic concerns but are also exceptionally exposed to global risks.

It follows, though, that the specific demand for such innovations represented by these groups and their ongoing *constitution* also feeds directly the logics of social polarization, into system winners and losers, but now with the added propulsion of a growing, self-propagating and system-supported *self-righteousness* – a powerful power/knowledge resource – amongst those who can and do buy such cleantech commodities over and against those who can't and don't. Finishing off the cycle, then, is the progressive development and marketization of such sustainability goods and services – again as further power/knowledge technologies – specifically tailored to, and so disproportionately empowering, their primary consumers, the Chinese middle classes.

Moreover, this emergent dynamic of positive feedback between division into green-virtuous middle class system winners and losers is fed by two further factors that seem all-but-inescapable regarding low-carbon and environmentally sustainable system transition in China: first, that it will necessarily progress relatively slowly (e.g. Zhang 2016; Kahn and Zheng 2016: 219, 223 regarding slow improvements in air quality at best, or regarding projected coal consumption), and probably alongside *worsening* exposure to some global risks in the meantime, with improvements limited to specific areas in China's massive territory and even then at a pace that feeds the impatience of the system winners; and, second, that given China's size and the inevi- tability that it will be both a rich *and* poor country, in global terms, for some time into the foreseeable future, it will also be home to disparities of exposure to global risk that will be evident and striking for all, in ways that, say, African poverty is not generally set alongside Californian wealth.

Such marked divergence, however, will be processed and translated by the *exist- ing power/knowledge system* with its sharp-elbowed hyper-competition, thereby likely *feeding* that very dynamic and, in turn, those of the society's deepening polarization. For instance, with no chance of any imminent equalization in cleaning up environ- ments and rebalancing urban/rural inequality, greening in the highly developed mega-city areas – perhaps slowly to world-class environments for work, home-life and leisure – are just as likely to catalyze the internal off-shoring of the heavily pol- luting industries to more remote, inland locations, out-of-sight but not entirely out- of-mind, thereby exacerbating the environmental *harm* in these regions. But, on the one hand, the clean environments will still be held up as shining beacons for the system as a whole, and achieve an impermeable moral gloss that preserves and per- forms their unique and incontestable universal moral legitimacy. While, on the other, the deepening exposure to global risks and the *cycles* of clustering dis- advantage these entail atop an individualized moral economy could well afford not necessarily growing sympathy and concern for those left behind (at least beyond their pre-adolescent childhood, perhaps as 'left behind children') but their increasing constitution in public discourse as the undeserving poor, whose failure in life is simply the inescapable corollary of their low *suzhi*.

The convergence of the middle risk-class and *suzhi* thus spells the progressive emergence of a new dynamic propelling the middle risk-class itself towards its qual- itative transformation. Here, global risks and a new risk-class moral economy specifi- cally of their production and distribution, benefits and burdens, are placed ever more prominently at the constitutive core of a new power/knowledge system, with the middle class cast from the outset as system heroes and saviours. In this regard, then, the final key aspect of this imminent climax is how the middle class themselves, in their very orientation to liveability, are currently being forced to confront the essential paradoxes of the clash of their growing material consumption and aspiration and the realizability of the clean, attractive, mobile city-living they increasingly desire. For it is this group itself and its growing demand for cars, (free-access to (see *The Economist* 2016c)) roads and parking space, warm or cool housing run on coal-generated electricity, cheap food or technology that has to-date been the main source of runaway demand for the production processes that generate poor air quality, burgeoning GHG emissions and/or the poor safety standards of many Chinese goods.

It is, therefore, again the middle class (in all its breadth and diversity) in particular that is confronted with this essential tension such that they are increasingly forced to take on system bads, and what thereby transpires to be the 'unnecessary' extent or profligacy of their production, as their *personal* problem. This, of course, thus feeds back into the dynamic above of conspicuous conservation and their increasing self-regard as agents of sustainability innovation and system transition. And it also further differentiates the middle class in their own estimation as against a decadent elite (both cadre-capitalist and Western) who benefit disproportionately from the current production of system bads while also escaping their consequences (Curran 2016), and an 'uncivilized mass' (again both Chinese and probably also across the global South, and even many in the global North amongst the erstwhile working class) who have 'system-irrational' habits and 'self-undermining' personal vices rather than living 'virtuous' lives of system-bad-mitigating consumption and system-good-*vs*-system-bad production (as in low-carbon innovation or high-skilled knowledge work jobs tackling the Four Challenges).

Altogether, then, (a), (b) and (c) are converging on self-propelling power/knowledge dynamics of the self-awareness of the Chinese 'middle class' as a specifically bourgeois risk-class of 2.0-networked and CP/KS-empowered green innovation-consuming and -producing individuals. Here, then, (a) risk-class provides the formal cause or novel category, (b) *suzhi* or inter-subjective assessment of character and personal quality the material cause or substantive defining characteristics and practices and (c) class-as-politics the efficient cause or empowerment.[3] The 'middle class' as prolepsis and power/knowledge technology is thus the key to this whole process, as a deliberately loose and even slippery concept – if definitely not a totally empty signifier. It is enabling and driving the ongoing constitution of a new moral economy that affords the productive containment and harnessing of contemporary China's multiple essential antagonisms and contradictions; and doing so specifically in ways conditioning transition of the system as a whole to a new, constitutively global-complexity-attentive middle-class-centric CP/KS.

In short, it signals the imminent emergence of the Chinese middle risk-class as a class-*for*-itself, in parallel co-production with novel power/knowledge technologies – or innovation-as-politics – for low-carbon system transition and systematic global risk reduction that is oriented specifically to the deepening strategic advantage of this group. Finally, then, we are back where we started. For these are thus the dynamics of the formation and shaping of the *demand* for sustainability-oriented, complexity-attentive innovation, as a new socio-political constituency with the immanent power momentum – as epi-phenomenon and unintended product of existing power/knowledge system dynamics, not in direct confrontation with them from without – sufficient to effect a progressive system transition in the medium term: a Gramscian historic bloc in-formation.

This potentially world-changing power momentum, however, resides primarily in the emergence of the new social categorization of risk-class itself. This is a new system of socio-political stratification *and* unification matching the new challenges of global knowledge capitalism, or complex risk-innovation-society, and the harbingers of its emergence that are the Four Challenges. Hence the middle risk-class as class-for-itself are also, as putative winners of this new socio-natural order, a class-*for-the-system-of-risk-class* itself. In this case, though, the very power momentum that makes possible a complexity system transition is also and indissolubly a dynamic of socio-economic polarization and division. And where that polarization itself propels, rather than undermines, the power/knowledge momentum of system transition even as it also feeds the dynamic of its essential contestation (Figure 8.5).

Finally, completing the picture is how this dynamic also promises to resituate the heightening tension driving the whole process in China of the IO vs UF. For the very emergence of the Chinese middle risk-class, together with its parallel sponsorship of complexity transition and its increasing systemic centrality, both within China and globally, also augurs the qualitative transformation of both immoveable object and unstoppable force – CCP Party-state and neoliberal financialized globalization – both in themselves and in their mutual interaction and inter-relational constitution (see Figure 8.6). We will discuss this final point in more detail below (see Chapter 11), but suffice to say at this stage that with the middle risk-class moved to hegemonic dominance and centrality in both China and the world this would involve both significant qualitative change and probably also significant continuity.

Conclusion

In these two chapters we have considered the demand side of innovation-as-politics in China towards a complexity- and sustainability-oriented system transition. As with the supply side regarding 'innovation' itself, here we have seen how 'class', as a key term shaping the emergent socio-political constituency that is driving demand for such innovation, must also be significantly redefined in light of the qualitative novelty and emergence of such a complex power/knowledge system. Likewise, in

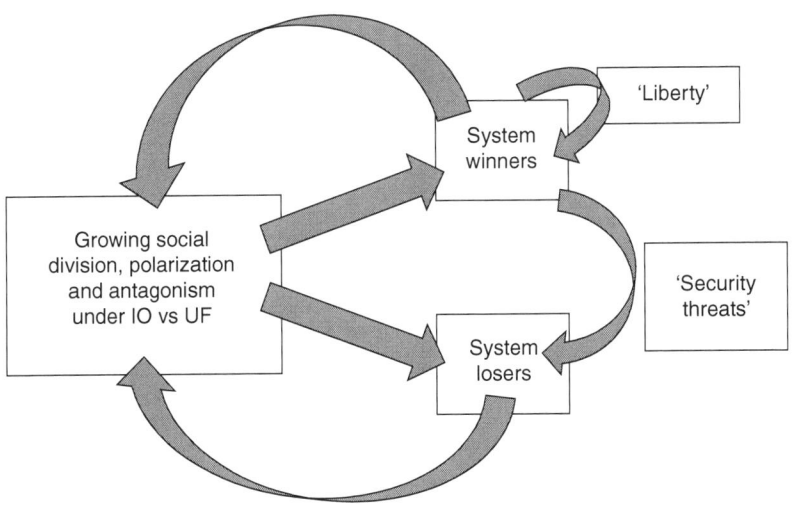

Figure 8.5 Polarization as system-productive not -destructive.

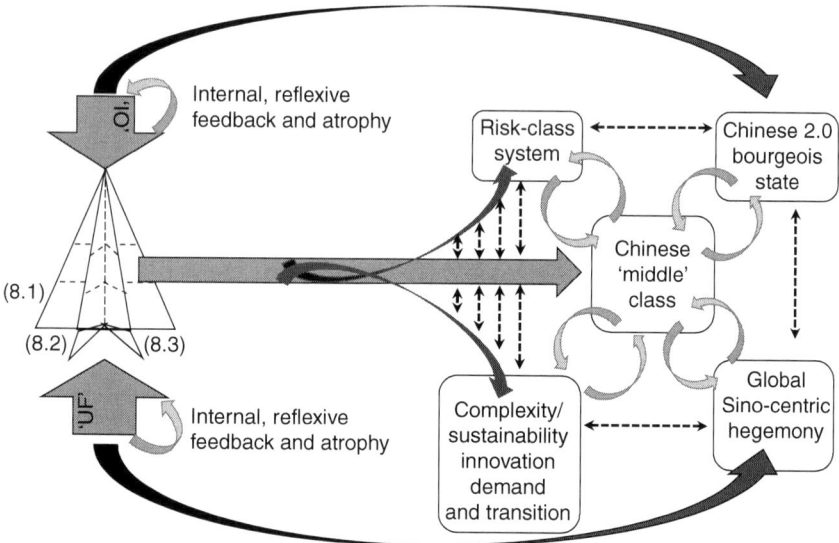

Figure 8.6 Towards the emergence of a historic bloc.

the impending convergence of major trends – trends that are making so momentous in China the dynamics of the emergence both of risk-class as a social category and of the middle risk-class driving that process – we find again a powerful impetus beyond even the existing system-functional role played by the Chinese middle class within the clash of immoveable object and unstoppable force. But so too the foregoing also sets out the capacity, agency, process and conditions for that process to be highly system productive rather than simply catastrophic – and in ways that precisely feed the parallel co-emergence of risk-class per se, the middle class and complexity-innovation transition, or what together add up to a(n urban, mega-city-based) Chinese(-centric) global capitalist resettlement (Table 8.1).

First, in terms of capacity, the emerging Chinese middle risk-class are marked out by several key characteristics: their massive collective consumer power and consumption-based aspiration amid slow global growth; their 2.0-enabled and networked-individually competitive venturesome consumption of gadgetry and (increasingly) system-bad-mitigating innovation; and their unique familiarity with and acclimatization to the pursuit of personalized advantage in the form of growing complexity-enabled autonomy and security from global risks amidst a particularly complex and illegible context of power/knowledge relations and practices. With over 200 million graduates, concentrated in STEM subjects, by 2020 (Tse 2016), this is also likely a highly scientifically literate, and so both hi-tech innovation-oriented and even *more* risk-attuned (Kahn 2002), group that is uniquely massive, notwithstanding what is probably the huge range in quality of their scientific educations.

In terms of agency, too, as this whole chapter has detailed, the Chinese middle risk-class are emerging as increasingly system-central and empowered. This is in terms of both their technical power/knowledges, propelled by their constitutive

Table 8.1 The quadrant of Chinese innovation-as-politics – demand-side

	Direct effects (at agent level)	*Indirect effects (at system level)*
Intended (*immoveable object*)	Continued high GDP growth feeding Chinese industrial and export-oriented 'middling' class as system winners and supporters	Emergent Chinese 'middle class' as entrepreneurial and private sector class • Essentially attached to socio-political order and *status quo*, pro Party-state and anti *nongmin* empowerment • Dependent on a strong state for professional opportunity (globally and nationally) • Foundationally adept at surfing complex global risk society at its most intense • Pragmatic and apolitical
Responded (*unstoppable force*)	• Yawning inequality from cadre-capitalist elite to subaltern *nongmin* migrant workers • Essential tension eliciting flight to safe harbour of 'middle class' status by urban majority, including many objectively outside this group • Dynamic of liberty/ security established regarding networked-individualized hyper-competition	• Increasing empowerment of a fundamentally self-serving, cosmopolitized and networked-individualistic critical mass only conditionally attached to Party-state regime and with multiple immanent (complexity-related) grievances • Support conditional on economic growth that is slowing amidst political economic reform that demands their further empowerment • Growing awareness of unnecessary abundance of regime-introduced complexity and frustration to everyday 'middle class' life • Growing self-conscious orientation to system bads and centrality in innovation efforts to mitigate them.... • ... Feeding growing pseudo-universalistic status as system heroes and saviours.

orientation to global system bads and their mitigation *and* the commercial opportunities thereof; and their power/knowledges of legitimation and quasi-universal moral rectitude, as archetypes of a new green, complexity-attentive virtue for the good of 'everyone' and the 'system as a whole'. Moreover, ongoing convergence with middle class expectations and living standards in the global North is underway. This has involved a prolonged stagnation underway in those countries, in the aftermath of debt-fuelled consumerist profligacy of privatized Keynesianism (Crouch 2011), and the concomitant sapping of these groups' power/knowledge enablement. In China, however, the middle class is ascendant, affording strong positive feedback loops of their progressive empowerment.

Regarding process, meanwhile, there is the oblique, non-linear and stuttering but relentless progress of the Chinese middle-class as a power/knowledge constituency within the informal, and increasingly also formal, architectures of contemporary government of that society. But, against the persistently disappointed expectations of many a Western observer, this is driven not by an increasingly strident and explicit political self-assertion against the CCP Party-state regime – precisely *not* a 'new class war' (Blau 2016). Rather it is in deliberately *a*political and pragmatic pursuit of a slowly accreting ambition and capacity for personal autonomy amidst global risk-society – liberty-security – through a hyper-competitive and highly pressurized everyday life.

Finally, in terms of necessary conditions, of crucial importance are the structures, institutions and infrastructures, the hard and soft power/knowledge technologies, of complexity-attentive, green knowledge capitalism. These are being deliberately incubated by the Chinese Party-state through industrial policies of uncommon ambition and financial clout. Whatever the future holds for the CCP itself, then, the legacies of these initiatives – themselves likely emergent in the non-linear processes described in Chapter 5 – will certainly remain standing and likely growing significantly, specifically enabling the continued growth of consumption and production capacity, demand *and* supply, of such innovation.

What remains for us to consider, then, is how these immanent, unfolding trends regarding innovation supply and demand in turn could come together, and what this means for our question at the start of this discussion of China: will China rule the world? And, of course, finally, how does this relate to and/or illuminate what comes after neoliberalism? As this is necessarily prospective and speculative, we can only proceed by zooming in yet further on to the meso-level dynamics of a specific, but increasingly pivotal, field of innovation and what is unfolding there in this regard. So it is to this we now turn, focusing on the key issue of e-mobility transition and what possible or plausible futures in this field tell us regarding these questions.

Notes

1 'Rule by law' connotes how formal laws increasingly exist as the key power/knowledge technologies of state Government but, implemented and enforced in the final instance still by the personal hierarchy of the Party-state, are simply tools at the disposal of the powerful, not relatively arm's-length structures that apply equally to them.
2 Normative grounds shared by this analysis, of course, lest this need be said.
3 And perhaps we could add, to finish off this Aristotelian scheme and jump to the conclusion in Chapter 11, liberalism 2.0 is the immanent, if not teleological, 'final' cause.

References

Anagnost, A. (2004) 'The corporeal politics of quality (*suzhi*)', *Public Culture* 16: 189–208.
Anagnost, A. (2008) 'From 'class' to 'social strata': Grasping the social totality in reform-era China', *Third World Quarterly* 29: 497–519.
Anderson, B. (1981) *Imagined Communities,* London: Verso.
Arrighi, G. (1994) *The Long Twentieth Century,* London: Verso.
Barton, D., Y. Chen and A. Jin (2013) 'Mapping China's middle class', *McKinsey Quarterly*, June.
Blau, R. (2016) 'The new class war', Special Report: 'Chinese Society', *The Economist*, 9 July.
Butollo, F. and B. Lüthje (2016) ' "Made in China 2025": Intelligent manufacturing and work': 42–61.
Cai, Y.S. (2005) 'China's modern middle class: the case of homeowner's resistance', *Asian Survey* 45(5): 777–799.

Calhoun, C. and G. Yang (2007) 'Media, civil society, and the rise of a green public sphere in China', *China Information* 21: 211–236.

Chan, J., N. Pun and M. Selden (2013) 'The politics of global production: Apple, Foxconn and China's new Working class', *New Technology, Work and Employment* 28(2): 100–115.

Chang, L.T. (2010) *Factory Girls: Voices from the Heart of Modern China*, London: Pan Macmillan.

Chen, J. and C. Lu (2011) 'Democratization and the middle class in China', *Political Research Quarterly* 64(3): 705–719.

Ci, J.W. (2009) 'The moral crisis in post-Mao China', *Diogenes* 56(1): 19–25.

Crabb, M.W. (2010) 'Governing the middle-class family in urban China: Educational reform and questions of choice', *Economy & Society* 39(3): 385–402.

Crouch, C. (2011) *The Strange Non-Death of Neoliberalism*, Cambridge: Polity.

Curran, D. (2016) *Risk, Power and Inequality in the 21st Century*, Basingstoke: Palgrave Macmillan.

Dickson, B. (2008) *Wealth into Power*, Cambridge: Cambridge University Press.

The Economist (2016a) 'From noodles to poodles', 2 July.

The Economist (2016b) 'Weibo warriors', 2 January.

The Economist (2016c) 'The great crawl', 18 June.

Ford, P. (2011) 'In China, toddler left for dead sparks heated debate about society's moral health', *Christian Science Monitor*, 19 October.

Fukuyama, F. (2012) Future of history: Can liberal democracy survive the decline of the middle class, *Foreign Affairs* 91: 53–61.

Geall, S. (ed.), (2013) *China and the Environment*, London: Zed Books.

Gold, T., D. Guthrie and D. Wank (eds), (2002) *Social Connections in China*, Cambridge: Cambridge University Press.

Goodman, D.S. (2008) 'Why China has no new middle class' in D.S. Goodman (ed.), *The New Rich in China: Future Rulers, Present Lives*, London: Routledge: 23–37.

Goodman, D. (2014) *Class in Contemporary China*, Cambridge: Polity.

Goodman, D. (2015) 'Locating China's middle classes: Social intermediaries and the party-state', *Journal of Contemporary China* 25(97): 1–13.

Gramsci, A. (1971) *Selections from the Prison Notebooks*, London: Lawrence and Wishart.

Griskevicius, V., J.M. Tybur and B. Van den Bergh (2010) 'Going green to be seen: Status, reputation and conspicuous conservation', *Journal of Personality and Social Psychology* 98(3): 392–404.

Guo, Y. (2008) 'Farewell to class, except the middle class: The politics of class analysis in contemporary China', *Asia-Pacific Journal* 26(2): 1–19.

Hanley, L. (2016) *Respectable: The Experience of Class*, London: Penguin.

Heberer, T. (2009) 'Evolvement of citizenship in urban China or authoritarian communitarianism? Neighborhood development, community participation, and autonomy' *Journal of Contemporary China* 18(61): 491–515.

Ho, W.C. (2013) The New 'Comprador Class': the re-emergence of bureaucratic capitalists in post-Deng China. *Journal of Contemporary China*, 22(83), 812–827.

Hui, E.S. (2016) 'The labour law system, capitalist hegemony and class politics in China', *The China Quarterly* 226: 431–455.

Hung, H.F. (2016) *The China Boom*, New York: Columbia University Press.

Jacka, T. (2009) 'Cultivating citizens: *Suzhi* (quality) discourse in the PRC', *Positions* 17(3): 523–535.

Kahn, M.E. (2002) 'Demographic change and the demand for environmental regulation', *Journal of Policy Analysis and Management* 21(1): 45–62.

Kahn, M.E. and S. Zheng (2016) *Blue Skies over Beijing*, Oxford and Princeton, NJ: Princeton University Press.

Kharas, H. (2010) *The Emerging Middle Class in Developing Countries*, Paris: OECD.

King, G., J. Pan and M.E. Roberts (2013) 'How censorship in China allows government criticism but silences collective expression', *American Political Science Review*, 107(02): 326–343.

Kipnis, A. (2006) 'Suzhi: A keyword approach', *The China Quarterly*, 186: 295–313.

Li, C. (ed.), (2010) *China's Emerging Middle Class* New York: Brookings.

Liechty, M. (2003) *Suitably Modern: Making Middle-Class Culture in a New Consumer Society*. Princeton RI: Princeton University Press.

Luo, Y. (2008) 'The changing Chinese culture and business behavior: The perspective of inter-twinement between guanxi and corruption', *International Business Review* 17: 188–193.

MacKinnon, R. (2013) *Consent of the Networked*, New York: Basic Books.

McGregor, J. (2007) *One Billion Customers: Lessons from the Front Lines of Doing Business in China*, New York: Simon and Schuster.

Mertha, A. (2009) ' "Fragmented authoritarianism 2.0": Political pluralization in the Chinese policy process', *The China Quarterly* 200: 995–1012.

Milanovic, B. (2013) 'Global income inequality in numbers: In history and now', *Global Policy* 4(2): 198–208.

Ngai, P. (2005) *Made in China: Women Factory Workers in a Global Workplace*, Durham, NC: Duke University Press.

Nyiri, P. (2010) *Mobility and Cultural Authority in Contemporary China* Seattle: University of Washington Press.

Ong, A. (2006) *Neoliberalism as Exception: Mutations in Citizenship and Sovereignty*. Durham, NC: Duke University Press.

Porter, K. (ed.) (2011) *Broke: How Debt Bankrupts the Middle Class*, Stanford: Stanford University Press.

Ravallion, M. (2010) 'The developing world's bulging (but vulnerable) middle class', *World Development* 38(4): 445–454.

Ren, H. (2013) *The Middle Class in Neoliberal China: Governing Risk, Life-building, and Themed Spaces*, Abingdon and New York: Routledge.

Rupert, M. (1993) 'Alienation, capitalism and the inter-state system: Toward a Marxian/Gramscian critique', in S. Gill (ed.), *Gramsci, Historical Materialism and International Relations*, Cambridge: Cambridge University Press.

So, A. (2003) 'The changing pattern of class and class conflict in China', *Journal of Contemporary Asia* 33(3): 363–376.

Solinger, D. (2012) 'The new urban underclass and its consciousness: is it a class?', *Journal of Contemporary China* 21(78): 1011–1028.

Sun, W. (2009) *Maid in China: Media, Morality, and the Cultural Politics of Boundaries*, London: Routledge.

Therborn, G. (2012) 'Class in the 21st century', *New Left Review* 78: 5–29.

Therborn, G. (2014) 'New masses?', *New Left Review* 85: 7–16.

Tomba, L. (2004) 'Creating an urban middle class: Social engineering in Beijing', *The China Journal* 51: 1–26.

Tomba, L. (2009) 'Of quality, harmony, and community: Civilization and the middle class in urban China', *Positions* 17(3): 591–616.

Tsai, K. (2007) *Capitalism Without Democracy: The Private Sector in Contemporary China*, Ithaca: Cornell University Press.

Tsang, S. (2010) 'Consultative Leninism: China's new political framework?', *China Policy Institute Briefing Papers* 58, Nottingham: University of Nottingham.

Tse, E. (2016) *China's Disruptors*, London: Portfolio Penguin.

Tyfield, D. (2017, forthcoming) 'Mobilizing the emergence of phronetic TechnoScienceSocieties: Low-carbon e-mobility in China', *Sociology of Sciences Yearbook 2017*.

Wang, R., and Q. Yuan (2013) 'Parking practices and policies under rapid motorization: The case of China', *Transport Policy* 30: 109–116.

Weber, M. (1978) 'The distribution of power within the political community: class, status, party', *Economy and society* 2: 926–940.

Whyte, M.K. (2010a) *Myth of the Social Volcano: Perceptions of Inequality and Distributive Injustice in Contemporary China*, Palo Alto, CA: Stanford University Press.

Whyte, M.C. (2010b) *One Country, Two Societies: Rural-Urban Inequality in Contemporary China*, Cambridge, MA: Harvard University Press.

Xiao, Q. (2011) 'The battle for the Chinese internet', *Journal of Democracy* 22(2): 47–61.

Yang, M.M.H. (2002) 'The resilience of guanxi and its new deployments: a critique of some new guanxi scholarship', *The China Quarterly*, 170: 459–476.

Yu, L. (2014) *Consumption in China*, Cambridge: Polity.

Zhan, J.V. (2012) 'Filling the gap of formal institutions: the effects of guanxi network on corruption in reform-era China', *Crime, Law and Social Change* 58: 93–109.

Zhang, C. (2016) 'Cleaner air is not clean air as ozone levels rise', *Chinadialogue.net*, 5 October www.chinadialogue.net/blog/9293-Clearer-air-is-not-clean-air-as-ozone-levels-rise/en.

Part III

Where are we going?

Sharing and haggling the long complex journey to green urban mobility systems transition in China

9 Electric vehicle innovation-as-politics in China

Introduction – urban e-mobility as key case

In the last part we explored in detail how the supply and demand sides of innovation in China, in the shadow of the Four Challenges, are taking shape and leading to strengths that are in unexpected and qualitatively novel forms. The question that emerges from that discussion – and gets us one step closer to our questions of 'will China rule the world? And what does this tell us about what follows neoliberalism?' – is how these twin tendencies are converging and where they, together, may be leading. This is the concern of this chapter, reconnecting the issues we had analytically separated above to explore their inter-relations and co-production. To do this we must zoom in to a more fine-grained and empirical analysis of a specific field of innovation, and preferably one that is systemically located such that, while not 'representative', it nonetheless affords particular insight into the system-level changes that Chinese innovation more generally is incubating.

There are many possible candidates for this domain of innovation, including especially the transformation of the production and consumption of food and of energy. We focus here, though, on the key issue of innovation transforming the system of urban mobility, beyond the dominant late twentieth century model centred on the 'steel-and-petroleum' auto-mobility of the privately-owned internal combustion engine (ICE) car (Dennis and Urry 2009; Paterson 2007) and all that is entailed for its easy, common-sense, everyday use as a *system* of mobility.

Urban mobility is a key case study for many reasons. First, it is crucial regarding each of the Four Challenges, and as above, with China particularly exposed and central to their global manifestation and efforts at overcoming them. Regarding the environmental challenge of planetary boundaries, transportation accounts for approximately one-quarter of global greenhouse gas emissions (GHGs) (IEA 2015), with road transport making up the lion's share, and is key to efforts to mitigate 'climate change'. Decarbonizing (the currently car-based) urban mobility, though, is not merely pressing but is also a 'wicked' set of intractable, huge and *system* problems (Marletto 2014). The system of automobility – 'automobile use and everything that makes it possible' (Rajan 1996) – is deeply locked-in and stabilized (Unruh 2000) in a complex and dynamic assemblage that involves a vast array of agents, infrastructures, and sedimented power/knowledge technologies ranging from the heights of military-industrial 'carbon capital' power of the global car and oil oligopolies (Urry 2013), via the high barriers-to-entry to the vehicle sector given massive manufacturing and R&D capabilities (Dicken 2014), down to the sedimented everyday common-senses, affective pleasures (and accepted frustrations) and practices of car use (or aspiration to it) by billions of drivers and their passengers (Sheller 2004). Altogether, then, a transition beyond ICE automobility is arguably the 'hardest case' (Geels *et al.* 2013: xiii; Tyfield 2013) of low-carbon transition, as against, say, a transition already imaginable to (big utility-provided) renewable energy.

Regarding post-human innovation and complex government, mobility and its innovation is also crucial. For the former, there is an increasingly clear emerging global common-sense regarding what will follow the 'car', namely the 'internet of things'. Here a 'sharing economy' of 'mobility-as-a-service' (MaaS) (*The Economist* 2016a; Flugge 2017) (for a critical analysis of the sharing city see McLaren and Agyeman 2015) or an 'internet of vehicles' is a key element and stepping stone, pushing the material, physical and industrial transformation of society by 2.0 digital technologies and 'Big Data' (Mayer-Schönberger and Cukier 2013) that is viewed by many in these sectors as the next frontier in a digital 'industrial revolution' (Rifkin 2011; Straw and Baxter 2014; Schwab 2016).

Regarding the latter, in the quintessentially capitalist and individualizing world of neoliberalism, mobility is both an essential freedom propagated, consumed and harnessed by that form of government and its entrepreneurial selves, and hence also a crucial locus of contemporary system control, ensuring responsible agents who can be trusted to move in ways that do not disrupt the constitutive flows of such a complex society. In a cosmopolitized globalizing world, this is now *global* mobility, constitutive of the emergent reality that is 'global' society (Beck 2009; Urry 2003) and hence of even greater system significance; witness the sheer inconceivability of the Chinese economic 'miracle' absent global flows. The car industry (and the digital firms nipping at its heels (Chapter 10)) too is quintessentially cosmopolitized, with most Big Car companies now having bigger sales and operations outside their home state than inside it.

Moreover, a key site in which all of the Four Challenges come together, and with the system-metabolic flows of systems of mobility – physical, but also virtual (Sheller and Urry 2006; Urry 2007) – as pivotal, is the key twenty-first century challenge of urbanization. This is centred in the burgeoning populations of the global South, especially south, east and southeast Asia and sub-Saharan Africa, with humanity's emergence as a predominately urban species around 2007 (Girardet 2008). Cities will thus be increasingly *the* key sites of both ecological footprint and innovation (to mitigate and adapt), and hence *themselves* key innovations in their form and system functioning (Glaeser 2011; Salat 2015). Mobility systems and their innovation thus sit at a crucial nodal point in terms of shaping and being shaped by, on the one hand, what broader innovation is enabled, through the co-presence they make possible, and on the other, what city forms as socio-technical-environmental assemblages arise in parallel (Figure 9.1).

So urban mobility transition is a crucial case study in its own right. But China is also again globally central to such a process, and in multiple ways. First, consider automobility itself and moves beyond it. On the one hand, there is embryonic evidence of 'peak car' in the global North, especially amongst younger people (Cohen 2012; Lyons and Goodwin 2014). In China we find to the contrary a titanic growth of car use to the world's largest car market and with fast rising mobility-related emissions (Schwanen *et al.* 2011), with no imminent peaking in sight.

From the perspective of Chinese society, the growth of (fossil-fuelled) urban mobility is a key feature of the immense changes since 1978. This automobilizing

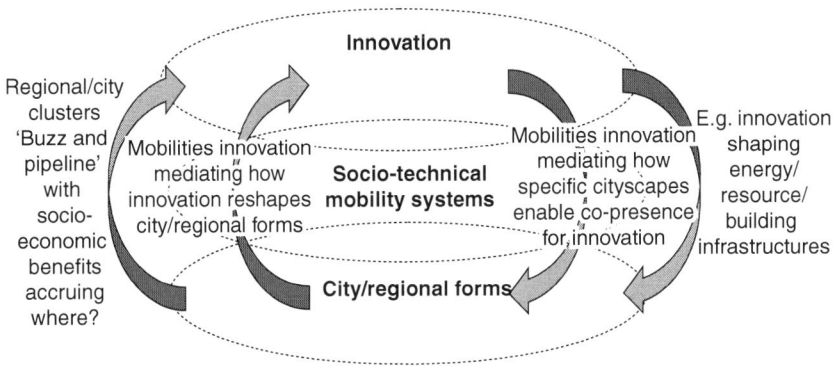

Figure 9.1 Urban mobility, innovation, cities.

transition in China is one of the fastest ever (Urry 2013). Cars in China increased more than six times between 2002 and 2013 to 137 million (OICA, 2013). Growth is expected to continue at 7/8 per cent p.a. in the medium term (Sperling and Gordon 2009: 209). China has become the top priority of car companies around the world. The diversity of brands and makes of cars is eclipsing even that of the USA (*Time* 2014), with a staggering 647 models available from 111 brands. One may ask, therefore, whether this forward momentum of steel-and-petroleum cars can be slowed down even to a limited degree.

Moreover, we find here again that manifestation of the multiple problems with the dominant ICE automobility system are particularly intense and experienced as such – as existential security threats, both to the CCP Party-state and to the corporeal being of its citizens. For instance, the key environmental issue regarding mobility is not so much the 'abstract' or 'global' issue of climate change but hazardous air pollution – a problem caused not just by transport emissions, but significantly worsened as traffic continues to increase (Qin 2015). But this interacts with and is compounded by other complex system dysfunctions. For example, as city rush hour traffic is increasingly faced with gridlock, this contributes further to the problems of air pollution as journeys take longer and the efficiency of emissions decreases. Worse air, however, encourages more drivers into their cars (Kahn and Zheng 2016: 10), in compounding feedback loops. This has been mitigated to some extent by various regulations, usually at municipal or provincial level, mandating improvement in emission standards (e.g. Winebrake *et al.* 2008; Gallagher 2006); as well as limiting the number of cars allowed on the road in various ways such as licence plate lotteries, auctions or daily restrictions (Wang *et al.* 2013). But rebound effects lead back to re-emergence of the same problems, only worse, generating treadmills of worsening congestion and air pollution (Wang *et al.* 2013). Add poor road safety (in a country where almost all drivers have been driving for 15 years at most), and stress and social isolation given long, uncomfortable commutes, and it is clear that the car is taking a terrible toll on China's health and well-being.

There is a great danger, though, that these high costs – economic not just human and environmental – could be literally set in concrete and locked-in via another epochal change in Chinese society and one inseparable from the political priority of economic growth that drives the rapid adoption of the car: its equally rapid urbanization, placing China squarely at front and centre of the surge of urbanization in the developing world. China's urban population now outnumbers its rural population for the first time.[1] By 2030, one billion people, or nearly 70 per cent, may live in cities, adding a further 400 million urban dwellers (*The Economist* 2014; cf. Brenner 2014; Angel 2012). The central Government has announced ambitious plans to build 400 new cities by 2030, while Premier Li Keqiang describes the challenge in terms of '3 problems of 100 million people each' (Kahn and Zheng 2016: 71).

In short 'urbanization is by far the biggest change China is undergoing' (Tse 2016: 114), and hence its biggest challenge for its massive, Party-state-sponsored drives for innovation. This includes the announcement of China's first ever national Urbanization Plan in 2014, most of which focuses on the issue of the 'floating population' of migrant workers but some of which makes significant noises regarding moves to liveable, attractive urban environments. It is very unclear, however, what these ideas will translate into on the ground and the record is poor to date. As with hi-tech innovation more generally, plans so far have focused on building car-free eco-cities but with almost total lack of success and great dependency on foreign expertise (de Jong *et al.* 2013; Baumler *et al.* 2012; Chien 2013).

The 'greening' of China's urban mobility is thus of urgent global significance across several dimensions. Even within the past decade there have been hopes that China could 'leapfrog' the ICE car to a cleaner and more efficient mode of urban mobility (Gallagher 2006; Altenburg *et al.* 2012; Rock *et al.* 2009). Yet this now looks increasingly improbable, even as efforts at low-carbon automobility are being loudly emphasized in China (see below). Instead, China is constructing, at exceptional pace, a system transition *to* ICE automobility. China's simultaneous transformation *beyond* the 'car' is thus particularly challenging. Yet this 'car-ing' of Chinese society also constructs China as a test-case in the challenges of decarbonizing urban mobility that face established 'car' societies in the global North, *as well as* those of rapidly developing and urbanizing countries. Mobility transition in *China* thus seems, in turn, to be the 'hardest case of the hardest case' *and* the key one globally, in terms not just of its massive quantitative significance regarding ICE-based emissions but also of leading a way 'beyond the car' that is of relevance to both already high-carbon global North and fast urbanizing and 'car-ing' global South alike.

Indeed, China is also the site of an unprecedented hubbub of innovation towards environmentally sustainable and digital urban mobility. If not 'leapfrog', the combination of the massive, ongoing urbanization process *within its own borders* and its general momentum of political economic ascendancy at least present unique opportunities for Chinese innovation leadership in the key and novel twenty-first century challenges of urbanization. This singular tumult of initiatives, of course, is not least spurred on by the very intensity of the challenges and the size of the stakes, as above. But the potential spoils for successful mobility innovation are also huge. In other words, the exceptional dynamism here is because of the uniquely dynamic pressures of security and liberty respectively.

Precisely as such, though, we again encounter in the field of urban mobility innovation *not* growing and incrementally accumulating evidence of China's linear progress towards its impending global leadership in the short-to-medium-term, as both defined by the criteria of the emerging and implicitly Silicon Valley imaginary of 'smart mobility' (Chapter 10) and in comparison with the existing strengths for such innovation in the US (or Japan or Europe or ...). Rather, we find evidence of much *greater*, but too easily overlooked, significance regarding ongoing dynamics of power/knowledge momentum of particular potency unique to China for formation, over the medium to long term, of a post-car mobility system the form of which still, in fact, remains largely opaque and uncertain to the observer in the late 2010s, but clearly manifesting the crucial processes of *socio-political* dynamism, constituting and empowering the new agencies that will drive that system transition in their own interests, at their core.

This reorientation of perspective – from 'technology' to the socio-political co-production of new innovation – is crucial though, because if the actual socio-technical innovations manifesting these prospective ideals are ever to present a viable challenge at the system level they will need considerable qualitative transformation and accommodation. This is the case even regarding the emerging dominant imaginary of 'mobility-as-a-service' (let alone the dominant one to date of the 'electric car' (see below)). And this in turn will need novel power/knowledge *agencies* behind these innovations to drive that change, who are themselves empowered in the process, over the medium term and through all the multiple, complex, knotty and even essentially-contested challenges they will face in the meantime on the road to fruition as a new system. Absent these conditions, it is highly likely that the proliferation of smart, electric, green mobility innovations will make little difference at the system level. Instead they could remain a miscellany of niche innovations (as they have so far (Geels 2012)) that ironically serve primarily to make the fundamentally dysfunctional dominant model of an ICE car-based auto-mobility system slightly less intolerable.

It is thus specifically innovation-*as-politics* and the *power/knowledge* dynamism that is the key here, *not* any particular company or hi-tech innovation nor any corporate or government vision. For, from this CP/KS perspective, we see, in reprise of the key themes and conclusions of the last section, that the severe *problems* of Chinese urban mobility innovation may yet prove medium-term processual strategic advantages. This is so given both the unique intensity of *that* dynamism in China, mediated by the IO vs UF dynamic on one hand; and, on the other, the supply and demand of innovation engendered by this situation that is intrinsically attentive to the Four Challenges and to complexity.

In what follows, we trace again the optimist, pessimist, disruptor and innovation-as-politics arguments regarding mobility system innovation in China. In this case, though, this involves several overlapping stages regarding an inescapably broadening purview of the relevant socio-technologies in our domain of interest, urban mobility. The starting point is the electric car, embraced by Chinese policy as the once-in-a-generation opportunity for China to break into the global oligopoly at the peak of contemporary hi-tech manufacturing that is the car industry. Exploring the liberty-security dynamics at the heart of the 'optimistic' EV policy and its 'pessimistic' reality, however, leads in two further steps beyond this hi-tech framing, to further, overlapping and interweaving dynamics of liberty/security – regarding a uniquely Chinese arena of disruptive innovation of electric vehicles and then even the emerging panorama of 'mobility-as-service' innovation.

Step 1: Optimism and pessimism about the electric car

The future of urban mobility, especially in China, undoubtedly requires significant low-carbon innovation and this is acknowledged at the highest levels of government, as well as within business, both private and state-owned enterprises (SOEs). Low-carbon mobility is understood as a significant, once-in-a-generation opportunity for China to develop innovations of global stature in this key economic sector. The electric car (or vehicle, EV) has been imagined as a 'national hero' (Tillemann 2015: 16) and as the route towards Chinese breakthrough – 'overtaking round the corner' in the language of Science Minister (and former Audi executive) Wan Gang – into the global automotive oligopoly. The EV has thus been invested with intense hopes in China to be a key pillar of the broader project of squaring globalized economic growth and continued one-party-state government through global innovation leadership (cf. Zhao 2010 on ICTs), and hence the key policy priority regarding China's programme of low-carbon mobility transition. This is particularly so given the present state of the automotive industry in China – which is highly fragmented, locally protected and dominated by joint ventures (JVs) between major Chinese State Owned Enterprises (SOEs) and foreign automotive transnational corporations (TNCs), with these JVs in turn dominated by the latter (see Table 9.1).

Given, then, the related political context of techno-nationalism (Jakobson 2007; Zhao 2010), and the lack of other countries with strong automotive sectors strongly committed to the EV, the apparent opportunities presented by the EV have proven decisive in setting the policy agenda (Yang 2015). The seeming need for significant Government support to develop a viable EV system also appears to play to another supposed strength of the Chinese political economy. For instance, a shift to the EV involves considerable challenges of coordination (Tyfield 2013), such as the construction of infrastructures for charging batteries, before there is any consumer demand for such vehicles. Absent such vehicles, there are few incentives for the private provision of charging services, in a classic 'public good' chicken-and-egg.

In 2010, EVs were declared a 'key strategic industry for the next 5 years', together with RMB100 billion (£10 billion) of Government support. Targets of producing 500,000 EVs by 2015 and five million by 2020 were announced. To encourage demand, a 0 per cent sales tax was introduced, along with subsidies of RMB 60,000 from the central Government, which was matched by some cities (notably Shenzhen, home of the EV/battery company, BYD) and even some districts. Furthermore, a programme focusing on the 'electrification' of mobility within 25 major pilot cities was also introduced.

All this support for an industrial and technological project would suggest, prima facie, that significant strides would follow towards Chinese global leadership in EV transition. Yet, there are numerous challenges. First, in terms of the EV as an agent of *low-carbon* transition, there are serious questions regarding its emissions, especially in China. Nationally, over 70 per cent of electricity is generated by coal. As much of this coal is, in turn, of low quality and burned in low efficiency power stations, the emissions associated with EV mobility even exceed those of conventional ICE mobility in some regions of China (iCET 2011).

Moreover, despite the favourable conditions listed above, sales of EVs so far have proven disappointing, notwithstanding a 'boom' in recent years. Against the target of 500,000 by 2015, fewer than 12,000 'alternative fuel vehicles' of any description (i.e. including HEVs) had been sold by end 2012. And while sales climbed significantly in 2014 and 2015, to approximately 220,000 vehicles (Reuters 2015), and 337,000 in the first 10 months of 2016 (Reuters 2016a), total sales are still likely to be approximately 26 per cent below targets. Moreover, while late 2015 saw sales surge, pushing the Chinese market into the global top spot, figures have been

Table 9.1 Chinese automobility vehicle innovation and liberty-security

	ICE automobility	EV automobility	Micro-EV/E2W automobility
Essentially adopted (as 'liberty') by	• Incumbent automobility system including Chinese carbon state-capital (supply side) • Elite and upper-middle echelons of society seeking conspicuous achievement (demand side)	• Hi-tech techno-nationalist Chinese party-state and private sector automotive corporations aiming to 'overtake around the corner' (supply side) • Elite Silicon Valley e-automobility (supply side)	• Chinese disruptor firms (supply side) • Lower-income urban and rural (and increasingly middle-class urban) drivers (demand side)
Essentially rejected (as 'security threat') by	• Hi-tech techno-nationalist Chinese party-state and private sector automotive corporations aiming to 'overtake around the corner' (supply side) Elite Silicon Valley e-automobility (supply side) • Emerging green public sphere (demand side)	• Incumbent automobility system (supply side) • Risk-averse non-elite middle-classes, and 'big, foreign car' demand (demand side)	• Government (supply and demand side) • Upper echelons of society concerned with low *suzhi* connotations (demand side)

brought into serious question as many of the sales were 'ghost cars'; accounting fictions used fraudulently to claim government subsidies (Yang 2016). This growth of EV sales has also been based on those subsidies, yet Government already planned to reduce these as quickly as possible (by the end of the 13th Five Year Plan in 2020) and will likely now do so even more quickly in light of the scale of the fraud. Increasing numbers tenfold to five million EVs (including plug-in hybrids) by 2020 thus seems a 'Herculean' task (Yang 2015). And, in terms of transition, such numbers must also be set alongside *annual sales* of approximately 20 million ICE cars (Johnson 2015). A rapid shift to system domination by the EV thus remains implausible even if these ambitious government targets are met and EV sales have finally, after over a decade of policy attempts, begun to take off.

Furthermore, up to 2015, EVs were also largely purchased in Government procurement for municipal taxis. Private purchases of EVs in China remain a major challenge and utterly dependent on the Government subsidies; for instance, EV manufacturer BYD's revenue would have been negative in 2014 and 2015 without these subsidies (Bloomberg 2015a), totalling RMB31 billion ($4.5 billion) in 2015 alone. Amongst automotive SOEs presented with Government targets for developing what are unprofitable EVs as against their profitable ICE businesses, the result has often been half-hearted engagement and positive foot-dragging, at best, (Wang 2013) – or apparently now subsidy fraud at worst – though private automotive companies are more aggressively pursuing an EV strategy (e.g. Bloomberg 2015b).

The usual explanations for this lack of progress concern issues of immature and hence expensive battery technology and inadequate charging infrastructures, which are undoubtedly important (Costa Maia *et al.* 2015). Such an analysis can draw straight from the 'pessimist' copy-book of arguments described above regarding the structural challenges and inadequacies for hi-tech innovation in China; problems that are all the clearer in an industry, such as the car, of exquisitely engineered, highly branded and stylized products, and their exceptionally hi-tech and globally-dispersed manufacturing processes, that must be extremely safe and hard-wearing for consumers given how they are used, as against, say the smart phone. Chinese car majors, not least through learning from their JV partners, now make cars of globally-acceptable quality, but their continued absence in global North markets signals how much of a stumbling block to Chinese innovation upgrade ambitions this sector has proven (Thun 2006; Brandt and Thun 2010, 2016).[2]

A pessimist analysis, therefore, could point to the continuing dependence of the Chinese car industry on its JV partners for technological upgrade, notwithstanding over two decades of partnership and policies aimed to force greater concessions and openness from those car majors (Winebrake *et al.* 2008). This is in part due to the structure of fragmented authoritarianism, which has consistently frustrated attempts by the central government to consolidate the Chinese car industry into several globally competitive industrial behemoths (Gallagher 2006, Thun 2006) attempting without success to follow Japan and South Korea's lead (Nolan 2004). Regarding the EV particularly, we also encounter in this regard a plethora of policies and initiatives that often add up to less than the sum of their parts given contending focuses and perverse incentives – familiar problems but compounded by the exceptional complexity of coordination needed for a transition to electric vehicles.

For instance, multiple ministries have inevitably had to be involved in EV policy, including not just the national planning agency (the National Development and Reform Commission, NDRC), but also the Ministry of Science and Technology

(MOST, overseeing the upstream research projects), and ministries regarding transportation, energy (e.g. for the electricity charging piles and infrastructure), the SOE automotive majors themselves etc … And this is true at multiple levels of Government, from the central state to provinces and municipalities, where the latter may well have their own local champions and industrial priorities. Yet each of these agencies has its own imaginary about what the as-yet-unformed system of future EV mobility will look like, with associated priorities. For instance, while MOST has focused exclusively on plug-in or battery electric vehicles (BEVs), the Ministry for Industry and Information Technology prefers to 'walk on two legs', supporting both BEVs and hybrid electric vehicles (HEVs) (Tyfield *et al.* 2015).

Similarly, Southern Grid, the massive SOE in charge of the electricity grid in south China, originally wanted to become a global leader in battery-swap technology and infrastructure, in partnership with Israeli company Better Life (now bankrupt). Yet instructions from the central government that it had to provide demonstration infrastructures for both battery-swap and fast-charging piles simply diluted efforts to the point of corporate lack of interest for Southern Grid. Instead of a real competition of charging options between global industrial behemoths (i.e. Southern Grid going for battery swap vs State Grid backing plug-in charging), therefore, policy simply incubated a culture of minimal and superficial matching of the letter of policy; a familiar trope in contemporary China. The result is inadequate charging infrastructures that are also poorly maintained and sparsely used, in positive feedback loops of negative interest in private ownership of an EV. As already noted, this has also been the result to date within the SOE car majors regarding commandments to invest in EV initiatives (Wang 2013).

In short, even notwithstanding something of a take-off – at long last – in EV sales that has received much adulation, both within China and globally in the business press, the result of the most ambitious industrial policy in the world regarding the EV is largely underwhelming stasis. And this is especially so when set against the parallel continued growth of the ICE (Cohen 2010) and the broader deepening of the *system* of automobility. For the latter includes processes of both planned building and redesign of urban areas around the 'car' and uncoordinated urbanization – driven overwhelmingly by local government land sales, building company speculation and highly uneven demand amongst the population for the financial security of bricks and mortar – that also instantiates the default model of wide roads, car-based construction.[3] If one chooses to be pessimistic about China's innovation upgrade, therefore, the EV will provide plenty of evidence to corroborate that position.

Going beyond the usual hi-tech supply-side focus of much analysis to a broader one of the socio-technical system of ICE vs EV automobility, there are multiple other reasons that this relative stasis seems deeply sedimented. Most notably, the political issues regarding EV transition in China are many and complex, involving issues of: industrial and innovation policy (for different industries, ICE cars, EVs, electronics and ICTs, oil/gas/(gasified) coal, infrastructure and construction, and so on); environmental politics and governance; and rapidly changing social power relations at the 'ground' level of society, including issues of consumerism, social distinction and intersubjective class and even gender definition.

On the supply or producer side, there is a tight interconnection between Chinese 'carbon capital' – of oil, coal, steel and cars – and the Party-state, epitomized in the strength of the Shanghai 'faction' and its specific model of heavy industry SOE-led development (Huang 2008). This puts these industries, all of which are currently very powerful and conversely profoundly threatened by a low-carbon mobility transition, in pivotal positions to resist and retard innovation away from the ICE car. Indeed, this power bloc is arguably particularly locked-in in China, notwithstanding the systemic empowerment of the car industry in Germany (where it is the largest industry, with turnover of €404 billion in 2015 (GTAI 2016)), Japan and the US. For in China such carbon capital giants *are* the state and Government itself, while they remain private companies, however well-connected and influential, in the other countries.

But it is user politics that are arguably of greatest significance since the most important issue overlooked by current hi-tech supply-side policy is consumer demand (cf. Wolf *et al.* 2015). An essential aspect of the extraordinarily rapid current *construction* of an automobility socio-technical system in China, the car has become the number one consumer aspiration (AC Nielsen 2011). This is especially so as cars are increasingly affordable for growing numbers of 'middle-class' Chinese with some disposable income, however modest. But, perfectly exemplifying the discussion above (see Chapter 8), this demand is socially complicated and complex, not simply a matter of consumption utility preferences. It is primarily through the consumption practices of oneself and one's family that contemporary Chinese citizens can exercise their individual freedom, cultivate a sense of individual and networked-collective identity and put it and their material success upon display, and hence claim a certain social status and personal quality (*suzhi*) (Yu 2014; Anagnost 2004). Hence demand depends not only on fit with existing social practices of (demand for) mobility (Shove *et al.* 2012; Pasaoglu *et al.* 2014), but also on how these are changing in ways charged with significant social importance.

Moreover, focusing on economic success or freedom, this conditions the adulation of a particularly unlimited and conspicuous form of wealth (Zavoretti 2013). In the circumstances of personal affective investment and the bodily experience of consumption of autonomous mobility (i.e. moving, possibly fast, under one's own control (Paterson 2007)), the car assumes an almost unrivalled position, arguably above even housing. This is compounded, not undermined, by the multiple novel risks, technological and financial, assumed in growing car ownership. Together these manifest as a strong preference for big, expensive, foreign (particularly German), gas-guzzling cars that epitomize the extravagant 'modern', 'technologically-perfect' and trust-worthy, 'Western' lifestyle to which many Chinese aspire as marks of high *suzhi*. As one contestant on a dating gameshow famously put it, she would 'rather cry in the back of a BMW than laugh on the back of a bicycle.'

Indeed, this connection between finding a spouse and car ownership is a particularly vivid and lived aspect of the contemporary cultural politics of the car in China, and manifest in the multiple connotations of 'BMW' itself. On the one hand, BMW is transliterated as '*Baoma*', which literally translates as 'precious horse'; while on the other, the joke goes that it stands for 'Be My Wife' – in English, adding to the cosmopolitan cachet of both joke and brand. In another extraordinary instance that links car culture with social media as a key mediation of networked individualization, when the popular Chinese messaging app WeChat began to use adverts in 2015 it sent out to every user one of three ads depending on the profile of their use of the platform. The outcry from its users that followed was not, however, that they would now be bombarded by unwanted commercials. Rather, many were outraged that they did *not* receive the BMW ad, while perhaps their friends and/or contacts did and made it *known* that they did (seamlessly forwarding it on), and the loss of 'face' this entailed.

Consider in this context the exceptionally competitive and individualized lives of the one-child generation of the urban young, now looking to partner up. Add in both a surplus of men and a culture of gender politics that makes marriage unattractive to many young women, since it still places the overwhelming burden of housework – and tending for the older generation, of both her own and *his* parents – on her shoulders. It is unsurprising, then, that a 'good match' to a successful young man has become a common expectation, of both bride and her parents. To the parents in particular, though, nothing signals such success, and achievement of security, like ownership of a flat and ... a (good = foreign) car. The car is thus the (literal) vehicle of both liberty, in growing autonomy and opportunity for oneself and one's family (or the very *making* of that family), and security, in terms of marks of financial security and as artefacts that can themselves be trusted both to be extremely well-made and to shelter oneself and one's family from terrible pollution and extreme weather. The conventional fossil-fuel car is thus currently a key element in the forming of the social identity of the Chinese 'middle class' (Zhou and Qin 2010). And given the importance of this stratum of society for broader regime legitimacy (Guo 2008; Goodman 2015), this suggests serious *political* obstacles to policies that would actively penalize the (petrol-powered) 'car' in support of, say, an (unpopular) EV alternative – as middle-class rejection of the congestion zones being suggested by the Beijing city government exemplifies (*The Economist* 2016b).

Meanwhile, the EV has been roundly rejected to date on purely pragmatic grounds also regarding security and liberty. On the one hand, as a novel technology the EV comes with several novel risks that put off potential owners. These include anxiety regarding charging and range, where estimates of distance left in the battery displayed digitally on the dashboard are often unreliable since they do not take into account unforeseen hills, use of other drains on electrical power in the car (e.g. heating or cooling, lights or wind-screen wipers – perhaps essential in the non-temperate climates of many parts of China) or congestion. The latter in particular renders EV use highly problematic in Chinese cities in which such congestion is all-but-guaranteed during rush-hour. On the other, there are also other security risks, associated with the uncertain but likely fast depreciation of the value of the car and its expensive battery, the upfront cost of these expensive vehicles (notwithstanding subsidies, now being phased out), and even widespread fears about the safety of moving around on top of a massive battery pack due to the risk of fire (with several high-profile explosions hitting the news) or 'radiation' (Zuev interviews).[4]

Against these considerable and unfamiliar risks, therefore, what the EV can offer to boost the owner's autonomy and opportunity *beyond* ownership of an ICE car is slight indeed and hardly adequate incentive. And meanwhile the supposedly green credentials of the EV still register little with the majority of Chinese consumers, though this is beginning to change: e.g. terrible air pollution in Beijing in autumn 2015 has been connected with increased interest in EVs. For the time being, though, the vast majority of electric cars on offer in China, including from Chinese car companies, are workaday and inconspicuous but still relatively expensive family cars just with an electric engine. What they are not, therefore, is highly attractive and visible cars that display superior status and 'conspicuous achievement' (Yu 2014).

There is, however, one notable exception to this that is informative regarding these very dynamics of liberty and security. This is the Tesla Model S: a striking and stylish sports car that appeals to the consumer on *this* basis of high-end display, not as a 'green' car that embodies self-sacrifice and sobriety. As itself a foreign brand,

and one that deliberately plays up its Silicon Valley associations, the Tesla is a car that directly challenges the luxury car competition by playing precisely to the tastes of the elite Chinese consumer. Of course, Tesla is not a Chinese company, nor even in a JV in China. This means all its cars are imported, are subject to high import tariffs and do not benefit from any purchase subsidies for Chinese-built EVs. And Tesla has had some problems in China, having to downsize its China staff considerably in 2015. Yet new showrooms continue to open in China's megacities, a national charging infrastructure from Beijing to Guangzhou has been constructed (at the expense of a private owner) and sales are robust … and conspicuous on the streets.

Whatever the future holds for Tesla itself in China, therefore, it is undeniable that its visible success to date has profoundly shaken up the EV sector and policy in China, spawning countless strategic rethinks in China's car companies and several copy-cat start-ups that loudly sport their combination of Chinese investment and Silicon Valley expertise (e.g. Faraday Future, Atieva or LeEco) (Bloomberg 2015c).[5] But it is equally obvious that the very success of this Tesla strategy has immanent limits, especially insofar as it aims (as Elon Musk, its CEO, explicitly says) to replace the ICE *tout court* with the electric vehicle. For in its current form, its very success depends on being targeted precisely to what is by definition a small fraction of demand.

Tesla has announced its plans for a mass market model, aiming to manufacture 400,000 a year by 2018 (Reuters 2016b). The challenge here is not just to manage the organizational challenges of such a quantum leap in manufacturing – and in competition with a fast-moving and extremely advanced manufacturing expertise elsewhere in the car industry, as its problems in this sphere have demonstrated (Fehrenbacher 2016). But also to manage successfully the dual challenge of selling many more cars to 'middle-class' owners looking for stylish dependability, not just statement-making flash, while not completely tarnishing the whiff of elite appeal that is their brand and primary appeal. There are multiple ways in which this could go wrong. Moving 'down' to the mass market demands not just the challenging scale-up of high-spec and economically competitive manufacturing, but also the building of an entire infrastructure of sales, servicing and repairs that can service the different and dynamic expectations of the (Chinese, networked-individualized) middle class, and with the potential for corporate blunder and (rapidly spread, 2.0-enabled) embarrassment in each case.

Together, then, the prospects of the widespread rapid shift to EVs – their production and consumption – do not appear promising, even as sales will likely continue to grow. Indeed, the last decade of efforts appears to evidence a landscape of largely separate spheres of evolving 'normal' consumer aspiration (market-pull) for mobility on the one hand, and of 'low-carbon' EV mobility innovation (technology-push) on the other. The prospects of an EV sociotechnical transition, by contrast, would need these to co-evolve and converge. Seeing how this may be possible, therefore, forces us beyond the hi-tech fetishism of analysis of mobility transition focused on the 'electric car', whether from optimist or pessimist perspective. The obvious first step here, in the context of discussions of China's innovation upgrade and the particular challenge of servicing but thereby transforming existing demand, is to look for what China's disruptive innovators are up to. The answer is, a great deal. But, true to form, perhaps not where we would first look.

Step 2: The electric mobility disrupters

In his seminal book, coining the phrase 'disruptive technology', Christensen (1997) actually devotes a whole chapter to the possibility of the electric car becoming the acme of this form of unexpected but transformative innovation. Indeed, in the likes of Tesla and especially China's lithium battery giant-turned-EV upstart, BYD, many – including Warren Buffett – have seen glimmers of a disruptive innovator in this sector. BYD, for instance, has published exceptionally ambitious targets to be China's leading car company by 2015 and the world's by 2020, when they only began manufacturing cars a decade or so ago. Of course, the first target has now been missed, as were its targets for sales of EVs, even though it occupies three of the top five slots in sales of EV models in China and, at 53,371 EVs sold in 2016 to July, it is by far the biggest EV company in China (Pontes 2016). And Tesla too is at best mobilizing, rather than 'disrupting', the car industry. So perhaps, while disruptive innovation can upend industries built upon relatively direct connection with end consumers and low-risk adoption that depends on no public infrastructures – e.g. think digital cameras or ICT – it is not so appropriate or effective as a strategy for a technology as systemically-situated and complex as the 'automobile'.

Yet in turning to China – and China uniquely – we do indeed find a massive, dynamic example of disruptive innovation in electric mobility that not only perfectly exemplifies that strategy and its advantages, and how they have been so successfully deployed in China especially, but also one that is at least as unexpected and easy-to-miss as Chinese innovation upgrade in multiple other sectors. This, however, is not in the four-wheel conventional electric *car*, but in a proliferating ecology of small, 'low-speed' electric vehicles on two, three or four wheels. With demand the singular gap in the current adoption of the EV-as-electric-car, the spectacular success of *these* micro electric vehicles is all the more significant.

The contrast between the disappointing uptake of the EV and that of micro-EVs in China could hardly be more striking. The former have 'boomed' in recent years … to a still minuscule and systemically-insignificant 1 per cent of annual sales, numbering in the hundreds of thousands at best. By contrast, there are now over 200 million electric two-wheelers alone in China (Timmons 2013), comparable with, or possibly more than, the total number of *ICE* cars on China's roads (Fishman and Cherry 2016; Wells and Lin 2015). And so-called 'low-speed' (or *disu*) three- or four-wheelers are also effectively ubiquitous in China's less developed cities, towns and villages and the peri-urban areas of its megacities.

This is all disruptive innovation servicing existing demand in classic fashion. Often based on (novel combinations of) existing and/or tinkered technology rather than new-to-the-world hi-tech cutting-edge innovation, these vehicles provide new functionalities that enable millions with cheap, easy-to-use but nippy mobility – much faster and/or easier than walking, bikes or public transport but also even cars in city traffic, as they weave through congestion. It is for this reason that these micro-EVs and E2Ws utterly dominate China's burgeoning logistics sector, which is so essential to its equally booming e-commerce. And, whereas for the EV the shift to electricity from gasoline is a key source of inconvenience and anxiety for the user and of need for expensive public charging infrastructure at system level, this shift is for micro-EVs a straight-forward advantage and direct enabler. E2Ws are cheap and easy to charge, with no need for new infrastructures but, to the contrary, taking advantage of existing ones for bicycles.

This disruptive e-mobility innovation thus perfectly exemplifies the characteristics of Chinese disruptors more generally, as described above: producing low-cost, good-enough products, starting from tinkering in thousands of relatively low-tech workshops but bringing together novel combinations of existing technologies and then constantly and incrementally improving them. The biggest brands of E2Ws are thus increasingly hi-tech and high-spec, with their own and growing R&D (and D&D) budgets built entirely upon profits, not Government subsidy or hi-tech upstream research funding. They are also, therefore, foundationally nimble innovators, responsive to their markets (which they are *creating* in the process), as fast-followers in a highly competitive market.

Moreover, this is a specifically *Chinese* sector, regarding both demand and supply, and one with potentially global significance and size, such as only a country as uniquely large (and socio-economically diverse) as China could offer. On the one hand, these companies have grown to significant industrial enterprises through servicing an almost entirely domestic market demand. This demand for low-cost but good-enough auto-mobility remains massive. In China's rapidly and chaotically individualizing market society, increasingly dependent on free and accelerating circulation, including of the poorly-paid masses driving China's economic miracle, these vehicles have enabled precisely such auto-mobility without which the country could grind to a halt. On the other, E2Ws/micro-EVs are almost entirely a Chinese industry, creating not just a set of potential national champions in a key industry of the twenty-first century but that *industry* with it – albeit, in familiar fashion, *despite* national policy in favour of completely different agents, not because of it. Going beyond the stealth rise to global dominance by China's disruptors in existing industries, such as container shipping, domestic appliances or pianos (Zeng and Williamson 2007), or even in deliberate incubation of new, emerging sectors such as wind, solar PV or LEDs in *ways* that are unexpected (Nahm and Steinfeld 2014; ten Brink and Butollo 2016), in e-mobility we find China's disruptors forging whole new and unexpected *sectors*.

Finally, we should also note how micro-EVs offer small, lightweight but nimble auto-mobility that is both much more appropriate for China's dense (re congestion) and high-rise (re parking and charging) megacities, and much lower in carbon footprint, since it involves powering only a small vehicle not several tonnes of metal with every journey. Absent the need to resolve the public good chicken-and-egg problem of EV adoption vs charging infrastructure this also could significantly expedite moves to a low-carbon mobility system. And this could be a model of 'liveable' automobility (alongside and supplementing, not replacing, increasingly sophisticated public transport infrastructures (Kopp *et al.* 2013)) based on low-cost low-carbon EVs that makes its relevance to the burgeoning mega-cities of the global South seem sublime as against the ridiculous inappropriateness of the massive, expensive, good road-, hi-tech-, and private charging pile-dependent Tesla; and with all the 'South-South' market opportunities for Chinese industry that would go with this. In all, therefore, it would seem that the micro-EV is an obvious national champion to be grasped and lauded by the Chinese Government; or, at the very least, a parallel development with the higher-tech and higher-profile EV efforts that should be a complementary innovation strength in China over the medium-to-long term (cf. Breznitz and Murphree 2011).

And yet this is not, in fact, at all what we find. Rather, regardless of the exceptional dynamism of micro-EVs as a sector and its clear advantages to China's own policy priorities and responses vis-à-vis the Four Challenges, far from embracing this uniquely Chinese opportunity for leadership in twenty-first century sustainable urban mobility, the micro-EV is neglected, proscribed and rejected.

On the one hand, the E2W as an industry has not only received no government support – as opposed to the lavish and wasteful support of the SOE car majors regarding EV projects – but is poorly connected even to local developmental states. This is exemplified by their extreme secrecy and the difficulty of even tracing E2W companies for interview, let alone arranging contact with them, in ways that are just not the case for EV companies. On the other, E2Ws and other micro-EVs are now formally banned from inner city areas in many of China's biggest cities, and with a coordinated national clampdown on dealers and users, confiscating vehicles across these large cities, in April 2016 (e.g. Beijing, Guangzhou, Shenzhen, Xiamen ...). In short, China's disrupters in urban mobility are only barely tolerated, as against their rise to national prominence and even official (if conditional) adulation in, say, ICT and home appliances.

The reason for this difference in treatment, however, is highly significant and illuminating. For it is precisely in terms of the 'security' threat – the term that is actually used – these vehicles pose that they are both formally and informally censured. And, again, this interpretation as a security threat, in multiple ways, is conditioned and heightened by the unique broader socio-political context of contemporary China regarding the immoveable object of the Party-state regime, the uncontrollable force of neoliberal global risk-society and their deepening inter-articulation.

This takes two forms primarily, at the level of formal Government and informal, everyday government respectively. Regarding the former, the E2W in particular is considered a security hazard by the Party-state because of how it is widely used: weaving swiftly, silently and often overloaded with stuff through traffic, ignoring traffic signals, onto and off pavements etc.... They are thus widely perceived as dangerous, causing multiple accidents at the cost of their drivers and those of the other, bigger vehicles or smaller pedestrians with which they collide. But is not just their straightforward challenge to the bodily health of Chinese citizens that makes them a security hazard. They are also a danger to the health of the Chinese body politic, and in several ways (Tyfield 2014).

First, the primary challenge for the Party-state, as discussed above, is to shape the Chinese citizenry such that they may be governed at a distance, and entrusted as 'responsible' individual agents who will push forward and not disrupt the accelerating circulation – of goods, capital, ideas and people – presupposed by the continuing market reform and economic growth that is the CCP's bedrock of continued legitimacy. While mobility and its everyday conduct – as 'governmobility' not just governmentality (Bærenholdt 2013) – is a key aspect of the government of advanced, complex market societies more generally (Rajan 2006; Foucault 2009), in China this has the added intensity of holding to ransom the existential continuation of the incumbent regime. What is needed, therefore, is 'responsible' auto-mobility that may be entrusted to move smoothly and with requisite self-discipline ... and that may be easily disciplined and governable by top-down authority when it fails to do so. From this perspective, however, the E2W is not only the very acme of *un*disciplined auto-mobility, ignoring rules of the road as a matter of course, but also, precisely as nippy and ubiquitous – a swarming 'mob' (Aradau 2016) not just a mobile demos – difficult to discipline. And causing multiple accidents and blockages to the smooth passage of the circulatory flows of its marketizing society (Usher 2014), E2Ws become a systemic security threat, not just a personal one.

But, second, a key aspect of the Party-state's legitimate oversight of China's socio-economic development concerns its stewardship of China into sites of global-leading industry and modern city-scapes, impressive to and consumable by the discerning twenty-first century mobile individual. This dynamic, however, is set within the tension of the immoveable object and unstoppable force and the hi-tech fetishist techno-nationalism it engenders. On the one hand, the invisible, lower-tech (if improving) micro-EV industry bears no resemblance to the gleaming, robotized factories and global brands of the car industry; while, on the other, these vehicles feature in images of 'backward' and undeveloped urban environments, not the gigantic, hi-tech, plate-glass sky-scrapers vision of the Government, elite car advertisements and the 'tourist gaze' alike (Urry 2002), notwithstanding advertisement campaigns in China for E2Ws featuring prominent young celebrities. And, of course, the very policies that threaten the micro-EV industries and exclude their vehicles to less developed peripheries simply translates this preconception into vivid and self-confirming fact.

Three- and four-wheeled micro-EVs have finally attained some official recognition in recent policy documents, perhaps due to concerted lobbying by the Governments of provinces and cities where these slightly-more-respectable vehicles have large industrial presence (such as the Yellow River provinces of Shandong and Henan – notably both provinces that are of middling development, heavily industrialized but also without a local car brand, SOE or private). Nevertheless, there seems little prospect on the horizon of the E2W receiving concerted official backing as the vehicle of twenty-first century urban mobility. But this ambivalence, at best, from Government is matched by similar misgivings amongst the population. In particular, amongst the urban middle-classes, considerations of *suzhi* not only make the car the quintessential item of display of one's high-quality but equally, vice versa, the E2W is strongly associated with low *suzhi* and the security threats this entails.

Crucial here is how the cultural politics of vehicle or mobility choice sit atop China's system of graduated citizenship (Ong 2006). The E2W not only appeals to those with less money, unable to buy a car but able to opt for the automobility on two- or three-wheels instead, but also to middle-class, urban residents it is especially amongst the rural migrant workers, in domestic service, construction or factory work, that they see widespread adoption of these vehicles. The E2W has thus become strongly associated with one side of this essential socio-political divide. Moreover, with a roaring trade in stolen E2Ws – their susceptibility to theft another security problem – many owners keep their rides deliberately battered and patched up, since the shiny new one is the most likely to be targeted. They thus become even less attractive in the popular imagination of the urban middle class. Finally, the very ease of use and charging that is so essential to their massive adoption itself feeds images of their 'irresponsible' and dangerous use. For instance, E2Ws are often charged by passing long electric extension leads from upper story windows, perhaps to existing bike sheds. This pragmatic, make-and-do tinkering, however, is not only 'unsightly' but has caused some terrible fires; in one case in Zhengzhou burning down an entire apartment building (China Youth Daily 2014).

Together, then, Government and government alike cast the dominant image of the E2W as a security threat. Yet they still persist in extraordinary abundance, and, if we include all micro-EVs, with markets that are still potentially all-but-limitless. For, on the other hand, these low-cost – and potentially low-carbon – vehicles unquestionably service the demand and structural imperatives for individualized auto-mobility that are just as strong amongst the poorer sections of Chinese society as they are amongst the wealthier. And, indeed, it is explicitly in terms of the greater 'autonomy' and 'freedom' afforded that users tend to describe their adoption of E2Ws, whether as commuters or as professional riders (e.g. in logistics or taxis) (Zuev interviews 2014/2015).

The totality of the dynamics shaping China's disruptive e-mobility innovation, therefore, take exactly the form of the interplay of 'liberty' and 'security', in a complex cat-and-mouse that has driven the extraordinary success story of their mass adoption *together with* their continual containment and neglect. Moreover, these dynamics are still playing out in an essentially-contested strategic manoeuvring of innovation and counter-innovation. This is being conducted, however, not in explicit political protest and debate in the public sphere, but in the everyday 'politics' of mobility and movement amongst over one billion Chinese citizens: ranging from big but everyday decisions about purchase of a vehicle or acceptance of a commute, to the micro-wrangling with other traffic over road space and rights-of-way as they drive along (Zuev *et al.* forthcoming). In other words, in a thoroughly pragmatic civilizational (and/or 'onto'-) politics regarding 'who gets to define what is "civilized" (urban conduct and mobility) in early 21st century China?' amidst the longstanding structure of Chinese society of mutual but asymmetric disciplining *and* disregard of Government and population (see Chapter 4) (Zuev *et al.* forthcoming).

Crucially, though, this dynamic and its output is building strong power/knowledge associations between such mobility and a specifically *middle-class* status and display that emerges as the key strategic goal. This is the case whether regarding the Party-state's Government of micro-EV mobility, the ongoing attempts to transform its connotations into those of *high*-suzhi or the forging of business strategies seeking to bridge these twin imperatives. In other words, just as the EV has to move 'down' from elite to middle-class appeal to be systemically disruptive, so too the E2W must move 'up' to occupy the same ground.

Moreover, there is emerging evidence not only of the explicit alignment with this strategy amongst micro-EV suppliers, but also of some growing success and momentum to their adoption by the Chinese urban middle-classes. For instance, regarding the former, recent years have seen considerable shifts in the micro-EV (including E2W) sectors. This has included the emergence of high-profile, celebrity-endorsed publicity campaigns that show a greater attention to intellectual property and branding, and with the urban, young and digital middle-classes as their specific target. E2Ws have thus begun to brand themselves as 'life-style' choices, to be associated with the *haute* attractions of European-style city-living, and, crucially, seeking to differentiate themselves from 'working' vehicles, i.e. as markers of 'conspicuous achievement' (Yu 2014) not *laodong* worker status. New brands have also emerged, either specifically targeting overseas markets in the global North, or with stylish design and digitally interconnected features explicitly aimed at the *xiaozi* (Bobo, bourgeois bohemian) urban young, such as the *Niu* – described by one dealer as 'my little Tesla' (Zuev interviews).

Regarding the latter, meanwhile, the tight connection between the E2W and associations of low *suzhi* is also being loosened through growing evidence of middle-class adoption, including on university campuses, and especially in Shanghai – China's second city but also one with a strongly European urban form in its inner city Puxi district. Here we find urban 'middle-class' residents deliberately choosing the E2W as the easiest and most efficient form of inner-city mobility precisely given (the security threats of) the worsening congestion and air pollution of the incumbent model. While out in newly built 'middle-class' suburbs connected to the city centre by subway, many have adopted the E2W as the best way to solve their 'last mile' problem, with subway station bike parks crammed full. In the latter case, then, the E2W, far from being a black mark on their status, is an essential means for these relatively low-wage but self-styled 'middle classes' to realize their ambition of a foot on the property ladder in the Shanghai metropolitan area with decent inner-city white-collar jobs.

This shift in public reasoning and culture of micro-EVs is even starker regarding those on three and four-wheeled vehicles. The biggest segment in the 'booming' growth in demand for electric 'cars' is actually amongst Chinese-branded small EVs, such as Ningbo's *Zhidou*. These vehicles are bought in general as useful 'second cars' by middle class families (often for 'her'), looking for enclosed, private auto-mobility that can take advantage of their ease of use for short peri-urban trips (shopping, transporting children) and the perks associated with EV ownership provided by national and local Governments (especially including quicker, cheaper access to licence plates and parking). At the boundary, then, these small EVs merge with bigger micro-EVs, suggesting this may be the key market.

In the meantime, however, micro-EVs, and especially E2Ws, remain insecure and politically sensitive, and hence essentially contested. For instance, notwithstanding its high-profile attempts to 'upgrade' the image of the E2W, *Niu* is a highly secretive company, while its management has been caught up in the Government anti-corruption drive, signalling the Party-state's disquiet with it as a company even as its impact to date is largely notional. Similarly, in interviews in Shandong in winter 2015/2016 with leading micro-EV companies, we encountered high corporate ambitions but also explicit nervousness about pushing their case too strongly in Beijing, lest it elicit a pushback from the much stronger forces of ICE carbon state-capital. Instead, these companies' strategy is to 'keep their own counsel' for the time being and continue to grow to a point where the central Government will simply have to recognize them as important market players.

In short, there is still a long way to go for China's e-mobility disrupters before they have securely claimed the (middle-class) mainstream – and with the turbulent and essentially contested dynamics of liberty/security playing out in the meantime. Far from retaining a relatively stable ground of persistent strategic advantage (cf. Breznitz and Murphree 2011), what makes out these disrupters is precisely their dynamism *atop* and *within* a systemic context that is *itself* one of rapidly shifting sands.

A tale of warring monsters … and efforts towards their (self-) domestication

Having traced both hi-tech optimist and pessimist *and* disrupter analyses, we can begin to discern the key dynamic of e-mobility innovation in early twenty-first century China; an exceptional case study that nonetheless, so we argue, illuminates the broader challenge of China's innovation upgrade in the context of the Four Challenges. As the 'hardest case of the hardest case', essential system dynamics – Chinese and global – are revealed that may otherwise be missed in less politically taut and contested and/or socio-technically complex domains. In particular, what we find in Chinese e-mobility is a context of a set of mutually contending socio-technical and power/knowledge systems *all* of which are currently unfit for purpose.

In other words, just as new innovations, and especially those aiming at radical and/ or system transition, may be described as 'hopeful monstrosities' (Mokyr 1991; Law 1991), in that they are both novel combinations of familiar features and awkward, unsightly and possibly dangerous vis-à-vis existing common-sense expectations, so too Chinese e-mobility is today a menagerie of monsters.

First, there is the electric car: a patchworked Frankenstein creature, in green lipstick, masquerading as a hideous parody of a robust and familiar beast (the 'car') while also trying to add the extra attraction of greenery, but in ways that largely fail to convince on either score. For, as we have seen, the electric *car* is trying, but largely struggling, to grow on the promise of familiar car-based automobility that is also 'green'. Second, there are the (diverse ecology of) micro-EVs and E2Ws: a vampiric monster, shady, alluring and successful but also extremely dangerous that must be guarded against by a vigilant guardian of respectability lest it spread its lethal and uncontrollable appetite for risky mobility. Here, the grey status of these vehicles, both legally and regarding industrial policy support, sits alongside their mushrooming, bottom-up demand and the multiple 'security' threats this generates. But, third, and crucially, there is also the dominant monster: a now gross but powerful zombie blob, insatiably and greedily consuming resources (environments, precious minerals, city-scapes and peri-urban land, time wasted in congestion, human lives in 'normal' accidents) and spewing its waste without regard to its consequences. And doing so in ways, under compressed modernity, that are both more intense and less sedimented than they are even in the neoliberal global North.

In these three contending systems, in other words, we see how Chinese urban mobility is manifestly confronted with three contending systems *none of which*, currently, is capable of furnishing mobility that is smooth, enjoyable and safe, either at the human level or that of society and planet. In urban mobility, thus, we encounter starkly and inescapably what is arguably the broader predicament regarding complexity-attentive system transition more generally in other domains as well, namely the absence of any *given* model to follow and hence the need instead to *build* it through processes of power/knowledge innovation.

Each of these three monsters is constituted and growing specifically *as* a monstrous system precisely through dynamics of liberty/security specific to that innovation that, today, render each as essentially contested. For instance, regarding the EV, its proponents in (Chinese) Government and business alike cast it as the epitome of liberty, as the latest, hi-tech incarnation of a pillar of twenty-first century global industry, offering green, individualized auto-mobility. Meanwhile, it is largely rejected on each of these scores as the manifestation of multiple novel security threats: to (car) businesses as risky and unprofitable ventures (especially vis-à-vis massive sunk investments in manufacturing and branding of ICE cars) while ICE emissions are also improving rapidly; to Governments as profound challenges of coordination for the building of massive new infrastructures and expensive, bottomless pits of subsidies and R&D funding; and to consumers as expensive, dangerous and unfamiliar vehicles. Yet, countering this counter in turn, proponents point to how the longer-term possibility of zero-carbon driving pertains to EVs alone, or to the emergence of stylish (elite) EVs with viable models of standalone use, or to the particular fit of EVs with novel socio-organizational innovations such as car-sharing (see below) etc....

Similarly, the E2W is the acme of liberty and autonomy to its hundreds of millions of adopters. But it is also, in other powerful minds, just as essentially associated with multiple security threats characteristic of the modern Chinese metropolis. And, finally, the ICE car is, of course, the archetype of conspicuous achievement, success and modern urban life for many, on the one hand; while it is increasingly rejected as the apex of selfish, gas-guzzling decadence and unhealthy, unequal urban environments, on the other. As such, to every innovation and association with liberty there is a negative association with novel security threats and ensuing counter-innovation, not least from the contending monsters.

From this perspective, then, we can also see anew why China's disruptive innovation in this domain remains so essentially troubled, and not *in spite of* but rather *because of* its very system dynamism. For the predicament here at its most fundamental is one of transforming socio-technical systems under pressure from the multiple novel Challenges of complexity, and this is essentially a matter of power/knowledge momentum that *necessarily* confronts the incumbent regime of power/knowledge relations. The very *existence* of these other, new monsters is thus 'political' and radical vis-à-vis given sedimented common-senses (Tyfield 2017). Yet the monstrousness of the *incumbent* system of urban mobility itself, which only ever grows more insupportable, propels a system dynamism from which novel alternative monsters and their experimentation and development relentlessly emerge. The growth of the incumbent monster, in other words, drives both increasingly enabled demand for alternative, monstrous systems of mobility that afford, to some, the individualized autonomy and liberty it itself is cultivating; *and* the deepening dynamics of existential concern and sensitivity about 'monster' innovations per se regarding mobility-related security risks.

On top of the liberty/security dynamics constitutive of each monster, then, is a key meta-systemic dynamic of liberty/security in urban mobility innovation more broadly. This imparts a further boost to the momentum of its power/knowledge system productiveness; a boost, moreover, needed to overcome the exceptionally locked-in dysfunctionalities of the incumbent system (see Figure 9.2). For monstrousness thus breeds monstrousness, in essentially contested feedback loops of both adoption and rejection of mobility innovations that together *make* these emergent vehicle-systems *both* irreducibly manifest and resilient *but also* system-incompatible and mutually wrestling, i.e. 'monstrous.'

The tendential outcome of this highly dynamic, and specifically Chinese, process in the medium-to-long term, then, is towards the strengthening of one or some of the (new) monsters as against the continuing dysfunctional mutation of the incumbent one. And the key emergent strategic goal here is the novel *domestication* of one of the monsters, such that what then incontrovertibly *exists* can also be actively *adopted* and *embraced* as not just system-compatible, but system-constitutive.

However, just as the domestication of the wolf, wild horse and ox profoundly transformed (the power/knowledge relations of) human society, and not just these animals, so too domestication of the monsters of e-mobility will do likewise; and will most likely be driven by those who stand to benefit the most from that transformation. In this instance, it is again the Chinese urban middle class that is the crucial agent, as both the segment of society whose allegiance would categorically stamp domesticated respectability upon one of the novel monsters and, vice versa, as the emerging power bloc who will likewise be not just further enabled and empowered, but specifically *constituted* and shaped, by that adoption. To see how this process is already unfolding, however, we must attend to one final sphere of e-mobility innovation, going beyond the default initial focus on vehicles per se:

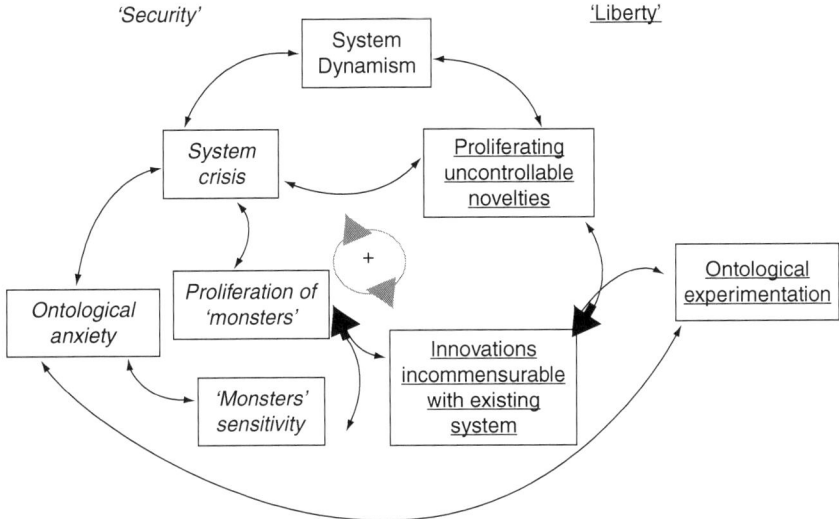

Figure 9.2 Liberty-security dynamics and monsters.

the ongoing convergence of mobility innovation with digital technologies and informationalization in the form of 'mobility-as-a-service' (MaaS).

Notes

1 See also data from the World Bank: http://data.worldbank.org/indicator/SP.URB. TOTL.IN.ZS.
2 As I write this, it seems that this threshold may have been broken with the first Chinese car company to enter Europe announcing its plans to do so: namely Geely, via its ownership of Volvo, which are together launching a new brand of SUVs called Lynk, aiming to compete with VW but at lower cost (Campbell 2016). This is further evidence of the (accelerating?) dynamism in mobility innovation in China today.
3 The unevenness here involves booming housing markets in top and second tier cities together with ghost cities and falling house prices in many smaller, newer and/or more remote locations (FT 2016) – a further mark of the current system dysfunction of the Chinese political economy.
4 This refers to interviews by my colleague Dennis Zuev conducted in China between December 2013 and March 2016 with funding from the UK's Economic & Social Research Council, grant ES/K006002/1, 'Low Carbon Innovation in China: Prospects, Politics and Practice'. I gratefully acknowledge the excellent fieldwork by Dr Zuev.
5 This combination of US cachet and engineering expertise with Chinese investment, consumer demand and engineering ambition seems particularly reminiscent of how a century before it was European cachet and engineering expertise (e.g. Daimler-Benz in Germany or Louis Chevrolet from Switzerland) and American investment, consumer demand and engineering ambition that took the ICE car from elite toy to mass market necessity.

References

AC Nielsen (2011) *A Snapshot of Today's Car Buyers in China*, www.cn.nielsen.com/documents/Autoreport.pdf.

Altenburg, T., D. Fischer and S. Bhasin (2012) 'Sustainability-oriented innovation in the automobile industry: Advancing electromobility in China, France, Germany and India', *Innovation and Development* 2(1): 67–85.

Anagnost, A. (2004) 'The corporeal politics of quality (*suzhi*)', *Public Culture* 16: 189–208.

Angel, S. (2012) *Planet of Cities*, Hollis, NH: Puritan Press.

Aradau, C. (2016) 'Political grammars of mobility, security and subjectivity', *Mobilities* 11(4): 564–574.

Bærenholdt, J.O. (2013) 'Governmobility: the power of mobilities', *Mobilities* 8(1): 20–34.

Baumler, A., E. Ijjasz-Vasquez and S. Mehndiratta (2012) *Sustainable Low-Carbon City Development in China*, Washington (DC): The World Bank.

Beck, U. (2009) *World at Risk*, Cambridge: Polity.

Bloomberg (2015a) 'China makes 100 billion Yuan bet on electric vehicles', *Automotive News China*, www.autonewschina.com/en/article.asp?id=14169.

Bloomberg (2015b) 'Geely aims for EVs, hybrids to account for 90% of sales by 2020', *Automotive News China*, www.autonewschina.com/en/article.asp?id=14055.

Bloomberg (2015c) 'Chinese-backed startup targets Tesla with $1 billion US plant', *Automotive News China*, www.autonewschina.com/en/article.asp?id=14002.

Brandt, L. and E. Thun (2010) 'The fight for the middle: Upgrading, competition, and industrial development in China', *World Development* 38(11): 1555–1574.

Brandt, L. and E. Thun (2016) 'Constructing a ladder for growth: Policy, markets, and industrial upgrading in China', *World Development* 80: 78–95.

Brenner, N. (2014) *Implosions/Explosions: Towards a Study of Planetary Urbanization*, Berlin: JOVIS Verlag.

Breznitz, D. and M. Murphree (2011) *Run of the Red Queen*, New Haven, CT: Yale University Press.

Campbell, P. (2016) 'Geely targets VW with Lynk launch in Europe', *Financial Times* 19 October.

Chien, S. (2013) 'Chinese eco-cities: A perspective of land-speculation-oriented local entrepreneurialism', *China Information* 27(2): 173–196.

China Youth Daily (2014) 'Electric bike safety is alarming' (in Chinese), 25 August, http://politics.people.com.cn/n/2014/0825/c70731-25528546.html.

Christensen, C. (1997) *The Innovator's Dilemma: When New Technologies Cause Great Firms to Fail*, Cambridge, MA: Harvard Business Review Press.

Cohen, M. (2010) 'Destination unknown: Pursuing sustainable mobility in the face of rival societal aspirations', *Research Policy* 39: 459–470.

Cohen, M. (2012) 'The future of automobile society: A socio-technical transitions perspective', *Technology Analysis & Strategic Management* 24(4): 377–390.

Costa Maia, S., H. Teicher and A. Meyboom (2015) 'Infrastructure as social catalyst: Electric vehicle station planning and deployment', *Technological Forecasting and Social Change* 100: 53–65.

Dennis, K. and J. Urry (2009) *After the Car*, Cambridge: Polity.

Dicken, P. (2014) *Global Shift* (7th edition), London and Thousand Oaks, CA: Sage.

The Economist (2014) "Building the dream" 19 April.

The Economist (2016a) 'It starts with a single app', 1 October.

The Economist (2016b) 'The great crawl', 18 June.

Fehrenbacher, K. (2016) 'Elon Musk: Tesla was in "production hell" this year', *Fortune*, 3 August, http://fortune.com/2016/08/03/elon-musk-production-hell/.

Financial Times (2016) 'Property risks loom in China's economy', 19 October, www.ft.com/content/563494ae-95aa-11e6-a80e-bcd69f323a8b.

Fishman, E. and C. Cherry (2016) 'E-bikes in the mainstream: Reviewing a decade of research', *Transport Reviews* 36(1): 72–91.

Flugge, B. (ed.) (2017) *Smart Mobility – Connecting Everyone*, Dordrecht: Springer Vieweg.

Foucault, M. (2009) *Security, Territory, Population: Lectures at the Collège de France 1977–1978*, Translated by Graham Burchell, Basingstoke: Palgrave Macmillan.

Gallagher, K.S. (2006) 'Limits to leapfrogging in energy technologies? Evidence from the Chinese automobile industry', *Energy Policy* 34: 383–394.

Geels, F. (2012) 'A socio-technical analysis of low-carbon transitions: Introducing the multi-level perspective into transport Studies', *Journal of Transport Geography* 24: 471–482.

Geels, F., R. Kemp, G. Dudley and G. Lyons (2013) 'Preface', in F. Geels, R. Kemp, G. Dudley and G. Lyons (eds), *Automobility in Transition?* Abingdon: Routledge.

German Trade and Invest (2016) *The Automotive Industry in Germany*, Berlin: GTAI.

Girardet, H. (2008) *Cities People Planet: Urban Development and Climate Change*. Hoboken, NJ: John Wiley & Sons Incorporated.

Glaeser, E. (2011) *Triumph of the City*, London: Pan Macmillan.

Goodman, D. (2015) 'Locating China's middle classes: Social intermediaries and the party-state', *Journal of Contemporary China* 25(97): 1–13.

Guo, Y. (2008) 'Farewell to class, except the middle class: The politics of class analysis in contemporary China', *Asia-Pacific Journal* 26(2): 1–19.

Huang, Y. (2008) *Capitalism with Chinese Characteristics*, Cambridge, MA: MIT Press.

iCET (2011) 'Electric vehicles in the context of sustainable development in China', *UN Commission on Sustainable Development*, 19th Session, Background Paper 9, 2–13 May, New York.

International Energy Association (IEA) (2015) *CO2 Emissions from Fuel Combustion*. Paris: IEA.

Jakobson, L. (ed.) (2007) *Innovation with Chinese Characteristics*, Basingstoke: Palgrave Macmillan.

Johnson, S. (2015) 'China surge drives global car sales to record high', *Financial Times*, 12 December, www.ft.com/intl/cms/s/3/569e90c0-a000-11e5-8613-08e211ea5317.html.

de Jong, M., D. Wang and C. Yu (2013) 'Exploring the relevance of the eco-city concept in China: The case of Shenzhen Sino-Dutch Low Carbon City', *Journal of Urban Technology* 20(1): 95–113.

Kahn, M.E. and S. Zheng (2016) *Blue Skies over Beijing*, Oxford and Princeton, NJ: Princeton University Press.

Kopp. A, R.I. Block and A. Iimi (2013) *Turning the Right Corner: Ensuring Development through a Low-Carbon Transport Sector*, Washington DC: World Bank.

Law, J. (ed.) (1991) *A Sociology of Monsters*, London: Routledge.

Lyons, G. and P. Goodwin (2014) 'Grow, peak or plateau – the outlook for car travel', Report of a Roundtable Discussion, New Zealand Ministry of Transport.

Marletto, G. (2014) 'Car and the city: Socio-technical transition pathways to 2030', *Technological Forecasting and Social Change* 87: 164–178.

Mayer-Schönberger, V., and K. Cukier (2013) *Big Data: A Revolution that Will Transform How We Live, Work, and Think*. New York: Houghton Mifflin Harcourt.

McLaren, D. and J. Agyeman (2015) *Sharing Cities: A Case for Truly Smart and Sustainable Cities*, Cambridge, MA: MIT Press.

Mokyr, J. (1991) *The Lever of Riches*, Oxford: Oxford University Press.

Nahm, J., and E.S. Steinfeld (2014) 'Scale-up nation: China's specialization in innovative manufacturing', *World Development*, 54: 288–300.

Nolan, P. (2004) *Transforming China: Globalization, Transition and Development*, London: Anthem.

OICA (International Organization of Motor Vehicles Manufacturers) (2013) '2013 Production Statistics', www.oica.net/category/production-statistics/.

Ong, A. (2006) *Neoliberalism as Exception: Mutations in Citizenship and Sovereignty*. Durham NC: Duke University Press.

Pasaoglu, G., D. Fiorello, A. Martino, L. Zani, A. Zubaryeva and C. Thiel (2014) 'Travel patterns and the potential use of electric cars – Results from a direct survey in six European countries', *Technological Forecasting and Social Change* 87: 51–59.

Paterson, M. (2007) *Automobile Politics*, Cambridge: Cambridge University Press.

Pontes, J. (2016) 'China electric cars sales up 188%, still dominated by BYD', *Cleantechnica*, 11 August, https://cleantechnica.com/2016/08/11/china-electric-car-sales-188-still-dominated-byd/.

Qin, L. (2015) 'Cars the main culprit for Beijing's smog: Govt figures', *Chinadialogue*, 2 April, www.chinadialogue.net/blog/7829-Cars-the-main-culprit-for-Beijing-s-smog-govt-figures/en.

Rajan, S.C. (1996) *The Enigma of Automobility*, Pittsburgh: University of Pittsburgh Press.

Rajan, S.C. (2006) 'Automobility and the liberal disposition', *Sociological Review* 54(1): 113–129.

Reuters (2015) 'Drivers scope out EVs in smog-choked cities', *Automotive News China*, 11 December, www.autonewschina.com/en/article.asp?id=14133.

Reuters (2016a) 'With tougher rules, China seeks fewer, but better, EV makers', *Automotive News China*, 25 November, www.autonewschina.com/en/article.asp?id=15483.

Reuters (2016b) 'Suppliers question Tesla's goals for Model 3 output', 20 May, http://uk.reuters.com/article/us-tesla-suppliers-idUKKCN0YB0CA.

Rifkin, J. (2011) *The Third Industrial Revolution: How Lateral Power is Transforming Energy, the Economy, and the World*, Basingstoke: Macmillan.

Rock, M., J.T. Murphy, R. Rasiah, P. van Seters and S. Managi (2009) 'A hard slog, not a leap frog: Globalization and sustainability transitions in developing Asia', *Technological Forecasting and Social Change* 76: 241–254.

Salat, S. (2015) 'Urban morphology, Spatial planning, spatial economics and climate change', Presentation to the *4th Nobel Laureates Symposium on Global Sustainability*, Hong Kong, 22–25 April.

Schwab, K. (2016) *The Fourth Industrial Revolution*, Geneva: WEF.

Schwanen, T., D. Banister and J. Anable (2011) 'Scientific research about climate change mitigation in transport: A critical review', *Transportation Research Part A* 45: 993–1006.

Sheller, M. (2004) 'Automotive emotions: Feeling the car', *Theory, Culture & Society* 21(4–5): 221–242.

Sheller, M.J. and J. Urry (2006) 'The new mobilities paradigm', *Environment and Planning A* 38(2): 207–226.

Shove, E., M. Pantzar and M. Watson (2012) *The Dynamics of Social Practice*, London: Sage.

Sperling, D. and D. Gordon (2009) *Two Billion Cars: Driving towards Sustainability*, New York: Oxford University Press.

Straw, J. and M. Baxter (2014) *iDisrupted*, New York: New Generation Publishing.

ten Brink, T. and F. Butollo (2016) 'A new upgrading paradigm? Innovation policy and domestic demand in strategic emerging industries', paper presented to the *International Symposium on Innovation-Driven Development*, Sun Yat Sen University, Guangzhou, 13–15 June.

Thun, E. (2006) *Changing Lanes in China*, Cambridge: Cambridge University Press.

Tillemann, L. (2015) *The Great Race: The Global Quest For The Car Of The Future*, New York: Simon and Schuster.

Time (2014) 'China's road show', 13 March.

Timmons, H. (2013) 'Consider the e-bike: Can 200 million Chinese be wrong?', *Quartz* 22 October, http://qz.com/137518/consumers-the-world-over-love-electric-bikes-so-why-do-us-lawmakers-hate-them/.

Tse, E. (2016) *China's Disruptors*, London: Portfolio Penguin.

Tyfield, D. (2013) 'Transportation and low-carbon development', in F. Urban and J. Nordensvard (eds), *Low-Carbon Development: Key Issues*. London: Earthscan.

Tyfield, D. (2014) 'Putting the power in "socio-technical regimes" – E-mobility transition in China as political process', *Mobilities* 9(4): 585–603.

Tyfield, D. (2017) 'Mobilizing the Emergence of Phronetic TechnoScienceSocieties: Low-Carbon E-Mobility in China', Sociology of the Sciences Yearbook 2017.

Tyfield, D, D. Zuev, P. Li and J. Urry (2015) 'Low carbon innovation in Chinese e-mobility: Prospects, politics and practices', *STEPS Working Paper 71*. Brighton: STEPS Centre.

Unruh, G. (2000) 'Escaping carbon lock-in', *Energy Policy* 30: 317–325.

Urry, J. (2002) *The Tourist Gaze*. London: Sage.

Urry, J. (2003) *Global Complexity*, Cambridge: Polity.

Urry, J. 2007 *Mobilities*. Cambridge: Polity.

Urry, J. (2013) *Societies Beyond Oil: Oil Dregs and Social Futures*, London: Zed Books.

Usher, M. (2014) 'Veins of concrete, cities of flow: Reasserting the centrality of circulation in Foucault's analytics of government', *Mobilities*, 9(4): 550–569.

Wang, L., J. Xu, X. Zheng and P. Qin (2013) 'Will a driving restriction policy reduce car trips? A case study of Beijing, China', *Environment for Development Discussion Paper Series*. 20–13.

Wang, T. (2013) *Carnegie Policy Outlook: China's Electric Vehicle Policy*, Beijing: Carnegie Institute.

Wells, P. and X. Lin (2015) 'Spontaneous emergence versus technology management in sustainable mobility transitions: Electric bicycles in China', *Transportation Research Part A: Policy and Practice* 78: 371–383.

Winebrake, J., S. Rothenberg, J. Luo E. and Green (2008) 'Automotive transportation in China: Technology, policy, market dynamics, and sustainability', *International Journal of Sustainable Transportation* 2: 213–238.

Wolf, I., T. Schröder, J. Neumann and G. de Haan (2015) 'Changing minds about electric cars: An empirically grounded agent-based modelling approach', *Technological Forecasting and Social Change* 94: 269–285.

Yang, J. (2015) 'Beijing gives another boost to EVs by setting sales targets for each province', *Automotive News China*, 18 December, www.autonewschina.com/en/article.asp?id=14166.

Yang, J. (2016) 'How Beijing's EV plan came unglued – and what to do about it' *Automotive News China*, 28 January, www.autonewschina.com/en/article.asp?id=14050.

Yu, L. (2014) *Consumption in China*, Cambridge: Polity.

Zavoretti, R. (2013) '"Be my Valentine": Bouquets, marriage, and middle class hegemony in urban China', *Max Planck Institute for Social Anthropology Working Papers 150*, www.eth.mpg.de/cms/en/publications/working_papers/wp0150.html.

Zeng, M. and P. Williamson (2007) *Dragons at your Door: How Chinese Cost Innovation is Disrupting Global Competition*, Cambridge, MA: Harvard Business School Press.

Zhao, Y. (2010) 'China's pursuits of indigenous innovations in information technology developments: Hopes, follies and uncertainties', *Chinese Journal of Communication* 3(3): 266–289.

Zhou, X. and C. Qin (2010) 'Globalization, social transformation and the construction of China's middle class', in C. Li (ed.), *China's Emerging Middle Class*, Washington DC: Brookings: 84–103.

Zuev, D., D. Tyfield and J. Urry (forthcoming) 'Where is the politics? E-bike mobility in urban China and civilizational government', *Environmental Innovations & Sustainable Transitions*.

10 Towards mobility-as-a-service

The new corporate imaginary of mobility-as-a-service

Search online for information about mobility disrupters today, and it is much more likely that you will find stories about Silicon Valley-based innovations aiming to 'disrupt' the model of private car ownership than about Chinese micro-EVs (e.g. *Economist Films* 2016, Straw and Baxter 2014). Of course, the two are not necessarily mutually incompatible (as we shall see). But the supreme significance of the former is as a sign of a powerful emerging common-sense (in 2016) amongst business and policy elites across the world: that the 'hardest case' of system transition – i.e. personal car-based automobility – is simply the next industrial summit that will (perhaps 'inevitably') be conquered by Tech innovation, with its mastery of consumer-friendly, stylish, individualized services via digital social media platforms. This is the imaginary of 'mobility-as-a-service', of which the archetype is the ride-hailing firm, Uber.

In its own estimation and that of a fawning international business press, Uber is not simply 'disrupting' the car industry but the very future of urban transport, ushering in an 'Uberworld' as *The Economist* put it on its front cover in September 2016. To be sure, Uber is an impressive corporate story. From its founding in 2009, it has set up in nearly 500 cities across the world and has around 30 million monthly users (*The Economist* 2016a). Through several large funding rounds and acquisitions (including of autonomous driving technology companies), it has accrued a valuation of close to $70 billion, making it by far the world's largest Tech start-up 'unicorn' and a third more valuable even than General Motors, despite the latter's annual global sales in 2015 of $152 billion (ibid.). And while initially targeting the elite market, *à la* Tesla, Uber has rolled-out a number of other services that afford P2P ride-sharing of varying degrees of comfort and exclusivity that are bringing their services to an increasingly large and low-cost market.

This spectacular growth is based upon a familiar Silicon Valley model, following Amazon or 'sharing economy' Apps like AirBnB, where the Tech company provides the online platform to bring together and mobilize existing assets with dispersed demand, thereby perfecting the market for these otherwise-underutilized goods and services. Ownership of the assets (here cars), however, remains with the drivers, who are also not employees but self-employed sole-traders offering their services on the platform and providing a cut of their earnings as payment. Such is the attraction of access to an abundant market of vetted and cashless customers that in many cities, including London, Uber drivers already now outnumber official taxi drivers, at around 25,000 (Knight 2016). Amassing enormous Big Data databases regarding services provided (here actual journeys) that are crunched for insights also then gives the Tech company a unique leverage in offering and innovating further services with a ready but untapped demand, exploiting the positive system externalities to the company's advantage as super-rents of innovation. The goal is maximum platform growth to achieve as dominant a position as possible, with maximized network effects and maximal control of these externalities, while profits and revenues remain a distant second in strategic importance.

It is also easy to understand why Uber has generated so much excitement amongst investors. The global market for individualized transport is estimated to be worth as much as $10 trillion *per annum* according to Morgan Stanley, as opposed to, say, $175 billion for online advertizing, $100 billion for taxis and $2.3 trillion for cars (*The Economist* 2016a, 2016b). Claiming a dominant stake in even a tiny fraction of this market thus promises extraordinary profits (at least in the medium-term), especially 'at a time when returns from other assets are widely disappointing' (*The Economist* 2016a: 17), i.e. amidst the stagnation of the terminal crisis of neoliberalism. And this is especially so when combined with the other key hi-tech aspect of this imaginary, namely autonomous vehicles (AV), in which, of course, Silicon Valley giants such as Google are also prominent. For together, absent even the need for drivers, MaaS and AV seem to offer the prospect of complete control by Big Tech companies of entire systems of urban mobility. In short, a growing manifest reality in many big cities of convenient, reliable, comfortable personal mobility through ride-sharing promises to make personal ownership of a car an increasingly expensive, unnecessary hassle. The declining rate of digitally native young people in the global North passing their driving test in recent years attests to these trends of 'peak car' (Lyons and Goodwin 2014).

These dynamics of 'disruption' of urban mobility by digital innovation are just as striking in China. Indeed, they are arguably greater and more concentrated. Key to this dynamic is the strength in digital innovation of China's own ICT giants: the acclaimed 'disrupters', discussed above, of the 'BATs' (Chapter 6). China's carbon state-capital players in the car industry are still technologically dependent on their foreign partners and are slowly, with Government funding and demands, shifting production capacities to the EV. Meanwhile, Chinese digital capital firms are simultaneously exploring opportunities for smart EV production using their own monies and much greater impetus. In particular, this is stimulating a tumult of corporate activity and deals in the digital MaaS domain.

For instance, in 2015, Tencent Holdings, the largest Chinese software giant, announced its agreement with Hon Hai (Foxconn) manufacturing and Harmony Auto dealer to enter electric vehicle production (Reuters 2015). Moreover, spurred by competition, several key representatives of Chinese digital capital announced their plans to participate in the smart EV race with each player setting ever bolder goals: Alibaba (e-commerce giant) has paired with SAIC motors and has bold plans to dominate urban mobility in China (Tse 2016), while Baidu (China's leading search engine) also announced independent entrance in the smart and driverless car production. It is noteworthy too how these initiatives in China are particularly focusing on the EV in ways that are not the priority for the Silicon Valley firms. Just as Uber is seen as the company to beat from Silicon Valley in this domain, however, it is in China's ride-hailing business too that we find perhaps China's most important mobility 'disrupter': Didi Chuxing, with a commanding domination of China's market and an increasingly essential app on every Chinese smartphone (see below).

Not only does all this feverish activity by corporate behemoths strike a notable contrast with the continuing plodding progress of the EV amongst the car majors, notwithstanding the new impetus Tesla has imparted to this sector. But, more importantly, all this activity seems to pose the question whether or not disruptive digital MaaS innovation is the key means and route towards the domestication of the various monstrous vehicle-systems discussed above. There are convincing reasons prima facie that it could be so. For instance, where private ownership is replaced by reliance instead upon accessible quasi-personalized, shared mobility, the number of cars on the roads could drop considerably, easing congestion. The wastefulness of parking, in terms of the time spent finding a space and the amount of land that must be devoted to it, and the under-utilization of the vehicles themselves as they sit idle for 95 per cent of the time, could also be substantially mitigated. This could also be transformatively enabling for electric cars in particular. For as the German Society for International Cooperation (GIZ) note, 'car-sharing does not need the EV, but the EV needs car-sharing' (GIZ 2014). This is because small fleets of shared cars concentrate the essential challenges of charging infrastructures and of coordinating time needed for charging and use in ways that significantly mitigate these issues; as the greater success with electric buses and taxis shows. (Autonomous) MaaS, in short, could make much more efficient use of resources and enable their quantum level decarbonization by alleviating bottlenecks in EV adoption and connection to 'smart' and green electricity grids.

Googliberal mobility-as-a-service innovation to the rescue?

In fact, there are also many reasons that the ride-hailing, Silicon Valley, Tech-heavy model of mobility-as-a-service is likely to remain, at best, a niche and/or pipedream. Indeed, far from being the heroic saviour of its own mythology – and perhaps thereby transmogrifying the vehicle monsters into systems friendly to humans and planet – it is perhaps most informative to see it as just another monster. To see how this is the case, we must characterize this dynamic of mobility system innovation in terms of its complex power/knowledge system dynamics, and then set these against the complex systems of urban mobility that it is aiming to 'disrupt'.

This comparison can be traced in several steps (see Figure 10.1 et seq.), that progressively reveal how Big Tech MaaS is no less essentially contested – and indeed, dynamic precisely as such. This is not, therefore, to argue that these imaginaries will come to nought, but that they will grow and develop only and insofar as they are profoundly transformed through similar dynamics of liberty and security that beset the other mobility monsters, and their interweaving with each other. To the extent that MaaS does indeed emerge as system-dominant over the medium-to-long term to the 2030s and beyond, therefore, the form of its reality remains fundamentally uncertain and will likely diverge considerably from its current Big Tech imaginings.

To start with, on the one hand, as described above, the model currently being pursued by Uber is one of digital social media platform-based exploitation of latent markets and incumbent infrastructures for maximal network effects and exclusive proprietorial control of those positive externalities; what we have above called a Googliberal innovation model (see Chapter 2; Tyfield 2013), connoting one that uses digital technology platforms for concentrated rentiership (Birch 2016) of existing assets in an essentially neoliberal, supremely proprietorial and parasitic form of innovation that nonetheless proclaims itself the nemesis of neoliberalism and the

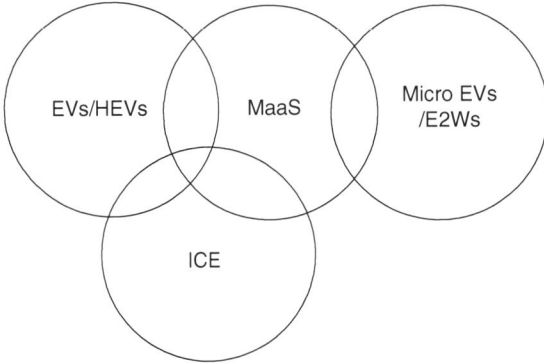

Figure 10.1 MaaS domesticating the 3 monsters?

'disruptive' hero of open-source, 'sharing economy' entrepreneurialism. This model thus presupposes a lot for the reaping of its massive monopoly profits, in terms of socio-technical conditions, infrastructures and power/knowledge relations, while deliberately excluding responsibility for tending and maintenance by the company of these conditions, as a necessary prerequisite of their unrelenting but profitless growth.

Moreover, on the other hand, the service Uber and other ride-hailing companies like Lyft provides is – and is explicitly framed in terms of – getting from 'A to B', albeit perhaps now 'enjoyably' (*Economist Films* 2016). They are thus providers of *journeys*, not mobility in its broader, systemic sense. Yet 'mobility' in this latter sense is both ontologically prior, in that structures of journey provision – and *demand* for those journeys (Shove and Walker 2014) – presuppose these broader systems of mobility, and thence emerges as increasingly pressing *strategically* as well, especially as innovations gun for transition at system level.

In both respects, these systematic oversights and constitutive blindnesses to a broader systemic perspective – as a badge of its very Silicon Valley 'disruptiveness' – can only engender deepening conflict with each step of its very success, as it pushes ever more insistently towards power/knowledge transition at the *system* level (see Figure 10.2). At its most basic, the Uber model presupposes forms of vehicle (perhaps autonomous) that must be economically, and – in a new CP/KS that is even relatively resettled and pushing at or beyond system transition – profitably, manufactured and distributed. To the extent these are manufactured by existing automotive majors, however, we immediately encounter the first potential clash. For the very promise of fewer vehicles being more efficiently utilized sets up this model in direct conflict with those powerful automotive companies, for whom such a drop in annual sales would be effectively lethal.

Indeed, seeking to disrupt the disrupters, big car companies are increasingly rebranding themselves as providers of 'elite mobility services' (e.g. BMW or Audi) and experimenting with models of shared mobility that will give them access to this $10 trillion market while also shaping new models of car use that do not presuppose a catastrophic decline in annual car sales. Ride-hailing as a security threat, therefore, elicits powerful counter-innovation. Yet, conversely, starting from a low-base and

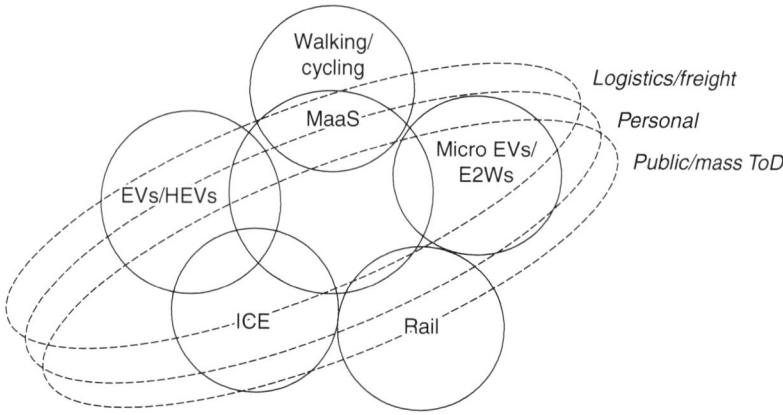

Figure 10.2 The interlocking systems of urban transport.

seeking only to grow bigger *than they are now*, a model of much reduced *total market* annual vehicle sales could still be a striking opportunity of much *increased* sales for start-up companies, notably in the EV sector. The Tech MaaS imaginary is thus essentially both security threat and opportunity for different existing industrial concerns, setting up some heated and highly dynamic contestation, not a straight march to its realization.

But beyond these early and already-manifest dynamics, from a CP/KS perspective it is hard to miss how profoundly contested the asset-light model of journey-sharing is likely to become the more it grows.[1] Uber is already met with controversy wherever it goes; and not just amongst aggrieved taxi drivers, frustrated by how these platforms are bypassing regulation, unsettling established local political economies of mobility services and undercutting the competition, or its own drivers regarding their status as 'employees' not 'independent contractors' (Osborne 2016). Passengers too have complained loudly (amplified by the social media on which, as digitally-enabled ride-sharing-users, they are prevalent) about being exposed to new security hazards of unlicensed and untraceable sexually predatory drivers, or fickle and sometimes extortionate pricing (as in Uber's infamous 'peak demand' pricing model), or privacy and surveillance of their journeys by these unaccountable companies, or simple unreliability and inaccessibility (as in many Chinese cities during Chinese New Year 2016, when many of the drivers had themselves left town).

The unique attractiveness to drivers is also dependent upon the viability of Uber's growth-at-all-costs business model. Uber can attract increasing financial investment and valuation on the basis of exponentially increasing revenues (if not necessarily profits, but with these *promised* 'in due course') while still charging low rates to participating drivers *and* low and attractive fares to customers, all of which thereby feeds the positive feedback loop of such rapid growth of the platform. In this upward spiral, the asset-light and low-cost model of sharing together with a ruthless focus on platform growth appears as a win all-round: for the firm itself, for self-employed drivers, for riders and for investors. Yet, in what resembles a Ponzi scheme – the characteristic of 'Googliberal' innovation more generally (cf. Mirowski 2012; Tyfield

2013) – as the growth of this model pushes at system saturation, as it inevitably must eventually, the positive feedback loop is likely to go swiftly into reverse.

Slowing system growth demands that revenue growth, relentlessly demanded by financial investors, would now have to be maintained by rising drivers' rates and fares alike. But the former directly subtracts from the profitability of the drivers' participation, exacerbating demands they be treated fairly as the de facto employees they are. While rising consumer prices dampens demand further, and so on.... At the limit where ride-hailing platforms 'provided' a significant percentage (let alone the majority) of vehicles, it is hard to imagine that this would not also involve demands for them to take financial and legal responsibility for (if not ownership of) its vast fleets, and elicit increasing attention from regulators and Governments to ensure the optimal provision of what would then be amongst *the* essential public goods of modern city life. Yet such developments would, no doubt, elicit further innovation-as-politics from the ride-hailing services. In these circumstances, then, the asset-light nimbleness of this model that is the key strategic asset to its rapid, 'disruptive' momentum becomes as well a growing source of contestation, *played out in innovation* not just in explicit protest. In short, again we see how the Tech MaaS model is productive to the extent it is essentially contested, i.e. precisely the dynamics that make it a monster. And wherever this leads in the medium-term, it fundamentally does not lead to a smooth, incremental embrace of Tech MaaS over personal automobility.

Second, even limited to the level of what modes of transit are necessary and must be coordinated in the complex and changing urban landscapes of contemporary (possibly 'world') cities, it is clear that the wholesale domination of mobility by such quasi-private digital ride-sharing as currently understood is impossible. This is quite obviously the case insofar as these journeys replace not just existing private journeys but also public or non-motorized transport. The latter would massively increase, not mitigate, congestion and emissions, and thereby frustrate at collective system level the growing enablement by these services at a personal level. Instead, mass transit-oriented development (ToD) systems are needed (Dittmar and Ohland 2004).

Quasi-personal motorized and enclosed mobility of ride-hailing may well be a key feature of such a system. But it is certainly just one aspect, and not necessarily the most important, even for replacement of the car. Liveable cities and the mobility systems that are their metabolism and circulation, with clean environments, green spaces, light and safe traffic and ready access, demand attention to a whole host of socio-technical concerns (Urry *et al.* 2016). For instance, this includes not just 'human-level' infrastructures of mobility, affording walking, cycling (and 'wheeling' in wheelchairs (Parent 2016)), and stasis or immobility (e.g. benches, cafes and restaurants, waiting rooms, terminals and shelters etc. ...). But also the interweaving with broader systems of mobility that afford the multiple forms of 'traffic' into, out of and through a space that allow it to *stay* clean, attractive and safe. This thus includes not only freight and logistics, but service workers, tourists/visitors and tourist services including high-quality entertainment, dust-carts, maintenance and construction vans, emergency vehicles, etc....

The top-down hi-tech imaginary of Uber, however, does not even register, if not actively excludes, such considerations. Moreover, such Googliberal business strategy and policy framed by the Tech imaginary of MaaS tends to frame other approaches, that may seek to *incorporate* but not be dominated by digital ride-sharing services (e.g. in Helsinki (*The Economist* 2016b)), as political opponents, including per se any form of regulation by the state. And while the complex coordination of equit-able, sustainable MaaS *systems* of ToD necessarily requires a pragmatic negotiation amongst multiple concrete agencies – Government, service providers, stakeholders – in specific locations, the Tech imaginary is specifically premised upon a one-size-fits-all bulldozing strategy that glories in its Tech-fetishistic entrepreneurial disdain for such effortful and painstaking cooperation-cum-competition. Yet the dynamics above of deepening liberty and security specifically concerning MaaS systems will ensure that such wrangling will come to Big Tech MaaS, and be constitutive dynamics of its successful emergence, whether it likes it or not.

But these dynamics of power/knowledge system contestation and emergence also extend way beyond the socio-technical architectures of transit to the transition of broader mobility systems. This incorporates not just physical movement of people and things, and systems *enabling* those mo*bilities* (and immobilities). But also parallel flows of services, ideas, data and images and the conditions for those circula-tions, and the conjunction of all this in the dynamic, developing assemblages that give rise to the possibility of and demand for specific movements. These complex power/knowledge systems, with their power/knowledge relations and technologies acting on and performing each other, are thus also characterized by specific, spatio-temporally located regimes of system conduct and government, enabling and har-nessing specific mobilities (or not) in relatively self-sustaining feedback loops (Tyfield 2014; Sheller 2016). In other words, Mobility-as-a-Service will only successfully emerge as system transition to the extent that all the power/knowledge relations and technologies it presupposes are also transformed around it.

To spell out something of the vast and unfathomable complexity of this challenge, we can consider how mobility systems sit in the middle, mediating the interactions and transformation of two key spheres that extend, again, way beyond the limited purview of the Tech MaaS imaginary. First, there is the transformation and redesign of the physical and infrastructural forms of the ('liveable', 'smart') city, and the power/knowledges involved in their day-to-day functional reproduction. This would include, for instance, not just the physical infrastructures of the smart city's vehicle fleet and charging infrastructures, but also those of production, circulation and consumption of (increasingly low-carbon, efficient, reliable) energy, data and communications, food and resources, and waste. And, second, the changing everyday practices of social life, including employment, provisioning and (e-)commerce, entertainment and leisure, maintenance, health and cleaning, and education, all of which must co-evolve in parallel with the changing institutions and architectures of these domains as they too are likely transformed – not least by *their* growing saturation with digital-based innova-tion of power/knowledge technologies in the medium-term.

This complex, multi-factorial co-evolution thus involves elements of contingent sequencing, path-dependence and bricolage that simply cannot be plotted in advance, but possibly can be worked through in practice. But this is a process of power/knowledge, not just 'technology', driven by strategizing and changing agen-cies and co-producing both new assemblages of socio-technologies *and* the practices and processes of their government from moment-to-moment, day-to-day, year-to-year. In short, the development of urban mobility systems *as a whole*, including but not limited to (quasi-)private passenger transport, is inextricably connected to the parallel and complex co-evolution of city-scapes and daily practices, where all of these are conceptualized as both discursive-material realities *and* aspects of system government (see Figure 10.3).

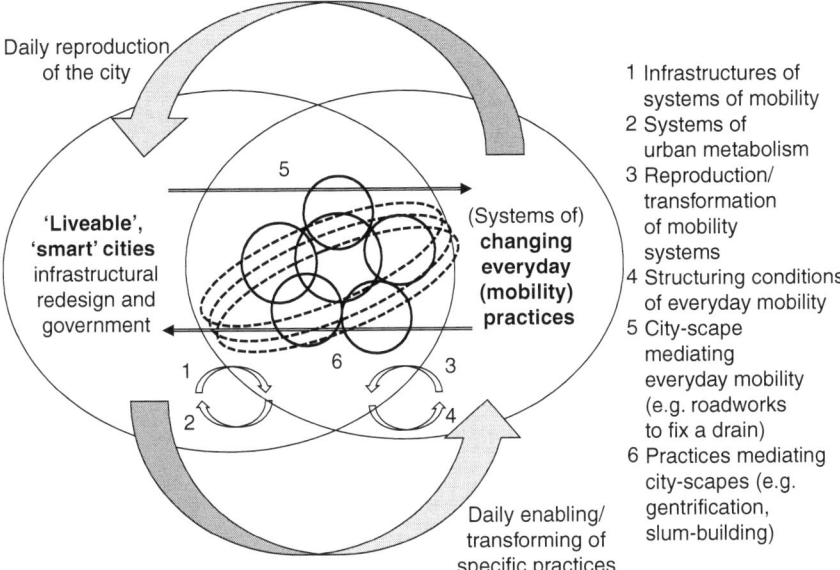

Daily reproduction of the city

'Liveable', 'smart' cities infrastructural redesign and government

5

(Systems of) **changing everyday (mobility) practices**

Daily enabling/ transforming of specific practices

1 Infrastructures of systems of mobility
2 Systems of urban metabolism
3 Reproduction/ transformation of mobility systems
4 Structuring conditions of everyday mobility
5 City-scape mediating everyday mobility (e.g. roadworks to fix a drain)
6 Practices mediating city-scapes (e.g. gentrification, slum-building)

Figure 10.3 Urban mobility, liveable cities, daily practice.

There is thus an irreducible socio-cultural richness and geographical specificity to the power/knowledge processes constitutive of mobility systems and their trans-formation that necessarily emerges with increasing force the more it is confronted with strategic action pushing at system transition. Yet the very constitution of the Tech MaaS trajectory is to be dismissive of all concerns beyond its own specific Pro-methean determination to disrupt existing industries and sectors and to profit for itself alone in its growing parasitism upon a system for which it categorically assumes no responsibility, with every push back encountered simply as further grist to that Randian me-against-the-world mentality.[2] As such, we see again how the more suc-cessful and expansive is the niche of Tech MaaS the more it will essentially encounter socio-political dynamics contesting it and/or attempting to mould it in ways expres-sive of the broader power/knowledge relations and constituencies of that place *and not* of the forces behind Tech MaaS itself. This is thus an escalation or power/know-ledge arms race of innovation and counter-innovation, liberty and security, as those newly enabled by MaaS grow increasingly enabled while the growing numbers it unsettles fight back with equal and opposite vigour.[3]

As such, we need also to see how these essentially contested dynamics *of* system emergence are also situated in, and in turn instantiating, the dynamics of innovation demand and supply, regarding the emerging power/knowledge constituencies dis-cussed above. Indeed, from the perspective of complex power/knowledge systems, it is only at this level that we reach an analysis that can begin to explore how and whether the system dynamics on display incorporate and are generating the novel *agents* of this systemic change and innovation. In other words, simply latching onto the explicit and self-serving corporate imaginaries of the incumbently powerful

Googliberal corporate giants as if they present a realistic vision of future urban mobility is to wish away precisely the self-propagating and turbulent *power/know-ledge* dynamics of innovation that will be constitutive of any success at system level that they do in fact achieve.

But while we must take these imaginaries with a blood pressure-bursting handful of salt, we must also acknowledge the persistent and pivotal importance of digital innovation regarding mobility systems. For, precisely *as* power/knowledge innovation, digital innovation occupies a uniquely privileged systemic position – at the reflexive crux of the innovation dynamics of power/knowledge technologies acting on themselves in particularly tight positive feedback loops, and controlling the key power/knowledge technologies for complexity innovation (including of mobility systems) of information and data management. The intensifying contestation of Tech MaaS, in other words, will most likely simply feed this monster, not kill it, and 2.0 digital innovations will indeed likely assume a particularly crucial role in these processes of innovation-as-politics of 'smart' mobility systems.

Beset by increasingly fundamental contestation and forced into novel coalitions – 'haggling' and 'sharing', respectively – the more they push to systemic dominance, therefore, digital mobility-as-a-service innovations and corporations are still a very long way from offering viable visions, let alone actualities, for complex, sustainable urban mobility systems. What is crucial, however, and the last piece of our puzzle here, is how these dynamics are, again, especially intense in China, turning current, synchronic strategic weaknesses into medium-term, processual strengths.

Step 3: Mobility system innovation alchemy in the Chinese pressure cooker

We have considered in the last section how digital MaaS innovations, though self-styled as the 'disrupters' of urban mobility, are, in fact, neither likely to 'disrupt' the incumbent system of personal automobility nor even really that concerned with 'mobility'. But it is in turning to China that we find an empirical contrast that illuminates how misleading is this self-serving misappropriation of 'disruptive innovation', effectively redefining it just as unsettling established forms of service provision with Tech social media platforms. In China, MaaS innovation really is taking place as disruptive innovation – i.e. cheaper, easier-to-use, novel combinations creating entirely new markets, technologies and assets, not merely mobilizing existing ones, and in which digital technology is undoubtedly important but not itself enough to merit the label, nor even its most significant aspect. And this disruptive MaaS innovation is being pushed forward by a blossoming ecosystem of entrepreneurial ventures, both start-up and already giant corporations, private and public.

But this is China, and so we cannot escape the multiple levels of reality of this process of innovation, through optimist, pessimist and disrupter analyses. An optimist, therefore, can point in usual fashion to the sheer size of some MaaS ventures in China and the growing focus of the uniquely enabled top-down power of the (Party-)state on such liveable urbanism and its mobility systems. For instance, the relevant innovations and policy initiatives here would include the largest bike-sharing scheme in the world, in Hangzhou, and multiple parallel initiatives across the country that together add up to an extraordinary process of experimental (policy) learning. Crucial here also is China's first ever Urbanization Plan, announced in 2014, which points clearly to the need for a new model of urban mobility, away from the car-based gigantism that has prevailed to date, towards smart, green, human-scale systems. This agenda has also been taken up with alacrity by China's mutually competing cities and regions, backing demonstration zones and eco-city projects hoping to make them the national, and possibly global, leader in digital, low-carbon cities – with all the commercial opportunities this entails for attracting inward investment and talent flows, partnership and commissions overseas.

To all this, the pessimist can, as ever, counter with compelling evidence of how much puffery is involved in these plans, meriting in turn profound scepticism that these fine words will turn into buttered organic, peri-urban-farmed parsnips. China's record in ecocity ventures over the past decade is particularly damning in this regard, producing a catalogue of, at best, high-end speculative real estate with a smattering of greenery to make it more aesthetically appealing and, at worst, total and ignominious failure (Zhuang 2015). This has certainly been a painful learning process for all involved. But, says the pessimist, the problem is that the forces structuring this failure are very much still in place, namely the systematic dis-incentives and perverse incentives of fragmented authoritarianism, pressure to sustain construction and real estate booms (and the growth they bring) and a Party-state veto on reforms towards meaningful inclusion of stakeholder voices in complexity- and/or sustainability-attentive innovation projects.

Given the heights of power already occupied within China by high-carbon (state-owned) industries (e.g. oil, cars, road construction, coal), it seems implausible indeed that such glacial political change could unsettle the juggernaut of car-based automobility. Moreover, on the demand side, the low-trust society incubated by these political structures and recent history makes the sharing economy a particularly hard sell in China, and especially regarding something as important for status display and as expensive as a car (Zuev and Tyfield 2015). In these circumstances, then, the chances of China (or, in other words, Chinese mega-cities, prosperous regions and leading companies) incubating a transition of 'the hardest case' that still eludes countries with much more advanced innovation systems seems implausible.

Yet, turning back to the disrupters, there is indeed plenty of evidence for a growing momentum of socio-technical innovation regarding MaaS. This includes a burgeoning set of initiatives in vehicle-sharing, often with overseas partners that contribute to their branding, and a landscape of massive corporate dynamism, that has filtered up even to new arenas of competition between SOEs. To be sure, many of these initiatives are making little impact, or are viewed, even by their managers, as experiments or learning experiences, testing the water and preparing a market that is still very immature while getting one's feet-in-the-door of this medium-term opportunity (Zuev interviews).

In parallel with Uber, though, perhaps the most influential and significant case for us to consider is China's own ride-hailing companies. Over the last few years, this sector has seen the rise of several massive firms almost immediately locked into a titanic struggle. In particular, three players – Kuaidi, Didi Chuxing and Uber – were quickly adopted by the BATs Chinese internet giants, turning the fight for dominance of China's ride-hailing market into a proxy battle between these new industrial titans. Through 2015 and into 2016 this played out in terms of massive price war, with Kuaidi and Didi having rivers of funding poured into them to outspend, and so crush, their competition with cheaper fares. Such loss-making market-claiming competition, however, was so expensive as to be potentially fatal to the industry as a whole, costing Uber $2 billion in two years and Didi far more (*The*

Economist 2016c), in a classic 'burning money' (ten Brink and Butollo 2016) 'market-quake' (Kirkegaard 2016) of Chinese non-linear disruptive innovation.

And, sure enough, with this surging wave exhausted, something has been left standing, namely Didi; now merged with Kuaidi, and with Uber exiting China for a 17.7 per cent stake and seats on the board (*The Economist* 2016c). Didi already commanded 80 per cent of China's ride-hailing market, against Uber's under 10 per cent. But with Uber out of the picture its command of this fast-growing market – revenues rising threefold just between Q1 and Q4 of 2015, up to RMB13 billion – seems now all but unassailable. As such, Didi is valued at $35 billion and has raised $8 billion in funding, placing it third and second respectively in the world to Uber amongst all start-up 'unicorns'. What is of particular significance for us here, however, is the intensity of the dynamics of liberty and security played out through this process, and still in place for further waves, and what this is uniquely *producing* in China.

On the one hand, there are multiple important constituencies in China for whom digital ride-hailing is a key form of an essentially-pursued liberty. This includes the BATs themselves who, notwithstanding the mergers of the ride-hailing companies, are already and remain intensely competitive over digital MaaS innovation in a way that is not (as yet) so evident amongst Silicon Valley giants. As adopted national champions, albeit private and overseas-funded, of China's techno-nationalist innovation policy and now pushing in globally-competitive ways at the emerging imaginary of twenty-first century digital MaaS, they are also recipients of considerable support from the Party-state, or at least significant sections thereof. But for drivers too, as in the West, ride-hailing has been embraced as an opportunity for a reasonable income, especially during the price wars. Rather than targeting taxis as competition with new drivers, both Kuaidi and Didi (but not Uber) included taxi drivers in their schemes, even allowing taxi drivers to participate in both platforms.

Finally, ride-hailing has grown so quickly in Chinese cities because of the demand it has been able to tap. Evidently, this includes in particular the networked-individual digital urban young who are both venturesome (Bhidé 2009) consumers in search of relatively personalized automobility and unable to afford a car, especially as city restrictions on licence plates have kicked in. This demand for personalized automobility is particularly marked in China too, as against the more developed cities in the West where Uber, in particular, has started its operations. Notwithstanding the booming (and even sometimes, unnecessary) growth of subway systems (*The Economist* 2013) and other public transport infrastructures that are provided at low cost, public transport remains a cramped and sometimes slow form of commute through rush-hour traffic. And against walking and (e-)biking, enclosed transport is often prioritized to escape extreme weather (cold, heat, rain) and/or terrible air pollution, or for reasons of *suzhi*. The taxi, always hailed on the street, has thus served as a key form of quasi-personal mobility since it is still relatively cheap – indeed, the cheap taxi is a singular hurdle to uptake of vehicle-sharing initiatives (Wang *et al.* 2012), especially atop the deep-seated scepticism of the 'sharing' economy in China (Zuev and Tyfield 2015).

Since ride-hailing is effectively a comparably expensive cab that you have called to your door and pre-reserved from the convenience of your smart phone, however, it has achieved rapid take-off. To the intrinsic attractions of quasi-personal mobility by taxi it adds the functionalities of not having to wait and compete with others on the street for a taxi to come by – an impossible competition when it rains, for instance –

while the layout of Chinese city streets, often with addresses getting you still only to the front gate or rough vicinity, not the exact location, is mitigated by (Chinese) online mapping and the need, with every trip, to speak to and direct the driver by phone. And, of course, in rapidly developing cities, a mobile individual may have to go to relatively remote and/or newly developed parts of the city that have sparse public transport and few cabs driving by. While in many Western cities, therefore, there exist infrastructures, mores and city lay-outs that afford relatively comfortable and 'respectable' use of existing forms of (quasi-)public transport, in China smart phone-platformed ride-hailing is particularly attractive.

But against these significant constituencies finding renewed empowerment and autonomy in ride-hailing, there are also multiple powerful groups for whom it is security threat. This includes, for instance, the car majors at the pinnacle of state power, keen to sell as many cars as possible. But the Party-state more generally also offers only capricious and conditional support of these innovations, since it is unsettled by moves that may either diminish its regulatory control over urban mobility and/or enable the 'wrong' kind of mobility or too much of it. For instance, at least during the price war, many of the new riders were replacing not private cars or taxi journeys but stepping off public transport (ITDP 2016). This relieved some pressure on overstretched public transport, but it also set an uncomfortable precedent for ride-hailing in China, in which its success conditions *worse* congestion, not better. This problem also seems a particularly important eventuality to consider given that car ownership (or, in ride-hailing, use) remains comparatively low and still a key aspiration of many millions in China's cities.

In another recent instance, the central Government has effectively stepped in to regulate the ride-hailing business and prevent any further money-burning price wars by banning subsidies, which were threatening even the solvency of their systemically-crucial backers. But it has also issued pronouncements that it will ban drivers from using these platforms unless they have the appropriate urban *hukou* (Clover and Ju 2016). This exemplary move of governmobility illustrates precisely how sensitive is the possibility that ride-hailing may encourage and enable the 'wrong' kind of mobility, in this case attracting migrant workers to the cities. But this latter case also illustrates the recursive and productive feedback loops of innovation *and counter-*innovation that this highly politically-sensitive pressure cooker of Chinese urban mobility transformation is propelling. For attempts to ban non-local *hukou* drivers from participation has been strongly resisted by the ride-sharing firms and their backers, since it will certainly exclude a large number of their effective workforce. The 'security' response of the Government is thus itself encountered as a security threat to the 'liberty' of those embracing MaaS-based opportunities for their enhanced autonomy, no doubt eliciting further innovation-as-politics on their part.

Similarly, the novel, monstrous security threats emergent with successful digital MaaS innovation will likely also, given the unique CP/KS context of China, emerge in ways that are particularly intense and productively contested. For instance, on some accounts, concerns about privacy and surveillance from MaaS, and online in general, are ironically lower in China than in the West at present (Minter 2016), with many Chinese citizens fully aware but unfazed by the intense scrutiny of their online activity by the Party-state. However, there is a significant difference between one's consumption of digital material being tracked and, perhaps, interfered with and one's autonomy itself, as expressed through the key issue of one's actual movement, mobility and access to it. As such, it is possible that surveillance of digital

MaaS may elicit particular ire and pushback from those, *without* a personal car, increasingly dependent on such services. Similarly, to the extent a fully autonomous fleet of MaaS became a viable prospect, in a country that has long depended upon government of a surplus of labour this will surely provoke a backlash that dwarfs that regarding the current policy of exclusion of *nongmin* drivers.

In short, digital MaaS in China is another domain of monstrous innovation, filtering and amplifying the dynamics of essential contestation we discussed above regarding vehicle systems. But *it* is not *just* another monster. First, because, its sits atop the Chinese monsters of vehicle innovation and potentially integrates them with the parallel reshaping of broader mobility systems by 2.0 digital innovation, including the transformation and replacement of demands of physical co-presence. And, second, because, as for the Googliberal dynamics of Uber *et al.*, digital innovation is uniquely positioned at the fulcrum or obligatory pass-point of the complex, iterative flows of power/knowledge innovation towards socio-technical systems that are constitutively complexity-attentive.

But also, and more importantly still, because, in China, the key dynamic of monstrous innovations and their interpenetration is not simply technological or corporate but precisely at the level of *power/knowledge systems and practices*. In China, in particular, we find complex power/knowledge system dynamics that both: (i) condition a particular intensity in that power/knowledge momentum; and (ii) are already shaping and giving birth to the crucial socio-political constituencies that can transform and harness that momentum for deepening system dysfunction into one of system productivity. In other words, it is in China specifically that we find how the *very dynamism of the mobility monsters*, concentrated by that of (Chinese) digital MaaS in particular, feeds the growth *not* of firms but of *socio-political constituencies* of deepening system significance – on both supply and demand sides, and increasingly converging with each other, as the Chinese mobility disrupters and the urban digital middle-classes respectively. The key achievement of domestication of the monsters is thus realized, but in the only way possible: through harnessing their irrepressible, protean vitality to the transformation and reconstitution by them, in turn, of the world around them, the complex power/knowledge system and its system-dominating regime of government (see Figure 10.4).

The exceptional power/knowledge momentum of Chinese mobility innovation

There are two key ways in which this socio-political dynamism is specifically Chinese. First, consider the growth of *agencies* particularly enabled by digital mobility innovation and, in turn, driving it forward in their specific interest and over the medium-term through all the complex and profound systemic wrangling. It is in the unique conjunction of dynamics of growing mutual empowerment and co-production of China's urban middle risk-classes, on the demand side, and the digital disrupters, regarding supply, that the crucial strategic advantage lies, harnessing China's boisterous mobility monsters. For they not only feed each other but also feed off the essential contestation of the monstrous emergent systems, drawing on the compounding dynamics of liberty and security in ways that are essentially productive of both.

The growth of the middle (risk-)class thus coproduces dynamics of digital(-mediated) mobility innovation. This would perhaps include innovations that are more than just the 'copycat-equivalent-in-China' of Silicon Valley's Big Tech MaaS

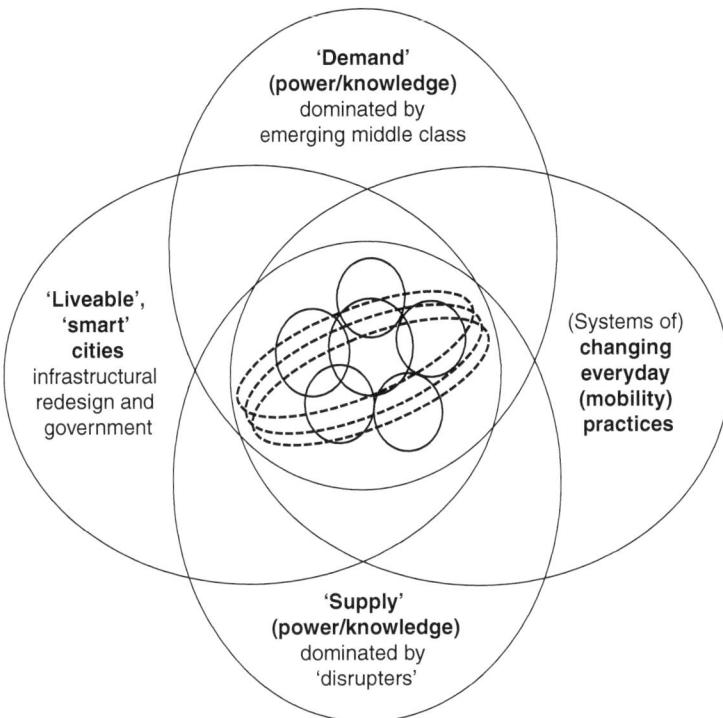

Figure 10.4 Urban mobility innovation-*as-politics.*

imaginary, such as micro-EVs or smart ToD initiatives or simple replacement of journeys with digital co-presence, with Chinese firms (including the BATs) major players also in virtual reality (*The Economist* 2016d). This venturesome consuming group essentially is just seeking out for themselves networked-individual autonomy and conspicuous achievement in their chosen forms of mobility, and flight from security hazards in 'liveable' urban environments. This then feeds and is fed by dynamics of new and deepening security threats and forms of autonomy; e.g. in demand for commercial mobility services that enable stylish, smooth mobile lives with mitigated exposure to (and, as innovation pushes at system level, increasingly conspicuous personal responsibility for *mitigation* of) congestion, poor air quality and/or dangerous movement, all of which are specifically middle class priorities and markets.

Meanwhile, the *Chinese* digital disrupter firms are specifically enabled regarding mobility innovation in being foundationally responsive experimenters. They are highly attuned to the complex, shifting field of *power/knowledge relations* that just *is* transition in urban mobility systems and 'smart', 'green' urbanization, and especially regarding serving their fast-changing demand with low-cost, good-enough novel combinations and service offerings. The very proliferation of new liberties and security threats from the mobility monsters thus particularly enables these Chinese *digital* disrupters, since their focus on digital technologies and data management yields more insights for further disruptive innovations with ready consumer appeal – not least, of course, by augmenting opportunities for autonomy and mitigating exposure to (mobility) system bads.

Together, then, the new and emerging digital mobility systems and smart city-scapes and practices, on the one hand, enable new industries hungry for 'high-skilled' knowledge work and offering productive investment opportunities to the middle-class; while conversely, on the other, the constitution and increasing empowerment of the middle-classes provides growing sources of demand for these digital mobility products and services, in positive feedback loops. It is thus in the building up of the specifically Chinese twinned power/knowledge agencies of its middle-risk-class and its digital disrupters that new agencies increasingly capable of asserting their own interests are co-produced with the crucible of innovation-as-politics that is an urban MaaS system transition.

Yet even more importantly, and inescapably, we must also note again how this entire emerging and self-propagating power/knowledge dynamic is all set within the uniquely Chinese context of the historical tension of the immoveable object and the unstoppable force, the CCP Party-state and/vs neoliberal globalization (in terminal crisis) (Figure 10.5). As we have seen, it is this macro context that particularly conditions both: (i) the dynamics of the parallel emergence of the digital disrupters and the Chinese middle risk-class as relatively enduring processes conditioning urban mobility innovation; *and* (ii) the intensity of the liberty/security dynamics of the Four Challenges that are compelling such concerted urban mobility innovation in China. Absent the IO vs UF meta-dynamic, in other words, we miss the fundamental reason why China is such a crucible of complexity-attentive innovation.

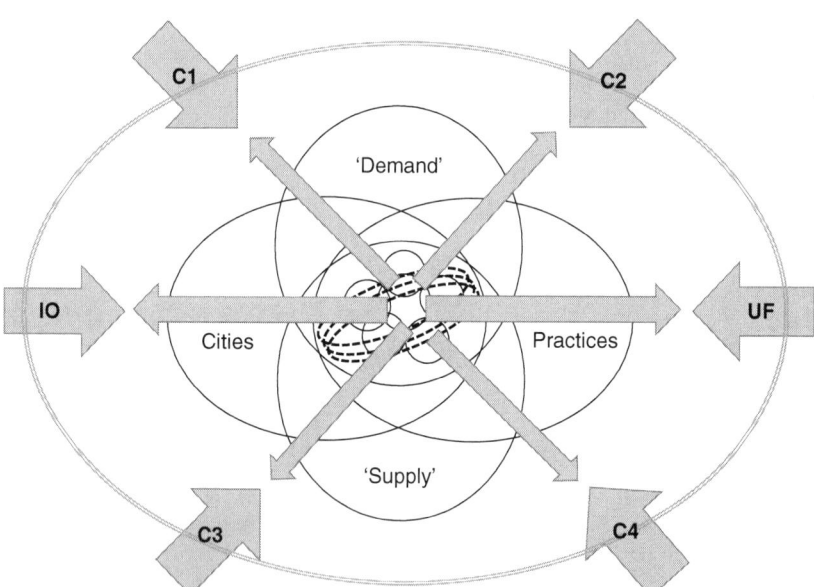

Figure 10.5 Urban mobility innovation-as-politics coproducing the 'IO vs UF' and the Four Challenges.

But the flipside also applies. Digital urban mobility systems innovation itself emerges as an arena of socio-political forces in which the clash of IO and UF is particularly heightened, and in which Government (not just government) of twenty-first century urban mobility promises to reach a new self-sustaining settlement. First, regarding the UF, in the dominant imaginary of Big Tech MaaS – itself increasingly central to neoliberalism's conception of its future as a global project, and how it will yet 'disrupt' the world and dig itself out of stagnation and implosion – we encounter a regime of innovation and socio-technical system change that is both the acme of Googliberalism and, as such, particularly empowered *and* contested. But, second, given the central systemic importance of government of mobility in twenty-first century complex, capitalist, individualizing societies *like China*, this also makes digital MaaS a key and emerging apparatus of contemporary system government. In China, however, the incumbent regime ensures that there will be a titanic effort devoted to ensuring its sustained control, with existential stakes in play. This is thus to set up a particularly fierce clash of IO and UF that, mediated by the emergence of digital disrupters and middle risk-class it also engenders, makes urban mobility transition in China, and China especially, a site of tectonic movement.

As this point highlights, however, these twin social forces are also shaping transition at the system level itself, in terms of the formation of new *processes of system government and Government* that *constitute* a new system. Both the middle risk-class and the digital disrupters are emerging, in the shadow of the IO vs UF, as key power/knowledge technologies for the government-at-a-distance that is both the most pressing challenge for the Chinese Party-state (Ren 2013) and an inescapable predicament of smart, complex, sustainable systems per se (i.e. the 4th Challenge of complex government of complex systems).

On the one hand, then, the progressive emergence of digital mobility innovations and novel electric vehicle forms alike is specifically enabling of *middle-class* subjectivities and aspirations. And these are also held up as, and increasingly aligned with, collective imperatives at the system level in the framing, empowerment and reinforcement of 'responsible' and 'respectable' mobility practices. This is thus to incubate a system specifically of middle-class govern-mobility as an emerging form of system government essential for and constitutive of transition at system level. For example, here aspirations to the premature universal status of 'middle-class' are increasingly given concrete substance as (a yet-to-be-specified) 'this' particular mode of low(er)-carbon, deprivatized, digital mobility – even as 'this' may presuppose levels of income, education, forms of work and living standards that are far from widely available. Indeed, it is precisely for the personal attainment of such 'liberties', and escape from the converse 'security threats', in setting up cycles of growing individualized advantage that 'middle class' status display and digital mobility choices are so fervently chased.

Yet the essentially divisive dynamic of risk-class, conditioned by the IO vs UF, also entails the flipside: that mobility absent these conditions becomes increasingly 'legitimately' sanctioned as 'irresponsible', 'disgraceful' and 'dangerous' (to self, others, society and/or planet), and expressive of the quality of the person. Crucially, however, this form of system government is *productive* precisely to the extent that it polarizes society into 'legitimate' and 'meritocratic' *winners … and losers*, according scarce and actively pursued status and empowerment to some *but not* others. The formation and constitution of the middle risk-class in coproduction with emergent forms of smart, green mobility system acts as a powerfully self-sustaining dynamic of system government precisely insofar as it *harnesses*, rather than retards or reverses, the monstrous production of both new liberties and their flipside of new security threats *and then* distributes these *unequally*, to emerging system winners and losers respectively. For only in this way and to this extent is the power/knowledge techno- logy of (middle) risk-class capable of translating amorphous, abstract system dynamics into the lived, anxious, sharp-elbowed competition between (networked-) individuals pursuing their own interests in individualizing, complex social systems. As such, just as middle risk-class emergence is a key condition of harnessing the liberty/security dynamism of mobility monsters towards system transition rather than ever-heightening system dysfunction, this is only system-productive to the extent it is also an essentially polarizing process.

Regarding the digital disruptive innovators, meanwhile, we find a parallel process as soon as we recall that 2.0 digital innovation (including regarding mobility-as-a- service) is in the business of creating new *power/knowledge* technologies. Such digital innovation is thus inextricably in ongoing negotiation with and reshaping of the apparatus of, not just government, but also the formal institutions of state Govern- ment; and with governmobility (Bærenholdt 2013) a particularly important aspect of system government of complex, twenty-first century mobile 2.0 systems. Digital MaaS innovation will inescapably, therefore, shape emergent forms of surveillance and anonymity, discipline and liberty, control and autonomy (Vanolo 2014). In China, however, as the Government of the internet by the Party-state already shows (McKinnon 2013), this is a particularly intense cat-and-mouse in which the giants of Chinese digital innovation and the massive online population both play along and resist, actively and passively, Government control, which then responds by increasing and targeting its efforts etc.... As mentioned above, though, as digital technologies increasingly stretch into the material sphere in an Internet of Things, including mobility systems, that is particularly attractive to the Tech-hungry Chinese con- sumer, this contestation is likely to become ever more intense.

What is again crucial, however, is the specifically Chinese combination of a strong (Party-)state determined to maintain social 'harmony' under its final oversight and its deepening asymmetric inter-dependence on forms of autonomy-hungry middle- class personal mobility. For these condition the emergence of digital (mobility) innovations that will serve the interests of *both* of these by distinguishing between responsible, trust-worthy (digital) mobility that the state can and should leave alone, as the 'legitimate' winners of society, and other forms of mobility, identified specifi- cally with disempowered system losers, that may just as legitimately be subject to the full force of the (now 2.0) apparatus of state surveillance and discipline. For instance, we can imagine how digital mobility innovation could converge with micro-EV innovation to create (quasi-)individualized middle-class low-carbon mobility that is both complex and autonomy-maximizing, and yet readily policeable with strong state oversight where the bounds of 'responsible' mobility are transgressed; in other words, finally *domesticating* the monsters of urban mobility through emergent power/knowledge relations and technologies of the middle class and disruptive digital innovations, as the soft and hard technologies of a uniquely harnessed power momentum of mobility system transition (Table 10.1).

Indeed, the publication of plans for a new digital-platformed 'social credit system' (*The Economist* 2016e) that will 'allow the trustworthy to roam everywhere under heaven while making it hard for the discredited to take a single step' formally expresses plans for precisely this goal. Of course, the actual shape of this system of

Table 10.1 The quadrant of Chinese urban mobility innovation

	Direct effects (at agent level)	Indirect effects (at system level)
Intended (immoveable object)	'Optimist' • Electric car 'overtaking around the corner' • World No. 1 in EV sales (by 2015)	'Disrupter' • E2W/micro-EV as specifically Chinese disruptive innovation • BUT neglected and proscribed by Government as 'security threat'
Responded (unstoppable force)	'Pessimist' • Slow EV sales, dependent on expensive and gamed Government subsidies (being phased out) and SOE disinterest • Relatively minuscule EV sales vs. deepening ICE automobility system	'Innovation-as-politics' • Evolving Chinese MaaS innovation-as-politics in co-productive parallel with middle-risk-class emergence

govern-mobility is another innovation still to be contested – and likely towards an enduring form that is 'not what the CCP wants'. Yet the key characteristic and promise remains: a model of urban mobility (innovation) that may not only domesticate the monsters of contemporary mobility innovation, but also thereby manifest a historical synthesis between IO and UF, squaring the circle of a Chinese yet liberal regime of system government.

Conclusion: not so '*Uber*' *alles* after all

Let us sum up. Chinese mobility innovation is the site not just of particularly intense problems associated with the incumbent system of urban mobility, and of an exceptionally ambitious industrial policy in response. It is also home to disruptive innovation regarding both a unique domain of micro electric vehicles and a crucible of digital mobility-as-a-service innovation. Yet each of these emergent mobility systems, still far from convergent and both in competition and cooperation with each other as against ICE automobility, is currently a hopeful monster – marked by both inescapable dynamism and profound incompatibility with existing power/knowledge systems.

What particularly sets Chinese mobility innovation apart, however, goes beyond all these colossal and exceptional socio-technical forces. Rather, the very dynamism of these monstrous mobility systems, mediated by the digital mobility innovation of China's disruptive ICT giants in particular, is coproducing the emergence of broader power/knowledge systems transition. This takes the form of the twinned construction of the Chinese middle risk-class and architectures and institutions of 2.0 (disruptive) digital innovation. Together these convert the seeming problems and weaknesses of Chinese mobility innovation into powerful drivers of further intense mobility system innovation, in self-sustaining feedback loops (of liberty-security) over the medium-to-long term; and based on novel agencies that are themselves increasingly *enabled* by these dynamics and are foundationally attuned to a pragmatic and restless responsiveness to complex, dynamic and so unmasterable systems. In short, the specific advantage of Chinese mobility innovation lies precisely in terms of innovation-*as-politics* of *power/knowledge* technologies and changing relations *at system level*, the arena of action necessary to achieve system transition. Chinese urban mobility innovation-as-politics thus promises to be as significant and influential a domain of twenty-first century global capitalism as was the steel-petroleum automobility of post-War American suburbia to US hegemonic capitalism in the twentieth century (Paterson 2007), and for similar reasons of system government and its domination.

From this perspective, there are also several respects in which Chinese mobility innovation seems better positioned than, say, an Uber or other Googliberal Silicon Valley poster boy to 'disrupt' urban mobility in the medium-term. This is so even as it is the latter that will likely continue to sport the impressive bling of the latest technology and biggest venture capital valuation and dominate the global business/policy imaginary. This is not just because the key element of disruptive *electric vehicle* innovation is a specific Chinese strength. Nor because Uber has effectively been beaten in China, ceding the ground to Didi – and if Silicon Valley giants can't win in China, it is hard to see how Chinese companies will not *have to be* globally significant companies in this domain – since Uber arguably now has the rest of the world, and is freed from the distracting and destructive focus on a competition in China it couldn't win (*The Economist* 2016c). Nor even is it just because in China there is both intense and manifest competition in the mobility domain between its digital giants, the BATs; e.g. their close connection with ride-hailing and other MaaS innovations, and/or their integration of Chinese MaaS innovation with the broader landscape of digital mobility that is these companies' home turf; none of which we see yet from Silicon Valley, even as it is often foreseen.

Rather, there are three key reasons why Chinese disruptive mobility innovation is strategically advantaged over Silicon Valley, all of which are specifically matters of power/knowledge system momentum. First, because in China on both supply and demand sides, given the intensity of the liberty/security dynamics in play, such mobility innovation is constitutively oriented to the Four Challenges, and is approached as an explicitly systemic challenge. Alibaba, for instance, amongst others, has explicitly stated that its interest in mobility innovation is as much about the profound social problem of cleaning China's environment, for which it wants to assume corporate social responsibility, as it is about claiming dominance in a key market of the twenty-first century (Tse 2016). The Chinese middle class dominating demand for these services is also increasingly attuned, as a matter of achieving their own 'liveable urbanism', to the *systemic* costs (and benefits) of personal choices. By contrast, the dynamism of Googliberal mobility 'disrupters' is premised upon the continued rapid expansion of its parasitism on existing (high-carbon) assets to the systematic *exclusion* of such considerations.

Second, the Chinese dynamic is built upon, and strategically enabled precisely because of, the emergence of new system-central power/knowledge constituencies across society, uniquely conditioned by the IO vs UF. Meanwhile, the Googliberal MaaS innovation trajectory tends only towards the escalating contestation of its asset-light, externality-appropriating business model. The latter thus essentially *is* 'just another monster', compounding system dynamics of growing turbulence and dysfunction. The Googliberal model is constitutively dismissive of incumbent power structures as 'obstacles' to its one-size-fits-all plan of world domination. This can only be a recipe for deepening rejection, even as it may powerfully feed such innovation in the medium-term. By contrast, quintessentially pragmatic and attuned to, and adept with, uncertainty in pursuit of their own fluid and developing self-interests, both the Chinese middle risk-classes and China's digital disruptors are much better equipped for the long hard slog ahead of negotiating and partnering, including with (unreadable) Government agencies, in the formation of mobility systems in particular places.

Finally, the Chinese innovation is taking place much more closely associated with, and with greater opportunities for, parallel development of changing and growing urban landscapes and other crucial infrastructures. Uber, meanwhile, is premised upon arriving and exploiting existing mobility infrastructures and vehicles in cities that are substantially set in layout and/or changing in ways that are simply not its corporate concern. Yet it is the former predicament that is of much greater relevance around the world in the burgeoning urbanization across the global South – with all the South-South trade and investment opportunities this entails (e.g. through China's massive 'One Belt, One Road' overseas infrastructure investment programme). Moreover, in these latter locations the institutional prerequisites of Googliberal innovation – such as strong property rights (let alone *intellectual* property rights), good existing mobility infrastructures that are well mapped (by Silicon Valley firms) etc. – could well be weak or lacking in ways that Chinese innovation will likely be more adapted to work with.

This is not to argue, of course, against the presumption of Googliberal innovation that an innovation does not have to be popular, whether amongst the majority or existing power structures, to 'win out' and become sedimented in the medium-term. Indeed, a (system) innovation may be plagued by controversy and even essentially contested and still prevail. This is, after all, the story of many totemic innovations through the neoliberal period. In this respect, we may, of course, soothe the inflated egos of the prophets of such Randian 'disruption'. What they miss, however, is that determined opposition to the existing system paired with individual-enabling digital technology and even oodles of venture capital support are not alone enough. Rather, genuine system innovation will only win out to the extent it is hitched to self-sustaining and self-referential power/knowledge momentum, including parallel co-produced emergence of specific socio-technical 'fixes' (Markusson *et al.* 2017). And this is all the more important and challenging the more profound the system lock-in that it is trying to dislodge. Regarding the 'hardest case' of urban auto-mobility, Chinese innovation displays just such dynamics, while Silicon Valley does not. It is the former, then, that at present is more likely to be central to global system innovation – and the resettlement of capitalism to which this would substantially contribute – than the latter. But how, and to what futures? It is to this, our final step in our genealogy of the emerging present, that we now turn.

Notes

1 Stop press: this chapter was written in autumn 2016, but in June 2017, with the sacking of Kalanick as Uber's CEO and its broader tribulations, the following discussion appears all-the-more germane.

2 For instance, Travis Kalanick, (recently and shockingly ex-) CEO of Uber, has been described by his colleague and CTO, Thuan Pham, as 'always see[ing] himself as an underdog' (*The Economist* 2016a).

3 And, to skip ahead a little, this is, of course, particularly the case in China.

L214   *Where are we going? The complex journey*

References

Bærenholdt, J.O. (2013) 'Governmobility: the power of mobilities', *Mobilities* 8(1): 20–34.

Bhidé, A. (2009) *The Venturesome Economy*. Princeton, NJ: Princeton University Press.

Birch, K. (2016) 'Rethinking value in the bio-economy: Finance, assetization, and the management of value', *Science, Technology & Human Values*, 0162243916661633.

Clover, C. and S.F. Ju (2016) 'Didi Chuxing to be hit by rules on migrant drivers', *Financial Times*, 16 October.

Dittmar, H. and G. Ohland, (eds) (2012) *The New Transit Town: Best Practices in Transit-Oriented Development*. Washington DC and London: Island Press.

The Economist (2013) 'Going underground', 27 April.

The Economist (2016a) 'From zero to seventy (billion)', 3 September.

The Economist (2016b) 'It starts with a single app', 1 October.

The Economist (2016c) 'Uber gives app', 6 August.

The Economist (2016d) 'Insanely virtual', 15 October.

The Economist (2016e) 'Creating a digital totalitarian state', 17 December.

The Economist Films (2016) 'The Disrupters – Gear Shift', http://eydisrupters.films.economist.com/.

GIZ (2014) 'Car-sharing: A contribution to sustainable urban transport in China?', presentation to *Sino-German Cooperation Project on Electro-Mobility and Climate Protection*, 21 September.

Institute for Transport & Development Policy (ITDP) (2016) 'Ride-sharing in China', presentation to Low Carbon Innovation in China, Mobilities Closing Workshop, Tsinghua University, Shenzhen, 13 March.

Kirkegaard, J.K. (2016) 'Rapid upgrading through experiment (self-)disruptive impasse: The case of China's wind turbine industry', paper presented to the *International Symposium on Innovation-Driven Development*, Sun Yat Sen University, Guangzhou, 13–15 June.

Knight, S. (2016) 'How Uber conquered London', *Guardian*, 27 April www.theguardian.com/technology/2016/apr/27/how-uber-conquered-london.

Lyons, G. and P. Goodwin (2014) 'Grow, peak or plateau – the outlook for car travel', Report of a roundtable discussion, New Zealand Ministry of Transport.

MacKinnon, R. (2013) *Consent of the Networked*, New York: Basic Books.

Markusson, N., M. Gjefssen, J. Stephens and D. Tyfield (2017, forthcoming) 'The political economy of technical fixes: the (mis)alignment of clean fossil and political regimes', *Energy Research and Social Science*.

Minter, A. (2016) 'Why China doesn't care about privacy', *BloombergView* 17 May, available at: www.bloomberg.com/view/articles/2016-05-17/why-china-doesn-t-care-about-privacy

Mirowski, P. (2012) 'The modern commercialization of science is a passel of Ponzi schemes', *Social Epistemology* 26(3/4): 285–310.

Osborne, H. (2016) 'Uber loses right to classify UK drivers as self-employed', *Guardian*, 28 October www.theguardian.com/technology/2016/oct/28/uber-uk-tribunal-self-employed-status.

Parent, L. (2016) 'The wheeling interview: mobile methods and disability', *Mobilities* 11(4): 521–532.

Paterson, M. (2007) *Automobile Politics*, Cambridge: Cambridge University Press.

Ren, H. (2013) *The Middle Class in Neoliberal China: Governing Risk, Life-building, and Themed Spaces*, Abingdon & New York: Routledge.

Reuters (2015) 'Hon Hai, Tencent partner in electric car business', 23 March, www.reuters.com/article/us-hon-hai-china-idUSKBN0MJ0V320150323.

Sheller, M. (2016) 'Uneven mobility futures: A Foucauldian approach', *Mobilities* 11(1): 15–31.

Shove, E. and Walker, G. (2014) 'What is energy for? Social practice and energy demand', *Theory, Culture & Society* 31(5): 41–58.

Straw, J. and M. Baxter (2014) *iDisrupted*, New York: New Generation Publishing.

ten Brink, T. and F. Butollo (2016) 'A new upgrading paradigm? Innovation policy and domestic demand in strategic emerging industries', paper presented to the *International Symposium on Innovation-Driven Development*, Sun Yat Sen University, Guangzhou, 13–15 June.

Tse, E. (2016) *China's Disruptors*, London: Portfolio Penguin.

Tyfield, D. (2013) 'Transition to science 2.0: 'Remoralizing' the economy of science', *Spontaneous Generations: Special Issue on 'The Economics of Science'*, September.

Tyfield, D. (2014) 'Putting the power in 'socio-technical regimes' – e-mobility transition in China as political process', *Mobilities* 9(4): 585–603.

Urry, J., J. Leach, C. Rogers, N. Dunn, A. Ortegon, C. Popan, K. Psarikidou, S. Pollastric, S. Lee, J. Ward, J. Hale and M. Dring (2016) 'Is car-free the future for cities?', Working Paper for EPSRC Project *Liveable Cities*, Lancaster University.

Vanolo, A. (2014) 'Smartmentality: the smart city as disciplinary strategy', *Urban Studies* 51(5): 883–898.

Wang, M., E. Martin and S. Shaheen (2012) 'Carsharing in Shanghai, China: analysis of behavioral response to local survey and potential competition', *Transportation Research Record: Journal of the Transportation Research Board* (2319): 86–95.

Zhuang, Y. (2015) 'Confucian ecological vision and the Chinese eco-city', *Cities* 45: 142–147.

Zuev, D. and D. Tyfield (2015) 'Car-sharing revs up as China's cities battle congestion', *Chinadialogue*, 19 February, www.chinadialogue.net/article/show/single/en/7736-Car-sharing-revs-up-as-China-s-cities-battle-congestion-.

Part IV

What can be done?

Conclusion

11 Liberalism 2.0 and beyond

The historic bloc of risk-innovation capitalism

This book has argued that the incumbent dominant power/knowledge regime of neoliberalism is not only itself in terminal crisis but that it is so precisely because it is catalyzing and deepening a series of four overlapping Great Challenges that confront humanity as a species and that must all be addressed simultaneously – something that remains far off in thinking let alone actual policy and action. At the centre of this dysfunctional ratchet of destructive creation is the collapse or 'death' of knowledge (in the post-Enlightenment sense) alongside and through the 'emergence' of the complex global *power/*knowledge system of global risk-innovation society. As such, neoliberalism has not only unwittingly brought the world together as never before, through its financialized model of globalization and the deepening of the four Great Challenges, but also led us to the limits of the peculiarly Euro-American world of literalist knowledge societies that has been the dominant episteme of the past 400 years (Duara 2014).[1] The fundamental challenge of the present interregnum and global turbulence, therefore, is to construct *new and more adequate relations to knowledge*, where this is effectively the challenge of constructing better, more reflexively-aware complex systems of power/knowledge relations and their government.

Drawing on a strategic, dynamic and complex systems approach to innovation, as a key lens on the socio-political and sociotechnical remaking of our 'societies', we have explored current tendencies in a key systemic location. The goal has been to see if – as starting hypothesis, being the continuation of existing systemic dynamics and rhythms – these tendencies augur the emergence of a new global regime that will meaningfully mitigate and address (or spatio-temporally and technologically 'fix' (Jessop 2006; Markusson *et al.* 2017)) many of the essential dysfunctions of the dying incumbent regime of neoliberalism; and perhaps thereby resettle capital accumulation into a newly restabilized upswing of productive capital and non-zero-sum socioeconomic development. The preceding chapters exploring innovation in China set out how this could indeed be the case, even as it will also, in turn, be only temporary and to the overwhelming benefit of only a fraction of humanity (albeit a bigger and different fraction to preceding regimes, especially neoliberalism).

For the key conclusion of our analysis of Chinese low-carbon innovation-as-politics is the unfolding emergence of a new *historic bloc* (Gramsci 1971; Arrighi 1994; Rupert 1993) of global knowledge capitalism. This new historic bloc consists of the twin and converging forces of a Chinese-led, high-skilled knowledge-working middle risk-class and digital, complexity-oriented 'disruptive innovator' firms and entrepreneurs. Crucially, these power/knowledge constituencies are strengthened and shaped by deepening challenges, imperatives and successes alike regarding such complexity-attentive system innovation, while they, in turn, are primary drivers of further such innovation.

The result is thus precisely a traceable, if still embryonic and defeasible, power momentum. And this process is uniquely dynamic in China given the intensity of the liberty-security pressures in that country under the darkening shadow of the historic confrontation (and possible synthesis) of (the complex power/knowledge systems of) global capitalism and China.

As Tse (2016: 226) puts it:

> China's entrepreneurs will be the key force driving the country forward through the coming decades, giving China the capabilities it needs.... In the process, they will remake the world – not because they want to remake the world, but because such is the interconnectedness of our world, and such is China's scale, that they cannot realize their potential without remaking China, and they cannot remake China without changing the world.

Not only does this spell the deepening asymmetric interdependence of Chinese state (currently as *Party*-state) and this historic bloc to the latter's strategic advantage. But it also places the latter, and the China it will increasingly fashion in its own image, at the core of a movement beyond the dysfunctional system feedback loops and innovation model of neoliberalism to one that is – and is constitutively and foundationally so – attentive to the Challenge of global complex system government. It is also noticeable that this historic bloc is being constituted in ways and through processes that resonate precisely with a renewed cycle of both of the analytically identifiable system rhythms of global capitalism to date described in Chapter 3, namely the Arrighi/Braudel cycles of political-cultural-institutional-military hegemony and the technoeconomic periodicity of Kondratiev waves. For, respectively, this historic bloc shapes both a plausible next hegemonic core to global capitalism of a transformed China (see below); and in intimate, interactive parallel with a new productivity in the ongoing capitalist revolutionizing of the means of production founded on a mature phase of digital innovation, now transforming the material and industrial economy, together with an ascendant clean energy revolution. And in which China plays a globally-leading, and proudly and loudly declared, role.

Such a regime is thus not only plausibly emergent from, and immanent within, the incumbent and still dynamically reproducing and increasingly turbulent conditions of global neoliberalism (in its many 'local' forms). But it would also clearly be able to boost its own power momentum further by increasingly drawing itself, in compelling concrete detail, in contradistinction to the culpable dysfunctions of neoliberal system government. This latter dynamic thus hinges on its progressive emergence into self-consciousness as a movement of national and global significance and strength, as a class-for-itself. And, indeed, in literature on Chinese innovation specifically (e.g. Tse 2016; Rein 2015, both notably diaspora Chinese) – but also, in parallel, on the new industrial revolution (Rifkin 2011) and the key global challenge of urbanization as a techno-economic challenge that exceeds the capacity of neoliberal market fundamentalist government (Glaeser 2011), both of which are literatures avidly consumed in China – we have already perhaps seen the publication of what will prove to be some of the key texts of this historic bloc's (political economic) self-assertion.

But how can we characterize this emergent system and so, finally, answer our opening question of 'what next after neoliberalism?' An emergent regime that is foundationally driven by both a new bourgeoisie and a reappraised relation to power/knowledge systems and complexity – as in the *middle … risk-class …* respectively – suggests we could do worse than the working label of a 'liberalism 2.0' or a 'complexity liberalism'. Here 'liberalism' connotes a break with neoliberalism to a revived nineteenth century classical liberal regime, but one reformulated and rejuvenated precisely by reframing its foundational credos and common-senses of system government and power/knowledge technologies, structures and institutions around complex systems. Thus 'liberalism *2.0*' seems particularly apt as it captures both the centrality of digital interactive media as key power/knowledge technologies and the *revival* of classical liberalism, in a new nineteenth century.

An emerging liberal 2.0 regime of system government

Classical liberalism is a power regime that deploys power-knowledge technologies to govern *through* (construction of) individual negative freedoms by way of complex dynamics of liberty-security. The central institution of this form of government is the market. But the market is crucially supplemented by multiple forms of power-knowledge technologies through which states, business institutions (firms and corporations) and individuals govern and are governed (and are constituted) by the strategic action of specific concrete people (i.e. including themselves). All of these processes of government are deemed legitimate to the extent they are 'rational', as defined by and co-productive of incumbent common-senses in a socio-political order that is essentially epistemic, founded in (specific) relations to knowledge.

Regarding the market, the key argument concerns the naturally spontaneous and self-correcting mechanism of matching individual desires to buy and sell, as seminally captured by Adam Smith and other eighteenth/nineteenth century liberal political economists at the birth of both that discipline and the classically liberal regime of (British) government of the global capitalist system. This presents the natural boundaries to the rational exercise of state power: the state can rest assured that, left to its own devices, the market will govern the otherwise unruly interaction of human individuals, *through and depending upon* individual freedoms, into outcomes that are *also collectively* 'optimal'.

Liberalism is thus the regime of *laissez faire* but where the market is thus entrusted because it is believed to be naturally and spontaneously arising, and is fundamentally rational in that it elicits what is the collectively optimal outcome for society. As such, it is also admitted that sometimes the market can fail to arise naturally and so realize this collectively optimal social order. This occurs, for instance, due to inappropriate conditions such as in the provision of public goods or the absence of appropriate rational individuals to take part. In such circumstances, the state then has its circumscribed legitimacy to step in, deploying rational, objective (including disciplinary) knowledges to plug this gap.

By contrast, recall, neoliberal market fundamentalism places faith in the market, as the infallible and supreme decision-maker, first and foremost, and so defines and subverts the very concept of 'collectively optimal' by defining it *just as* that which 'maximizes gains for market-based entrepreneurship'. Moreover, for neoliberalism, the market is not spontaneously emergent – indeed, it is often vehemently rejected – but must be forcibly constructed. On the one hand, then, the state has no real role to play in the political economy, since the market always does it better;[2] but, on the other, the *unlimited* power of the state is legitimate and necessary insofar as it is deployed in the reconstruction of society and all things *as markets*. While both classical and neo- liberalisms are thus variants of a broader liberal-capitalist system of government, built on trust in the market as the optimal allocator, the latter is a radical mutation, underlying a highly productive but also destructive disregard for integrity and order at collective or system level and for accountability in terms intelligible to the thinking individual, as we discussed in Chapter 1. The market simply always knows best and that is all *we* need to, or can, understand.

The keys to the entire power regime of a (classical) liberalism, by contrast, may be seen as a twofold distinction: first, the privileging of 'rational' processes of '*knowledge*' (and its progress) – attributed to the autonomous individual, who thereby becomes the key agent of this power regime – as against a constitutively opposed '*politics*' (with cognate dichotomies of fact vs value, truth vs power etc ... and market vs state, private vs public); and, second, the '*natural*' and spontaneous emergence of such collective rationality if 'left alone' set alongside the possibility of establishing 'rational' governmental institutions to supplement any unfortunate failures or shortcomings of market rationality by 'rational' policing, exclusion and management of the '*unnatural*'. 'Rationality' is thus that which is in accordance with 'nature' as unarguable but demonstrable reality, while 'nature' is that which is revealed in 'rational' science and system government, in iterative feedback loops.

Regarding both founding distinctions, an emerging CP/KS regime specifically attentive to, and enabled by, system complexity and its understanding in new power/knowledge technologies offers potent redeployment and updating of precisely these arguments. In other words, growth of a middle risk-class historic bloc and its co-produced complexity-attentive power/knowledge technologies is not only de facto a unifying collection of newly innovated and self-propagating complex system liberties. But it is also actively driving a co-production process with a transformed *liberalism* as both socio-political *and parallel epistemic* paradigm shift, where the two are bridged and mediated by the – specifically complexity-attentive – sociotechnical innovation-as-politics we have explored above.

Consider, first, 'rationality'. The emerging complexity paradigm would certainly effect a radical transformation of the definition of the 'rationality' that underpinned the original argument of classical liberalism. But in doing so it could nevertheless revitalize the specifically rational and epistemic mode of justification of political economic order – at collective, system level – that is precisely the mark of a classical liberalism. This would thereby meticulously separate anew rationality and politics, siding always (in discourse, if not, of course, in practice) with the former. Hence, for classical liberalism, rationality was understood as the reasoned thought of the free, disinterested individual – i.e. tacitly, the propertied, white (north-west) Euro-American bourgeois, straight, Christian male, with the leisure needed for such 'rational' thinking – that could then be explained in (persuasive) prose and rationally debated and defended in public (amongst similarly 'rational' or reasonable fellows) (e.g. Habermas 1989; Shapin 1995). The classic case of just such rational argument, of course, is the emergence of (British and French) political economy as a power/knowledge technology of liberal government (Foucault 2010), showing – rationally and 'beyond reasonable doubt' – the rationality of *laissez faire*.

Today, for complexity, however, rational conclusions are no longer computable by individual rational thinkers sitting thoughtfully at their desks and/or argued in town halls or salons. Instead, they are the unpredictable 'emergent' outcomes of 'complex' networked processes characterized by positive and negative feedback. 'Rationality' thus demands acknowledgement of, first, the limits of individual thought in the face of systemic complexity and irreducible uncertainty, though outcomes may be summarized and understood *post hoc*; and, second, the crucial emergent epistemic powers (of calculation) of dispersed human-machine networks and their massive computing power. Similarly, the key agent of rationality is no longer the isolated disinterested individual deploying their autonomous reason, but the digitally inter-connected, networked individual (Rainie and Wellman 2014) deploying their spontaneous, pragmatic, depoliticized and voluntary will to participate in the complex-*and-therefore-rational* calculation of important conclusions and societal orders.

This emergent epistemic paradigm, therefore, affords three crucial steps for the rejuvenation of a classical liberalism:

1   It reconceptualizes the basis of 'rational' calculation in ways that not only accord with the apparent 'new frontier' of knowledge production but also, thereby, appear to afford a relegitimation – at the relatively autonomous level of epistemic justification itself – of the *very possibility* of 'rational' appraisal and assessment or faithful representational 'capture' of (not just complicated but) *complex* realities. This thus directly rebuts the assault on such possibility and epistemic authority by the neoliberal 'marketplace of ideas' and agnotology.
2   It resituates and redefines the role of the individual in this process, in ways that resonate strongly with the affective experience of actual strategic agents – and particularly system winners of highly-skilled knowledge-working networkers – involved in this inchoate political project of assemblage, but also in ways that essentially preserve the source of both its self-presented epistemic and ethical goodness in the *voluntary participation of self-directing individuals*.
3   Finally, taking the previous two issues together, it reinstitutes a contemporarily credible distinction between knowledge and politics upon which a liberal power regime constitutionally depends: as complex rationality vs heavy-handed ideological top-down decisions; and uncoerced, spontaneous networked individuality vs action enforced for the supposedly but dubiously *ex ante* and top-down defined good of others, regarding issues 1) and 2) respectively.

Second, the 2.0 complexity paradigm redefines 'nature' *as spontaneous, complex systems* and, again, this affords several crucial steps for an emergent liberal power regime. It sets up a new site of 'smart' control of 'nature' while working *with* the increasingly undeniable unpredictability and complex systematicity of 'natural' processes. This agenda may be witnessed in acknowledging the essentially ineliminable and unpredictable danger of natural processes (especially in an age of the Great Challenge of climate crisis and other planetary boundaries) as challenges for system government. But it is also apparent in the emergent effects of nature per se – which are real and remarkable but still to-be-explained and (it is presumed) scientifically explic*able* – that are the research questions at the forefront of contemporary sciences. For instance, the latter would include issues of the (embodied) brain and consciousness, climate and (socio-natural) ecology, (epi)genomics and disease, urban systems and urban-rural problems, or socio-technical change itself.

But complex systems (analysis) also blur supposedly ontological boundaries between 'nature' and 'society'. 'Nature' is thus no longer – as it was for classical liberalism – the pre-social and pristine 'sphere' of reality that commands special epistemic and moral force, demanding, as a matter of fact not value, our submission. Rather, a complex conception of 'nature' demands we admit a reality that is inherently unstable and potentially hostile, while also being the necessary presuppositions of valuable human life. This thus compels 'responsible' management of this complexity to render (socio-)'nature' hospitable, and in diverse 'heres' and 'nows'.[3] It is also, notably, an approach to nature that is strikingly resonant with Confucianism (Tse 2016; Elvin 1993) in terms of privileging government that aims for human-nature harmony but through intelligent *intervention* and *control* of the latter by the former.

In this case, though, any and all 'variables' may – indeed, *should* – be incorporated in complex models of increasing sophistication if we are to be able to understand and intervene in these 'natural' processes 'rationally' and 'responsibly'. In a move equivalent to redefinition of the rational individual above, therefore, this redefinition of 'nature' resituates the normative force of the claim regarding the 'naturalness' of that which is revealed by 'rational' science while essentially preserving its privileged status; in this case, as that which ends all arguments (and, per Latour 2009, controls all mobs), the brute unarguable fact. The 'natural' is now anything that is spontaneously emergent, that exceeds the essentially limited (if always progressing) knowledge of humanity – whether as individual or institutional understanding and planning – and thus (rationally!) merits our supreme respect … but with a view to its mastery for 'our' optimized benefit now with 'smart' power/knowledge technologies, not for its own flourishing (whatever that may mean).

The emergence of this new 2.0 complexity episteme, therefore – coproduced with complexity-attentive *innovation* – reframes and rejuvenates an essentially classical liberal power regime that posits the *natural* superiority of a form of government that works through the *rational* agency of free *individuals*, with each of the italicized terms suitably redefined. As a classical liberalism, there is no clearer demonstration of this, and of how these redefinitions come together, than regarding a revitalization of the argument for the market as the crucial, *as epistemic*, institution of a free and rational society. For with 'natural', 'rational' and 'individual' redefined in terms of 'complex' and 'spontaneously emergent', 'intelligent/smart', and 'networked' respectively, the key argument of liberalism no longer seems transparently false, as it has for nearly a century since the collapse of liberal British hegemony in the 1914–1945 interregnum. Let us express the classical liberal common-sense thus (with apologies for the repetitions, which are included to highlight the argument in its fullness):

> There is a natural tendency for self-regulating markets to emerge from the natural interaction of negatively free and rational individuals expressing the natural propensity to truck, barter and exchange. And this, if rationally left to itself, constructs a rational social order, characterized by *collectively* optimal outcomes [of prosperity and its allocation], that also maximizes individual negative freedom and thereby also presents the limits to rational exercise of state power.

I write this out again in full to show how this absurd (and obviously, historically *disproven* (e.g. Polanyi 1944/1957)) argument – one that even libertarians and neoliberals do not believe today – is now rendered curiously compelling (if inescapably prolix):

> There is a complex, spontaneous bottom-up tendency for self-regulating markets to emerge from the complex, spontaneous interaction of negatively free and complexity-attentive, responsible networked-individuals expressing the complex, spontaneous propensity to truck, barter and exchange [and network]. And this, if responsibly left to itself, constructs a complexity-attentive and responsible social order, characterized by *collectively* optimal outcomes [of resilience and optimal innovation], that also maximizes networked-individual negative freedom and thereby also presents the limits to complexity-attentive, responsible exercise of state power.

One crucial element of this redefinition – and the promising system-productiveness of that liberalism, breaking with neoliberalism – is the way in which, through a reframing of the composition of knowledge and of 'nature', it reconstructs a credible connection between the normative basis of liberal power in both the free individual *and* its production of *collectively* optimal outcomes. Credible, we must add, but not 'true', and hence still inseparable from projects of asymmetric empowerment (see below).

The admitted possibility of market failure is another crucial element of reconstructing the normative basis of a classical liberalism *against neoliberalism*. For it resituates the normative justification of the market not any longer in itself, as for neoliberalism's epistemic market fundamentalism. Rather the argument is rebased in a supposedly rational case (i.e. of reasoned means-end argument) regarding what of value it can usually be expected to achieve, *especially in terms of collective or systemic-level outcomes as appraised through an epistemic lens.* Market failure, therefore, can be readmitted by an essentially liberal-capitalist episteme precisely because it now once again consists of *specific* and rectifiable instances of market-based interaction in which this test was failed.

The necessary response, therefore, is either to repair the *epistemic* shortcomings of the networked-individual participants that triggered the failure of that market; or, where 'rational' complexity-attentive knowledge newly tells us that the conditions are such that provision of this specific good by the market will be collectively suboptimal, then the state or other public institution can legitimately provide them instead – especially if the state *itself* is reconstituted as complexity-attentive (Goldstein and Tyfield 2017) – such that it avails itself of complexity-informed judgement and processes of policy-making. In both instances, however, this is an open invitation to digital and complexity-oriented innovation to provide platforms that will plug these gaps in complex system government (and Government) – both of which are projects already being pursued with vim, albeit at present predominantly framed by self-destructive Googliberal strategies (e.g. Morozov 2014; Micklethwait and Wooldridge 2015). Of course, a renewed legitimation for state power as *central* in constructing complexity-attentive political economies also resonates strongly with the possibility of contemporary *Chinese* global leadership and hegemony (see below).

In short, in the (brief and inescapably speculative) outlines of this liberalism 2.0 or complexity liberalism we can discern self-propagating power/knowledge dynamics that would feed both themselves and the immanent triumph over neoliberalism –

thereby *allowing* this zombie regime finally to fail, power systems abhorring a vacuum – based on a profound reorientation to new complexity-attentive relations to knowledge and system government. And in ways that bear strong *ex ante* resonance with trends and characteristics of contemporary China and its innovation-as-politics, and hence with its emerging historic bloc and leading agency.

To be sure, while identifying the historic bloc merely affords the tendential conclusion that a new resettled regime of global capitalism is in prospect, the openness and uncertainty of the future – and especially a future beset by the Four Challenges – offers no concrete guarantees of actual outcomes. For instance, we have illustrated this elsewhere regarding a set of (at least) four plausible (Wilkinson *et al.* 2013) scenarios for urban mobility futures in China (Tyfield, Zuev *et al.* 2016, summarized in Table 11.1). While primarily formulated in order to stimulate strategic responses by stakeholders in that domain of innovation, the four scenarios also illuminate four of the ways in which the interaction of current trends regarding urban mobility could co-produce broader regimes of system government. As Table 11.1 summarizes, however, two of these four scenarios would involve a liberalism 2.0 that is thwarted or stalled in some profound way. But even in such eventualities, a liberalism 2.0 remains pre-eminently influential just as the historic bloc would still continue to shape such futures profoundly, and likely to their systematic advantage.

Our task was to trace a plausible and empirically observable continuation of the incumbent dominant system logics of global capitalism, beyond and *through* the terminal crises of neoliberalism and its Four Great Challenges. Here, therefore, it seems that this has been reached. In other words, not only do we now have evidence against the imminent implosion of capitalism, but we also have strong evidence across multiple dimensions of its tendency towards resettlement and resurgence, perhaps to historically unprecedented heights. And, crucially, we have this in the outlines of *how* it will do so and the substantive, meso-level, *qualitative* power/knowledge dynamics of this regime of system government.

One cheer for liberalism 2.0 – how 'China' may 'rule' 'the world'

It would be churlish in the extreme to deny the multiple advantages and benefits of such a new capitalist regime regarding each of the Four Challenges, especially over the extraordinary destructiveness of an exterminist (Biel 2012, citing Thompson 1980) neoliberalism and the dark forces of outright fear and hatred its demise is unleashing. Climate change would become an unquestioned and uncontroversial priority of Government across the polluting world, 'developed' and 'developing', 'North' and 'South'. For instance, China and India could increasingly accept their inescapable global responsibility for mitigation and adaptation (Harris 2011; Duara 2014), thereby also shaming the US into action and/or empowering intra-national American forces of cleantech innovation. While the growing success in low-carbon innovations and industries could finally make decisive moves to sustainable system transition as these sectors become self-propelling, commercially attractive and highly competitive.

Global system complexity may also become something that is worked with, not denied, in governmental practice of increasing adeptness and in ways that harness, rather than seek to control and order, human individual (and global risk-mediated) liberty, including through digital platforms. Techno-economic innovation and

Table 11.1 Summary of the 4 scenarios for future Chinese urban mobility

Scenario		Slow people-centric mobility	Hi-tech elite mobility	'BAU' digitized mobility	Secure, splintered e-mobility
Equivalent CP/KS Regime		Neo-Confucian liberalism 2.0	Hi-tech liberalism 2.0	Stalled liberalism 2.0	Post-catastrophic liberalism 2.0
Key System Elements	(Daily) mobility patterns and demand	Mobility is slower, smoother, less hurried and simply less	Always on the move	Highly individualized lives, competing for networked opportunities	Personal responsibility for local (i.e. metropolitan) system resilience
	Vehicle forms and use	A shift to diverse ecologies of small electric vehicles and transit-oriented development	Elite EV systems … with a highly proscribed but still massive E2W niche	Hybrid cars trump plug-in battery EVs	Shift from cars to EV-*disu* small electric cars … and bikes
	Industries, innovations and employment	Chinese mobility innovation, including social innovation, leads the world	Global corporate knowledge-economy concentration, centred in China	Innovation is concentrated in digitized information systems, not in mobility systems	Global-local champions of security innovation
	Infrastructures and energy and resource systems	Significant new infrastructures effecting a major decarbonization	Unlimited growth of high-speed, high-tech, smart mobility infrastructures	Infrastructures for smart interconnectivity … and cars, not EVs	New local public-private infrastructures of distributed provision and surveillance
	Regulations and (local) governance	Cars and acceleration of life are strictly limited	Green but socially regressive regulation	Personal carbon budgets, monitored at every step	Strict regulation by devolved city governments
	Aspirations, values and status competition	The modern Chinese sage	The fast-living high-mobility innovator	The influential entrepreneur celebrity	The responsible custodian of the complex urban system
	Environment	Deep decarbonisation and local environmental sustainability measures	Environmental improvements, but erratic and unequally distributed	Rebound effects, slow environmental improvements and worsening environmental risks	Accelerating but highly urban-localized environmental improvements

Source: Tyfield, Zuev et al. (2016).

employment may become re-regulated and re-energized, not least through taking into account the systemic costs of mass un- and/or under-employment for *capitalism itself* and for individual corporations, especially massive transnational and knowledge-intensive ones like the auto sector (Hofmann and Sun 2016). Similarly, disruptive, low-cost innovation that focuses on *capital*-substituting and (cheap) labour-*using*, not labour-substituting, socio-technical change could spell new productive and profitable industries creating low-cost system-bad-mitigating innovations and new jobs, even for 'low-skilled' workers, that support growing wages and demand, in positive feedback loops (see Figure 11.1).

And the combination of a rebased and more self-confident and empowered complexity liberalism and non-elite (i.e. 'middle-class') cosmopolitanism – across not just the global North but also in China, India and many other populous 'rising' nations with massive and influential diasporas – with an ever-more empowered China (and,

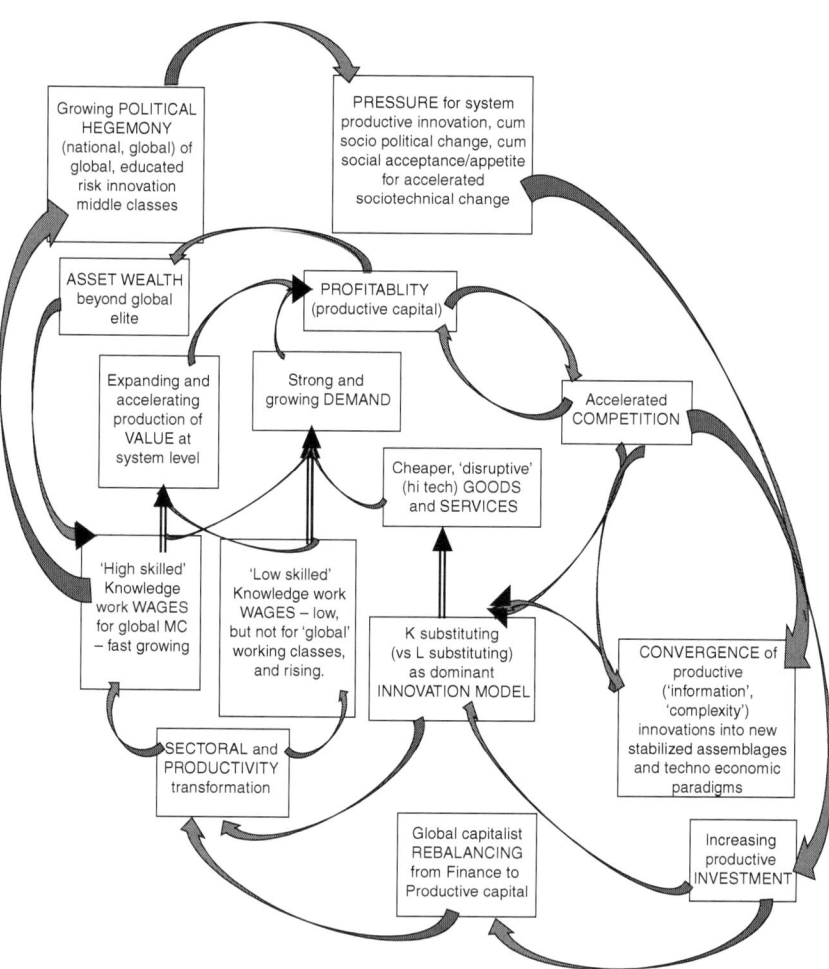

Figure 11.1 Re-established positive cycles of increased production of value and wages.

possibly, (parts of the) global South (Starrs 2014; Hung 2016)) could reach towards a new common global geopolitical settlement and a new Pax Sinica. Regarding simple cycles of geopolitical dominance, for instance, it seems at least plausible that even the military security threat of fundamentalist militant Islam will be defused more quickly where China is ultimately brought into the 'GWOT' than by an over-reaching (Johnson 2006) US alone.

For instance, on the one hand, the new routes to the Middle East and Central Asia that China itself is building under its 'new Silk Road' policy could facilitate both a new traffic of jihadism into its own Islamic-separatist challenge in Xinjiang, and conversely new dependence on this region's fossil fuels. And then China's ongoing massive growth in military projection capabilities, conventional and cyber (Austin 2014), may prove decisive in tipping the balance. But, on the other hand, China's promise of a new non-neoliberal, non-oil-based and simply non-*American* global peace and prosperity may be essential to the ultimate restabilization of this volatile region; just as supersession of the colonial Pax Britannica by American-sponsored European peace and global decolonialization was crucial in the late 1940s (Arrighi 1994).

In the midst of the whirlwind of collapsing neoliberalism – its bloodied hands in almost every major news story or catastrophe today, from Trump (and Clinton!), Brexit and a new populist illiberalism now ascendant even at the heart of global capitalism, to Syria and the ensuing migration crisis, to Hong Kong, to yet another warmest year on record ... – this preview of a *more* settled world surely merits at least one cheer; and so too, therefore, does the regime of government that could make it happen. And this is true from the perspective of both China itself and the world system as a whole.

Liberalism 2.0 – emergent from current Chinese developments – would, in turn, be highly advantageous for China in multiple ways through growing productive, low-carbon digital innovation firms and middle-class consumer demand. This would rebalance the economy towards growing wages and consumption (at least of the white-collar middle classes) and substantially transform the growth model to the officially hoped for 'new normal' of increasingly globally-competitive knowledge-intensive services, goods and jobs. And it would do this through obliquely harnessing the singular intensity of the Four Challenges to their progressive domestication in driving further growth and power of the historic bloc. Indeed, even as that domestication is likely to prove slow and unevenly distributed, this too is simply more fuel to the dynamic, presenting the permanent stick of individual exposure to *worsening* security threats *and* their deepening unequal distribution that drive the dynamics of liberty-security.

But the supertanker of socio-political change in China would also be highly productive for the world (system). For it would increasingly furnish a singular (i.e. singularly large, rich-*and*-poor, still fast-developing, and exceptionally state-backed) agency for global capitalist transformation into a new upswing of productive ('MC' *per* Arrighi 1994) non-zero-sum growth, again harnessing the Four Challenges towards socio-technical system transition. Here, then, we may also finally return to our question from Part II and see *how* 'China' could well come to 'rule' 'the world' in the next 15–20 years – but where this obfuscating but ubiquitous phrase actually connotes and conceals profound qualitative transformation regarding each of these terms.

'China' now (i.e. following such redefinition) means a quintessentially capitalist nation-state dominated by the globally-connected, characteristically pragmatic and networked-opportunity-seeking historic bloc; and specifically the urban mega-city politics and political structures they dominate. Moreover, the increasing asymmetric interdependence of CCP Party-state on this historic bloc does not necessarily license any rash projection of the former's imminent collapse. For this is a *liberal*, not a *democratic* (e.g. Dean 2003; Losurdo 2010), China that we are imagining here, which could well be compatible with continuation of a one-party-state system, albeit with aspects of meaningful empowerment of the historic bloc. After all, nineteenth century Britain was still very far from universal suffrage. And, indeed, given the intensity of the Four Challenges in China and the scale of resources needed to tackle them, a strong state (if not necessarily a *party*-state) apparatus would remain significantly to the historic bloc's advantage, nationally and globally. But, conversely, it could well involve profound constitutional change in China, as corroborated by a recent surge in prognostications regarding the particularly intense problems today for the CCP that are epitomized, not disproven, by the ongoing moves to concentrate state power in the hands of Xi Jinping (e.g. Pei 2016; Shambaugh 2016; Brown 2016).[4] In either case, as we saw in Chapter 10, digital complexity could well be harnessed to afford a recognisably *Chinese* strong, authoritarian state that is nonetheless quintessentially *liberal* in systematically enabling and being co-produced with the autonomy-seeking networked-individuals of the urban middle risk-class, squaring this historical circle.

Similarly 'rule' is here profoundly redefined from the incumbent common-sense of the China ascendancy literature. This is not 'rule' in the sense that the US did even after 1945, through the Cold War with its nuclear military supremacy and leadership of the 'free world' in all spheres; let alone since its unipolar moment as the USSR collapsed through the 1980s/1990s. Rather, like British hegemony in the nineteenth century, it is 'rule' as *primus inter pares* and medium-term process; as global coordinator, dynamo and increasingly undeniable exemplar (in key Chinese locations for the global business and tourist gaze, if not, by far, everywhere) of world-city cosmo-networked complexity capitalism. And this could be increasingly set against the high-carbon (Steffen 2016), Government-disparaging, booming inequality and political and military dysfunction of the US. This new form of global 'rule' will also likely be compounded by the characteristically oblique and non-linear development of China's capacity for innovation and system government. For this will make China a hidden and unexpected hegemon that creeps into global centrality and will likely still furnish powerful evidence-based denials of its dominance right up to the moment when the RMB rivals the dollar (cf. McNally 2012; Hung 2016).

For instance, an ever-increasing number of Chinese firms will likely scale the heights of global leadership, combining their unique, disruptive complexity-adeptness and flexibility with a strong state that is increasingly oriented to assisting them in tackling the 'hardest cases' of system transition. Conversely, not only will these firms and industries develop non-linearly, especially as they negotiate the broad-based cross-societal liberty-security dynamics such grand projects cannot *but* elicit; but there will also remain for any foreseeable future a massive cohort of Chinese firms that are still cheap manufacturers of poorly-made goods and services. The development of Chinese innovation capacity thus takes on the profile of an iceberg with a tiny but soaring peak but also a huge rump below the surface threshold of global competitiveness that affords representative truth to almost every assessment in comparison with, say, the US: both more impressive and significantly weaker, the 'best' and the 'worst' place to do business etc. ... and where these double truths are, in fact, mutually interdependent (see Chapter 5) (Figure 11.2).

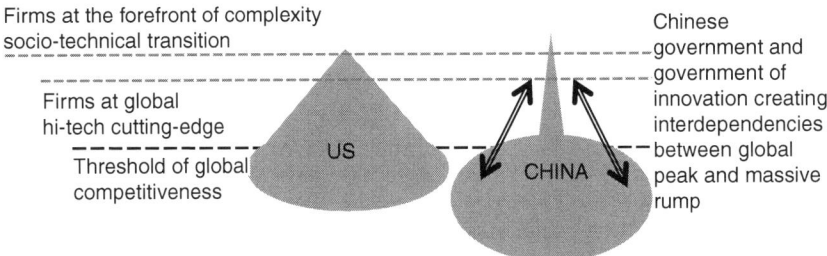

Figure 11.2 The 'iceberg' of Chinese innovation competitiveness.

Of course, these dynamics will also then drive the ongoing mutual learning process between China and the West that will characterize Chinese hegemony as a *process*; and a dynamic, quintessentially cosmopolitized process at that. Finally, then, it should also be noted that this 'rule' is thus to no-one's liking, and certainly not the incumbent parties dominating discussion of China's rise, namely the CCP, 'democracy'-exporting neoconservatives and neoliberals and orientalist Western left-wingers.

Finally, though, the 'world' too is recast here as a changing global system increasingly emergent beyond the self-destructive cycles of neoliberalism and its divisive model of globalization to a new dominant model of innovation-as-politics and concomitant global business organization, with all this entails: transforming global regulatory architectures and institutions, an international division of labour of innovation, flows of capital, investment opportunities and business or policy common-senses over the coming decades. Crucial in this regard, for instance, will be the re-entry of the state described above, itself transformed into the complexity-aware 'entrepreneurial' state (Goldstein and Tyfield 2017 cf. Mazzucato 2011), and the form of globalization that this coproduces. This could well take shape, for instance, as a two-speed globalization of *deeper* global mobility and interdependence for the knowledge-rich and 'skilled' alongside new *barriers* for knowledge-poor and 'unskilled', as is already apparently the May Government's approach for the UK in its Brexit negotiations. In short, a form of globalization transformed from that of a one-size-fits-all, borderless US-dominated neoliberal financialization to Chinese (urban)-dominated complexity-attentive globalization built upon productively reshaping (world) cities in light of their specific advantages and system situatedness. And where the more this embryonic regime grows, the more it will be able to harness complexity that incumbent neoliberal dynamics can only exacerbate, towards the eventual tipping point of the former against the latter.

The new nineteenth century – an age of liberalism, an age of revolution

One cheer, then, for liberalism 2.0. But only one. For that is just enough to orient strategically towards working *with* its power momentum and the present opportunities for shaping it (see below), but not enough to hail it without profound qualification. First, consider how even this brighter prospect (than neoliberal-induced system collapse) still has a long way to go, both in initial formation and emergence and then in broader global 'roll-out' (cf. Peck and Tickell 2002). There is no sense in the foregoing argument, then, that we – humanity as a whole, let alone the planet – are out of the proverbial woods yet regarding the Four Challenges, only that a glimmer of light has appeared on the horizon, faintly illuminating a way 'forward'. Moreover, crucial to the very power momentum and strategic plausibility of liberalism 2.0 is that essential contestation, as liberty-security, is the primary and constant source of its dynamism throughout.

Hence, liberalism 2.0 may be a resettled global capitalist regime, founded in non-zero-sum productive capital, and yet it is not a new *stable* and *harmonious* order.[5] But this would simply be another way in which it is a revived classical liberalism since this nineteenth century regime was also characterized by its parallel and mutually productive 'light' and 'dark' sides (Foucault 2004, 2009, 2010; Losurdo 2010), where the manifest and undeniable former – an age of unprecedented socio-economic and scientific 'progress' and adventure, this time found not in Africa and 'the Orient' but in space and undersea (*The Economist* 2016a; Vermeylen 2014) – also serves to occlude and overrule the latter, even as it itself co-produces it.

These dynamics would be compounded by a complexity episteme that studiously cleaves to its purely 'epistemic' status. The greater mastery of complexity and claims to 'reason' – albeit a complex, strategic reason that is always in practice dominated by particular sectional interests and tacit, self-serving common-senses – not only supports liberalism 2.0 CP/KS dominance, triumphing over and actively replacing a neoliberalism constitutively uninterested in emergent systems (except for the Market). This 'rational' progress, though, is actually driven by the arational, but highly affective and system-productive, vying of liberty and security, ambition and fear, individual opportunity and sharp-elbowed defence of it, that systematically benefits the system winners in positive feedback loops – not least through power/knowledge dynamics of innovation-as-politics. It thus also affords, and even compels, a growing self-righteousness in lock-step with its growing self-consciousness, where being a system winner is read as both meritocratic and good for the system itself. The flipside of this belief in the universal rational legitimacy of one's success, however, and with the support of all the emergent sophistication of complexity sciences at the forefront of human knowledge, is a deepening constitutive blindness to the concerns of subaltern groups – as invisible, illegitimate, outlandish and/or selfish from that hegemonic systems perspective – that will then likely underpin its most egregious injustices. And at the heart of this dynamic is the emergence, driven by its' winners, of a new category of capitalist social stratification – risk-innovation-class – just as the more familiar class system of industrial capitalism was constructed in the first instance in the eighteenth/nineteenth centuries by its bourgeoisie; hence a new form of 'global knowledge' *class* for the new form of 'global knowledge' *capitalism* (Figure 11.3).[6]

This process may be sharpest and most techno-economically productive where the stakes of liberty-security dynamics and exposure to the Four Challenges are highest, as in China with its pragmatic, hierarchical politics. But this is also a process that may be most turbulent and loudest – and hence most *politically* and *culturally* productive – in the global North, with its strong (if, at present, essentially contested) institutions of free speech in the public sphere, rule of law and free assembly.[7] Indeed, we can already begin to see this latter process at work in the pivotal instance of social media and the digital public sphere as it is taking shape in the West especially.

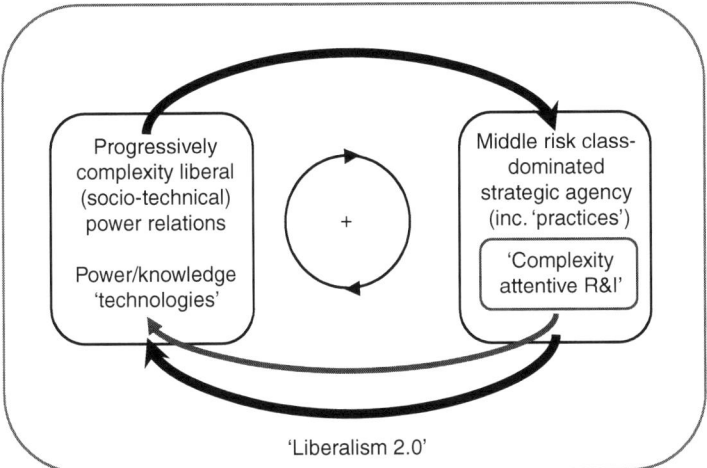

Figure 11.3 The complex power/knowledge system of liberalism 2.0.

To the great disappointment of libertarian prophets of the internet, social media has not unleashed a new dawn of free, rational and evidence-based argument that is leading to new and greater rational consensus – or at least not this alone and primarily. Rather, conditioned by the Googliberal innovations of its platforms and neo-liberalized subjectivities of users, it has enabled an efflorescence in the purely strategic and quickly heated-to-boiling-point battling of claims. This is generating both polarized, almost tribal culture-war division while normalizing, on all sides, the common-sense strategic education that knowledge claims are simply another tool in their strategic arsenal.[8] Of course, the archetypes of this public sphere and its effect on the broader process of government (and, indeed, Government) are the 2016 campaigns for Brexit and a Trump presidency in a 'post-truth politics' (Pinto 2017; *The Economist* 2016b).

This is, therefore, neoliberal agnotology and 'truthiness' gone mainstream, to the point that it threatens to infect fatally the democratic political process. But the systemic result of that process is the combination of: (i) the contingent and capricious shifting of the ground of the particular argument at stake (e.g. Brexit, (trans-)gender politics, fracking, immigration … and even free speech itself, regarding 'safe spaces', no platforming and 'trigger warnings', particularly in higher education settings (Bromwich 2016)); and (ii) the mass rejection of such 'polarized extremists' by a newly defined and *constituted* 'reasonable majority' of the 'normal' middle (class) as 'power majority'.[9] These arational dynamics of power/knowledge reconstitution thus not only engender (increasingly influential calls for) an emergent liberalism 2.0 digital public sphere and transformed, re-energized Western democracy (Flinders 2013). That is, by putting the established gains of liberal democracy in *question*, they rouse a new political movement in its defence, including ways to revitalize it vis-à-vis the new (epistemic/government) challenges of modern, cosmopolitized, complex 'knowledge' societies, as in citizens juries and 'sortition' (Van Reybrouck 2016). But they also *empower* an emergent liberalism 2.0 and its common-senses, which are quintessentially constituted in what they are *against* – i.e. the novel system-threatening political security threats emergent with the ongoing implosion of the incumbent neoliberal regime – in turn setting up the constitutive exclusion of perspectives and voices outside this emergent mainstream 'reasonableness'.

This dynamic can then be paired with that of liberty-security-responsive liberal 2.0 innovation-as-politics more generally. For this innovation merely productively *exploits* system complexity and dynamics of liberty-security for the benefit of the emerging system winners, in feedback loops, round and round. To be sure, this must involve taming at system level some of the most threatening system bads unleashed by neoliberalism, as systemic existential security threats not merely 'risks'; even if the elite can escape climate change on their personal yachts, the global middle-class cannot. But it also inevitably involves doing so slowly and unevenly, hence with the system bads continuing as matters of system metabolism and necessity in the medium term at least, and with them increasingly being moved (spatially and abstractly) where they are least offensive to the historic bloc. To be sure, this will almost certainly *not* be enough to establish anything resembling effective elimination of the system threats. But it will be demonstrable 'progress' in that direction, while the continuation of such system threats will continue to shape the deepening 'common-sense' that they must be tackled. In both respects, then – success and relative failure regarding socio-technical system transition – the complexity liberal power bloc is specifically empowered as icon of the newly emergent 'universal' good of 'us all'.

Together, then, the current common-sense strategic education on top of and shaping such innovation-as-politics elicits a fundamentally divided and unequal future, that *continues* the neoliberal process of systematically producing novel risks and then exporting them on to the shoulders of the regime's losers but with the significant difference that it does so 'better' (cf. Foucault 1991) given its foundational attendance to systemic risks and dangers. The very system dynamism, *as* dynamism of *system winners* within a fundamentally unequal incumbent socio-political order, thus spells a regime equally marked by huge, and indeed likely unprecedented and unconscionable, costs of this new settlement on the majority of humanity (as *not* members of the new global middle class) (e.g. OECD 2016; McGranahan *et al.* 2016), and non-humanity to boot. For, as a liberalism, such a regime cannot but be essentially divisive, exploitative and blinkeredly self-righteous – empowered *only insofar as it is such*. And armed with the historically unprecedented power/knowledge technologies emergent from its growing mastery of complexity and systems government, this will also likely come at a price of unprecedented ontological intensity to those newly bearing its costs, not least through liberalism 2.0's harnessing of, but not 'solutions' to, the Four Challenges.

Of course, though, this simply fires further the essential contestation, in turn to the systemic benefit of the emergent historic bloc over the medium-term. The combination of incumbent neoliberal regime breakdown – increasingly *driven* by liberal 2.0 regime emergence – with the failure of the latter to alleviate the uncertainty and insecurity of many (a global and intra-national majority), and even compounding it through new power/knowledge technologies, on the one hand, serves to enable newly radical and populist political movements, of Left and Right. Yet, on the other, this growing revolutionary fervour and break with established political common-senses both directly informs a redoubled determination by the historic bloc to remain pragmatically 'depolitical' in defence of the status quo of what they have won so far (including, in the West, liberal democratic constitutions), hence defining the 'reasonable' middle and power majority increasingly *against* such movements, in direct proportion to their popular appeal. And, indirectly, the dynamics of liberty-security and the pragmatic, enabled focus on liberty of the system winners in particular – their defining characteristic *as* system winners – continues to power their growing system enablement.

A new liberalism, in short, augurs a new nineteenth century of extraordinary advance – in the twenty-first century, across the world not just in the Northern trans-Atlantic – *and* neo-Dickensian intra-national inequality, exploitation and squalor. But this time it is not premised on the overweening Enlightenment epistemic confidence in the gospel of 'Progress' of a Macaulay, but the bleaker and more 'resilient', seemingly humbler but actually no less arrogant – and hence possibly much more dangerous – prospect of managing a global system accepted as untameably complex and with irreducible 'downsides' to one's (middle-risk-class networked-individual and collective) advantage.

The collapse of twentieth century social democracy

These cruel system ironies are particularly barbed for a progressive, radical Left. On the one hand, the continual vilification (e.g. in the UK, including by a hostile and virulent right-wing press; in the US, by the alt-right media) and frustration of these movements simply engenders their further radicalism. At its worst this pushes to self-defeating self-parody and certainly waves of excess, perpetually threatening wilful self-exclusion from mainstream politics, thereby proving the 'reasonable majority' right. On the other, adding insult to injury, their very impassioned and rejuvenated political imagination serves the crucial purpose of the emergent liberalism of raising issues and posing questions of (complex system) government that must be addressed but which the latter and its pragmatic, individualistic agents simply cannot imagine, much less resolve, by themselves. The radical movements simply shape the system-necessary 'progressive' gloss of the emergent liberalism, shifting the 'reasonable middle' in their direction but never gaining the initiative or political credit, in an essential parasitism of the emerging liberal regime. Radicalism thus strengthens and shapes a new self-styled liberal reasonableness both directly (in opposition) and indirectly (through the backdoor) to the latter's continual advantage. Like the early nineteenth century, then, the age of liberalism is also and inseparably an age of revolution (cf. Hobsbawm 1962), where the latter, unwittingly and self-defeatingly, fuels the former.

These dynamics thus compound the profound strategic weakness of a Left politics that could challenge liberalism 2.0's tendency to a chilling but highly system-productive heartlessness. First, against the incumbent regime, the singular goal of the neoliberal project has been to destroy the political power of the industrial working class and the public sector in the global North (Mason 2015). This political constituency was broadly system-empowered by the turn of the nineteenth/twentieth century and became increasingly so through the Keynesian welfare-warfare state regime, controlling key nodes of system government, such as energy flows (e.g. Freese 2003; Mitchell 2011, Nye 1998). But neoliberalism has achieved the dismantling and delegitimization (Jones 2012; Frank 2005) of this power bloc with extraordinary success, sundering both the complex power/knowledge system of twentieth century social democracy and with it the social relations that enabled social democratic parties to form large and powerful coalitions across their multiple and profound differences.

Moreover, the post-industrial (Western) working class is now facing the renewed assaults of the 'gig economy' and the continued rise of much cheaper *white* collar labour too in the global South. And nothing has – yet – been built in its place. The (Western) Left today is thus *necessarily* disunited, lacking any settled core constituency on which (what must inescapably be) a *coalition* of a self-confident democratic socialism can be based. While the end of 'real existing socialism' in 1989 and the 'end of history' (Fukuyama 2006) was then lamented as the nadir for the (Western) Left, we see today that it is the crisis of neoliberalism that is paradoxically the real nadir for twentieth century Left politics in the death of *social democracy*, now evident across the whole of Europe (*The Economist* 2016c).

But, second, to this we must add the evident alignment of the right and left of the British Labour and American Democratic parties, say, with the *real* dictates of the present as an age of emergent liberalism and revolution respectively, but where both ignore the crucial *interdependence* of these two essential characteristics of the present. Hence the social democratic right see only – though they are justified to see – a society (or least, reachable 'electoral middle') that is overwhelmingly uninterested in, if not hostile to, (or simply, beset by austerity, just too busy for) 'radical', organized (and especially work-based) politics and social mass movements. Pragmatic accommodation to this reality is thus the only possible starting point for a Left party to enter Downing Street or the White House. But this goal is pursued in its own right, at the cost of further evisceration of any popular support for such party machineries, now even from their taken-for-granted working class base. While conversely, embracing the revolutionary zeitgeist, the social democratic left see clearly the undeniable system need and potential for a 'new politics', but become intoxicated by the renewed sense of political possibility and idealism in ways that tend increasingly to speak only to the like-minded and thereby *reject* the broader public, catapulting the movement towards ever more strident rejectionism.

The ultimate victory of the age of liberalism thus is precisely that it *uses* the revolutionary fervour, and particularly of the Left, as unwitting drone in the construction of an essentially Liberal regime, not least through enfeebling the Left by turning it on itself. And it does this by presenting the Left with a Hobson's choice: accommodate and capitulate (namely, Blair's declaration to Bush that he was 'with you whatever' (*Guardian* 2016), or the Clinton campaign), or oppose in ways that feed the Dionysian beast underlying liberalism's cool Apollonian façade.[10] **But we do not have to accept this false choice.** Indeed, from this perspective, the challenge for the Left today is precisely to do everything possible to straddle these twin perspectives, and to work together so as to use the existing situation in crafting a brighter, progressive future … or face being used by it to entirely inimical ends.

New relations to knowledge: a phronetic ethics of complexity

Lest it need be said, therefore, this book is no argument *for* liberalism 2.0, even as it does argue both that liberalism 2.0 is as plausible a working hypothesis for contemporary strategic shaping of the future as we have and, as such, should be embraced to some extent. This is now clearly the case in late 2016 when we can see a resurgent liberalism as the best hope of defeating the new clear-and-present political danger of an insurgent *il*liberal, neo-fascistic (McDougall 2016; Müller 2016) populism that signals not just a Braudelian 'autumn' for capitalism but the darkness of a looming 'winter'. Rather, and this is the point of such embrace and engagement, the aim of this analysis is to reckon with fully and in advance the terrible violence and destruction that may yet be unleashed by the very *defeat and supersession* of neoliberalism, so that we may, strategically, mitigate it as much as possible and instead actively redirect trajectories of socio-technical and political change to brighter futures. A key aspect to this is precisely the challenge of working with the *power/knowledge* dynamics of an emergent complexity episteme (including complexity innovation-as-politics) that will likely prove particularly productive and, conversely, strategically enabling for those who link most closely to and dominate its development.

Given the still-early stage of its emergence and its openness to strategic intervention *before* it becomes sedimented, the question of 'what next?' – or rather '*which* new relation to knowledge? And to what "good society"?' – remains open to strategic action to a singular, once-in-a-lifetime degree. Hence it is a question for us – you! – to answer, in strategic practice (including of an innovation-as-politics as key arena of twenty-first century politics (Tyfield, Lave *et al.* 2016; Callon *et al.* 2009)), as the axiological imperative of the age, working *with* (the projection of) the emergence of liberalism 2.0.

How can we do this, and forge new relations to power/knowledge that promise more egalitarian, sustainable and equitable futures? We must start by acknowledging that understanding these system dynamics is itself an important strategic insight, since it counsels a politics that takes avoiding that Hobson's choice as a priority. This approach must first draw on, and continually learn from, the power/knowledge momentum of liberalism 2.0 to manifest crucial characteristics that must be in evidence in any proposed way forward and that together add up to a paradigm shift that can offer complex system government and productive responses to the other 3 Challenges. This perspective is fundamentally practice-oriented and strategic, processual, constitutively relational, complex systemic and essentially productive (see Chapter 3). Or to put this in a more easily recalled way, it involves a shift from dualistic and literalistic thinking, in terms of means and ends, to thinking in terms of practices and deepening capacities. 'Practices' are simply the everyday processual counterparts that build up 'capacities', but also presuppose them in turn.

Against a familiar Left politics, such a change in perspective rules out any detailed concrete formulation of ends or means, and hence any positing of fixed, future goal or of the favoured agencies and means (and/or power/knowledge technologies) to produce it. This is a radical turning about of perspective regarding knowledge itself, implicit in acknowledging that 'knowledge' is always and irreducibly *power*/knowledge situated in mutually constitutive but dynamic relations with other elements of broader systems. This counsels a broad strategic or 'political education' regarding knowledge – building on the one that is clearly well underway in global society under the twin pressures of neoliberal agnotology and internet 2.0-mediated production and circulation of knowledge and argument (discussed above). Indeed, it is in harnessing this political education, and the networked-individual and system-attentive responsibility it will slowly engender through cycles of liberty-security, that liberalism 2.0 will itself be able to re-establish a new political economic settlement.

To go *beyond* liberalism 2.0 involves two further key steps. First, whereas liberalism 2.0 works with the ongoing political education to reinstantiate a new boundary between politics and knowledge, a progressive response demands the *explicit* acknowledgement of *power*/knowledge, reformulating our implicit but presupposed ontologies and epistemologies around this. This immediately politicizes (or makes explicit the irreducible politics already therein) not just 'knowledge' and 'innovation', but also the worlds thereby produced. This is something that liberalism – constitutively tied to and empowered by the distinction of power *vs* knowledge, politics *vs* reason – cannot and will not admit, even as it pragmatically *deploys* power/knowledge. Let us call this the Foucauldian moment.

This first step, however, is not enough and risks being self-defeating. On its own, it would tend simply to compound a radical politicization of all things and knowledge claims that fuels the revolution-liberalism dynamic; to the extent it is not *directly* self-consuming as a virulent epistemic relativism. *Using* complex power/knowledge system dynamics, without having to declare this process – and, indeed, actively denying it – thus seems to be the secret weapon of liberalism 2.0 and its episteme. To escape this trap, then, we must turn to a new *ethical* vision that addresses what is potentially so troubling in liberalism 2.0 while guiding actual normative decision-making, in personal or public life in ways that temper the radical politicization of all things. This ethical vision is also most likely to be missing without explicit attention to it, as systemic blind-spot of liberalism 2.0.

Certainly, with this new ontological vision of complex strategic systems we can immediately see the sheer inadequacy of both of the (highly epistemic) frames of modern ethics that have increasingly dominated Western philosophy since the sixteenth/seventeenth centuries. Consequentialism becomes hopeless in complex systems since causation is so unsettled and consequences so uncertain and multiply produced. While deontology is simply ridiculous even in the most straightforward of cases (*Never* kill? *Never* lie?), dependent as it is on (an ontology of) timeless abstract truths and states as against the irreducible complexity and concreteness of actual lived predicaments, and particularly those of modern complex social life that *most* cry out for ethical guidance. Both approaches thus fade even further from practical usefulness for ethical decision-making, not even any longer coherent perspectives for academic philosophizing.

But taking this complexity turn, we find that we already have the answer at hand: in the phronesis (or situated practical wisdom (see Chapter 3)) that has framed this whole book, we already have the new *ethical* relation to knowledge for which we are looking, incubating a new relation, in turn, to each other, our 'societies', non-humans and even ourselves, that is (over time and *through* future practice) co-constitutive of all of them.

A CP/KS perspective entails that any thinking subject (including the professional (social) scientist) is situated *within* the world – of dynamic power/knowledge relations – they are trying to understand and thereby to influence to their advantage in action on and through power/knowledge technologies. The ethical predicament here, therefore, is recast in terms of the questions of how one can do this 'better', and what 'better' itself means; and, of no less importance, how we may come to *learn* in each case (cf. Harari 2015). But phronesis is actually the answer to all of these questions.

First, as a means and learning process, we start by noting that phronesis in the first instance involves epistemic practices that actively engage with concrete and/or meso-level realities so as to illuminate for as many stakeholders as possible (including oneself) the dynamic strategic landscape – of power/knowledge relations and technologies – of the issue in question, and how they are situated within it. The goal here is thereby to stimulate insightful strategic thinking in others *for themselves* – all of whom contribute to and condition their complex power/knowledge systems – in ongoing, never-ending contestation and (re-)constitution of such systems so that they become more responsive to and expressive of the irreducible diversity of agencies co-producing them. This book has aimed to contribute in just this way, illuminating a key emerging meso-level dynamic that will profoundly shape the twenty-first century, but which leaves open – and ethically calls urgently for – significant steering (see Table 11.1 in this light).

But, second, the ethical good implicit in this form of epistemic initiative goes beyond simply enabling as broad a (direct, practical, strategic) participation in shaping future worlds as possible. Rather, as a goal this itself presupposes a broader meta-epistemic, ethical vista implicit in ontologies of complex power/knowledge systems with their dynamic and constitutive relationality, namely regarding: (i) the compassion and generosity of spirit implicit in seeing the constitutive interdependence of oneself with countless others (human and non-human) who are essentially and irreducibly different to oneself; (ii) the patience and tenacity implicit in acknowledging the impossibility of worlds that are perfectly and durably to the liking of oneself or any of the interdependent others; and, bringing these together with the political education above, (iii) the centrality of an ethics of power/*knowledge* technologies and relations – a *wisdom* – as of supreme importance in actually transforming one's action in the world – as a thinking, ethical subject – and, in turn, the world thereby conditioned into actuality. The *goal* of an ethics of complex systems, therefore, is also (the cultivation of) phronesis as situated practical wisdom, where this is precisely a deepening *capacity* of selves and systems for ethically-aware and skilful judgement – responding to the complex, changing and unique situation at hand – in strategic, productive action.

But as such, we may return to phronesis as 'means' since this is now likewise recast as the *practice* through which that capacity is itself cultivated, and thus an orientation that guides the ongoing process through which one can come to deeper awareness and skilfulness, learning how to act strategically in the world 'better' *and* what 'better' means as two sides of the same coin. While phronesis thus can be formalized in acts of scientific analysis – such as this one, and our the starting point above – it is not limited to that domain and is, in fact, a much broader practice or epistemic *virtue* (MacIntyre 1997, Wright 2011); an open invitation and unshakeable imperative to everybody to engage strategically and ethically with the mutually conditioning processes of continually constituting themselves and their worlds, in an ongoing and deepening process of cultivating wisdom.

Together, this reveals a new ethical vision, for relations to knowledge and inseparably for the good society, as the 'phronetic civilization' and its corollary and coproduct(ion) of (not just 'responsible' (Stilgoe *et al.* 2013; von Schomberg 2013) but) '*virtuous* innovation'. The former connotes associations of reasoning, but also strategic and ethical, beings who by continually practicing the virtue of phronesis, in both their personal and public government, are actively constructing a civilization that just *is* learning (by doing) to live together well – a 'con-viviality'. While the latter attempts to capture the crucial process by which power/knowledge technologies mediating and coproduced with social agency and everyday practice are not just 'better' regulated and controlled (as in liberalism 2.0) but actively construct and instantiate *ethically* superior complex power/knowledge systems.

Here, as a practical and complexity-attentive epistemic virtue, phronesis becomes something that we – that is, each and everyone one of us – can, and must, begin to work on immediately in (any given, or 'our') 'here' and 'now'. This would include, not least, phronetic exercises of democratically holding innovation trajectories to account and actively shaping them, as producers, regulators, consumers and stakeholding publics. In both respects, it is always practice *in the present* that is the locus of concern, as opposed to formulation of 'real utopias' supposedly then to be realized (somehow) in the future (Wright 2010; Mason 2015; Bregman 2016). Rather, phronetic practice in the rolling present then necessarily and actively *shapes* unfolding futures that are understood to be unknowable and unspecifiable (vs forlornly attempting to manage the uncertain future (Nordmann 2014)) but now slightly different and, hopefully, significantly improved for the strategic practical wisdom addressed to them. Importantly for the Challenge of cosmopolitizing globalism, this dialogical, processual and pragmatic ethics is also much more strongly resonant with non-Western traditions, not least in a rising East and South Asia (Duara 2014). Hence the more enlightened elements of the historic bloc itself, the Chinese/global middle-risk-class, also become crucial protagonists in this project, not just its opponents.[11]

Transforming our relations to knowledge also thus includes transforming the substantive power/knowledge technologies of our episteme as well. But while a complexity turn can certainly be harnessed to regressive ends (Chandler 2014, Tyfield 2014) there is also much to celebrate on this score in terms of a raft of work that is engaging with complexity and systems thinking but in ways that are both analytically fascinating and ethically enlivening. This impeccably interdisciplinary work, opening a veritable Pandora's Box of modern heresies and surely amongst the most exciting intellectual frontiers of the moment, would include work regarding, for instance: Big History (Spier 2015; Christian and McNeill 2011); *longue durée* and cosmopolitan histories of civilizations (Duara 2014; Sachsenmaier 2011; Han and Park 2014; Morris 2010), geo-politics (Clark 2014; Yusoff 2013), the Anthropocene and Gaia (Bonneuil and Fressoz 2016; Szerszynski 2015; Palsson *et al.* 2013; Hamilton *et al.* 2015), the AQAL work of Integral Theory (Esbjörn-Hagens 2010) and, turning inward, the increasingly intuitionist psychology of judgement (Haidt 2012; McGilchrist 2009). Together this work may be seen to be incubating a new, anti-utopian and non-Romantic holism; another echo, of course, of the nineteenth century and the emergence of the classical liberal age albeit, again, with a crucial 'complexity' twist. Here, in other words, phronesis *as* practice of 'science' becomes the counterfoil to a renewed *liberal* episteme, as not a 'critical' complexity episteme – indeed, moving beyond the politically self-defeating attachment to the 'critical' label – but (much more hopefully and productively) an engaged, strategic and ethical one.

In all of this (power/)knowledge practice – of both 'science' and innovation-as-politics – it is the practice of phronesis *itself*, not the outcome of any single exercise of phronetic engagement with concrete innovation trajectories, that is most important, given the CP/KS perspective and the priority therefrom of what knowledge practices *do* and *coproduce* rather than what they ideationally conclude, agree or explicitly state. Engagement in exercises of phronesis to some small but never insignificant degree serves to cultivate phronetic *citizens* and *selves*. These agents are (ever more) adept with and attuned to not just the openness and complexity of socio-technical futures and the concomitant importance *and* limitations of (prospective) knowledges, but also their own specific situatedness at present within dynamic complex systems affording strategic openings that may otherwise be missed; and inseparably, to the inescapable predicament of *essentially* imperfectible, dynamic power/knowledge-strategic realities, and systemic interdependence. This is thus also a knowledge politics that both works with and celebrates irreducible diversity and multiplicity of agents constitutive of the world while also furnishing grounds for their coming *together* (not least *against* the as-yet unclear but deepening global challenges unleashed by emergent liberalism 2.0 in turn, through the coming century, e.g. of a trans-human global capitalist imperialism of a global elite and their androids vs 'mere' humans) and *common empowerment* in new power/knowledge *closures*, not just endless opening up, differentiation, fragmentation and weakening.

In other words, phronesis cultivates thinking active beings embodying and practising (progressively deeper) new relations to knowledge and who thereby *themselves* instantiate the only way societies of skilful and ethical complex government of complex systems could possibly be realized: not in any great and one-off 'transition' (or 'revolution'), from (bad) 'here' to (good) 'there', but a never-ending practice of deeper capacity to themselves assume responsibility for the essentially unknowable – and so ethically confounding (Harari 2015) – world-productivity of our power/knowledge technologies and the (mediated) creation of *new* ones. Indeed, regarding this latter point, through phronesis we can begin to imagine also how actively to shape specifically *virtuous* innovations. These power/knowledge technologies would in turn serve to mediate and co-produce socio-technical political futures, power/knowledge relations and selves that may be both more strategically enabled – thereby actually *constructing* better societies – and again more attuned to complex systems interdependence, irreducible diversity and system openness, and the limits of knowledge – thereby constructing *better* societies. And so on, round and round (see Figure 11.4).

Precisely, therefore, as practices and capacities, hence *both* means *and* ends in iterative feedback loops in the rolling present, phronesis offers a new vision: deepening cultivated complex power/knowledge systems of (the innovation of new) strategically-situated wise institutions, technologies and living, practising selves that

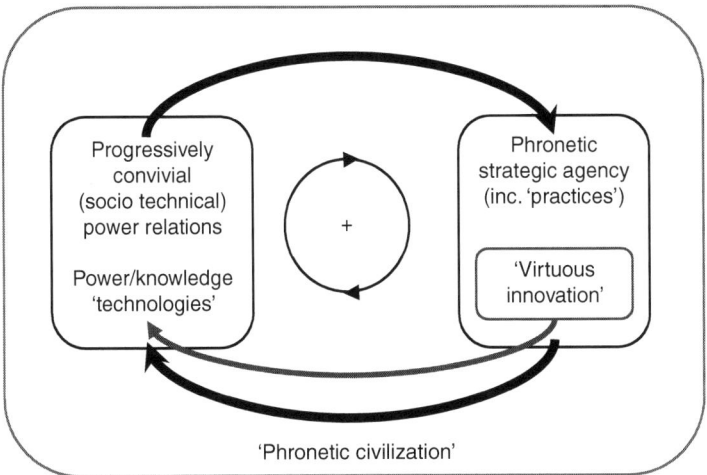

Figure 11.4 The complex power/knowledge system of phronetic civilization.

not only *can* work with the complexity of socio-tech-natural systems (including the likely intense and growing complexity of liberalism 2.0) but *do* so with growing skilfulness and attentiveness to others in a never-ending, strategic and ethical learning process across that society as a whole – and not just among the middle risk-class system winners.

Left liberalism 2.0 towards a mid-century neo-socialism

Having set out this ethical vision of new complex relations to knowledge, however, we can see the significant *strategic opportunities* of the present conjuncture for such a Left. For, on the one hand, neoliberalism has definitively destroyed not only the socio-political, but also the conceptual-philosophical, bases of post-Enlightenment secular Leftism in the crisis of knowledge (Chapter 1) to which it has led. We may add that there is little to lament about this death of nineteenth/twentieth century social democracy (let alone Stalinist or Maoist communism), given the enormous damage it reaped upon the world in both the Great Acceleration of fossil fuel consumption (Steffen *et al.* 2015) and the division of the world in First, Second and Third Worlds and cycles of under-development upon which the Keynesian welfare-warfare state systematically depended (Huber 2013; Urry 2013). To the extent we are prepared to look clearly at the Four Great Challenges, it is evidently impossible today simply to recreate a new New Deal as grand plan, albeit now a global and 'green' one (Perez 2016), that does not tackle one of the Great Challenges without exacerbating another; and this is primarily due to the total absence of socio-political constituencies, especially with their hands on the reins of Governmental power – and at what scale(s)? (e.g. Barber 2014; cf. Beck and Grande 2007; cf. Cabrera 2012) – capable of creating one and then shaping and implementing it over time (Goldstein and Tyfield 2017).

But, on the other hand, it is also the case that collapsing neoliberalism is simply a whirlwind of global destruction, destroying our prior relations to knowledge but putting nothing, other than the further destructiveness of unconstrained 'market forces' and 'entrepreneurialism' (subverted into its antithesis of parasitic asset rentiership (Birch 2016) or now flagrantly chauvinistic populist untruth), in their place. As such, the moment is ripe for a Left that takes this challenge on to have significant and lasting effect in shaping global futures in the twenty-first century and beyond, while it has nothing left to lose by doing so.

Moreover, to the extent that it is (the family of geographically diverse projects that will be) liberalism 2.0 that will most likely dominate this new relation to knowledge initially, there remains significant strategic opportunity for an engaged, strategic (and equally diverse (Duara 2014)) global Left to shape *it*, pulling it in more egalitarian and equitable directions, with significant effect in the medium/long-term. And this is so precisely because of the crucial strategic lacuna in the project of a secular liberalism 2.0 itself, namely its lack of deliberate (rather than *post hoc* and rationalizing) engagement with the reformulation of the ethics implicit in that emergent complex power/knowledge system; while this is *precisely* the ground on which a revitalized Left must and can be based.

Furthermore, this opportunity resonates with a particular strategic opening implicit in a resurgent classical liberalism. The latter, at the very least, enables both an ascendant economic liberalism and, inseparably, new horizons of *political* liberalism. The former is likely to be the preserve of a new Right, lobbying for the capitalist-conservative demands of the middle risk-class at their most self-preserving. But the latter opens up significant opportunities for a new Left, capturing the cultural politics of the zeitgeist, if not (immediately) the levers of hard, techno-economic power.

For instance, new but (in the first instance) essentially liberal discourses of rights are likely to be a highly enabling political agenda; and especially to the extent they are self-consciously phronetic rather than 'radical', as practical experiments infused with a new ethics of complexity and systemic inter-dependence. For instance, consider rights regarding individual access to systemic goods as commons (e.g. the 'city' (Harvey 2008), clean air (Walker 2012), well-being (Gay 2013), a universal basic income or social dividend (Stern 2016; Painter and Thoung 2015; Reed and Lansley 2016)) or rights *more* universally applied (across human intersectionality, non-Western cultures and/or non-human living beings, ecological systems and (possibly sacred) landscapes).

But the socio-economic issues can hardly be ignored. On the one hand, therefore, complex systems insights should be used to emphasize systemic interdependence and uncertainty in economic policies, e.g.: in the importance of high-quality and enabling education and training for all and at all ages, universal health and social care, creating complex system-adept citizens and 'innovators' not just workers; and in the cultivation of (knowledge and cultural) commons. For instance, noting how the essential liberal concern of social mobility and meritocracy in fact correlates with greater socio-economic equality (Calder 2016), a robust, resilient and innovative society thus becomes one that stewards this *systemic* health, not just maximal enabling of individual choice. Strategically, therefore, this approach would provisionally work with and beyond liberalism 2.0 as a *proximal* 'left liberalism 2.0' that is defined and distinguished by its *distal* vision of a trans-capitalist 'neo-socialism' later in the century. Such policies would contribute to the growth of the grounded, socio-economic forces for the latter's realization as they take shape through the emerging age of liberalism 2.0, just as the industrial working classes and its socialism crystallized through the nineteenth century (see Table 11.2). This is thus a strategy that is always framed along these two temporal strategic horizons, hence also thereby affording a constant dialogue between 'reform' and 'revolution' that sidesteps the self-defeating Hobson's choice described above.

The label 'neo-socialism' captures two essential aspects to this distal project. On the one hand, it is essentially 'neo-' in ways that bear striking resemblance to both the *goal* and the *means* of the renewal (and thence also repudiation) of the classical liberal project that was neo-liberalism. In the seminal discussions of neo-liberalism at the nadir for (classical) liberalism in its collapse of the early 1930s and '40s, the goal was the strategic revival of the liberal project (at that moment against not just the extremes of communism and fascism, but also the emergent hegemony of social democracy) while the means for this was the profound epistemic, and thereby ontological, rethinking of liberalism (see Chapter 1) (Mirowski and Plehwe 2009).

More specifically still, this involved the repudiation of the ill-founded epistemic self-confidence of classical liberalism in 'natural' categories of the Enlightenment – of human reason etc. (see above) – which actually turned out to be based on specific and, by the early twentieth century, passing socio-political conditions. These foundational categories were replaced through their wholesale rejection, instead placing the epistemic faith of *neo*-liberalism firmly in the supra-human decision-maker of the Market itself and its Promethean capacity to *create* the humanly-incalculable

Table 11.2 Left liberalism 2.0 towards neo-socialism

Four Challenges	Key Dimensions of Strategy and Policy	Proximal (strategic opportunities emergent from a CPKS analysis of the present, here as those of liberalism 2.0)	Distal (rolling, provisional extrapolation and longer-distance social goal: the slogans and open ideas of the future hence necessarily more abstract and possibly novel terminology)
	Overarching Political Project	Left Liberalism 2.0 towards …	"Neo-Socialism" as productive placeholder for Phronetic Civilization
	Innovation and Knowledge production	Experiments in phronetic government of major innovation/transition policies and bottom-up initiatives. Innovation/technology democracy beyond and disciplining RRI Experiments with different forms of knowledge ownership and benefit-sharing, with a view to systemic productivity of innovation	Institutionalized 'innovation democracy' for phronetic government of socio-technical creativity within an established 2.0 public sphere Multi-scalar regulatory architectures for Intellectual Monopoly Privileges (not Intellectual Property Rights) and growing global and local cultural commons
Cosmopolitized Globalism	Living together globally	Liberal Left cosmopolitan global 2.0 risk communities, particularly attentive to and held to account by multiple bottom-up, subaltern experiments of dialogical transcendent civic activism	Deepening con-vivial practices, institutions and power relations
Complex (Ethical Self-) Government of Complex Systems	Political communities	Beyond the 'entrepreneurial state' (taking incumbent neoliberal definition of 'entrepreneurial') with innovation itself redefined as power/knowledge Experiments with state as multi-levelled and -scaled Political parties as 'thought leaders/stewards'	Phronetic, engaged, public government and economic/innovation 2.0 trans*human democracy Established institutions of managing onto-political disagreements and difference Reinstitutionalized 'states' for maximum global-local self-government
	Government	Experiments in sortition (including regarding use and production of new future-illuminating power/knowledge technologies e.g. scenarios) Experiments in city and regional government and transnational cooperation Campaigns for new 'rights' e.g. 'to the city', to clean air, of 'higher' non-human animals and ecologies.... Reinvigorating 2.0 public sphere including new rights to non-trolling and new mores of social media practice and misuse....	New institutions bridging local and global democracy New institutions of accountable decision-making building on intuitionist insights regarding how to make public/private reasoning 'exploratory' not 'confirmatory'

Socio-technical Post-Human Emergence			
	Fiscal and monetary policy and institutions	Fiscal stimulus for green innovation and systemic debt-reduction, perhaps with simultaneous monetary policy tightening to return to 'normal' climate for investment/saving Experiments in living wage rates and universal basic income (UBI), together with revival of 'meritocratic' contributory welfare reform (against 'universal' 'entitlements') A new deal for global finance towards rebalancing in prudent 'boring' banking and financial innovation focused on stimulating socio-technical low-carbon transition Reduced working weeks and tax incentives for part-time work (focusing on 50–80% FTE)	Citizen wages or social dividends, perhaps now with global (or globally harmonized) elements Growth to systemic 'tipping point' of non-monetary forms of payment and trans-capitalist digital economies surrounding small islands of 2.0-public capitalist financial institutions Degrowth of resource-intensive capitalism and its positive replacement by 'development' measured with new metrics of creativity and well-being (vs. GDP)
	Class	Left/Green wings of liberal 2.0 global middle class maximally expanded and enabled through attention to 'left behind' of the emerging knowledge subaltern class and in both North and South, as systemic conditions of possibility and inter-dependence *for* productive liberal 2.0 innovation	A loose but resilient global networked coalition of trans-(capitalist)-labour ['class'] as *virtuoso* vanguard of neo-socialism for planetary post-humanity (cyborg and non-human) vs. elitist trans-humanists (and androids)
Planetary and Locally-diverse Ecologies	*Ecological sustainability and repair*	Climate change and planetary boundary mitigation and adaptation with low-carbon innovations that support and drive maximal demotic enablement 'Rights' to clean environments and environmental rights Growing complex system-attentive R&I focused on growing demand, not just supply New modes of valuation of ecologies, landscapes etc....	Global and local socio-natural self-government and ecological civilization

'optimal' socio-technical future. Just as neo-liberalism was thus a project of reframing the political project of *liberalism* through the lens of an explicitly post-Enlightenment strategic-productive epistemology, so too for neo-*socialism* in its Foucauldian moment.

Furthermore, and as corollary of this, neo-socialism is also thereby an *explicit* project of slowly and surely building the practical-cum-intellectual resources and resilience over the medium-to-long term such that, when the *next* crisis of capitalism comes along – as surely it must, and probably as soon as in the early part of the second half of this century at the exhaustion of the Chinese digital cleantech MC cycle – it will be there, ready-to-hand and ripe for its meaningful storming of the citadels of (global) state government, just as the neoliberals explicitly strategized in the twentieth century.

But, of course, neo-socialism is also significantly different from neo-liberalism. And given the intra-implication of 'means' and 'ends' in complex systems, just discussed, these differences speak not just to the clearly different political goals and visions, but also to the 'means', or practices, of their construction. This thus takes us to the second aspect captured in this label. For, on the other hand, it remains as essentially 'socialist', just as neo-liberalism is 'liberal', in its fundamental *ethical* rejection (not acceptance) of the essentially divisive CP/KS regime of liberalism in favour of the eudaimonistic vision of interdependent flourishing of all things (Therborn 2016); the key target of a necessary political agonism and a revitalization, if recasting, of 'politics' itself. Indeed, it must also be stressed that a turn to phronesis is to be set *on top of* and productively to *resituate*, not to replace or abandon, a more familiar 'critical' concern to defend those oppressed and exploited, not least by global capitalism – including regarding the as-yet unclear, but likely grave, injustices and wounds liberalism 2.0 will inflict on humanity (both the majority and the species as a whole) and non-humanity.

Yet this goal too has thereby been entirely recast, as discussed above, where this reframing may be captured in the crucial difference between the *conviviality* of neo-socialism against the *collectivism* and *solidarity* of socialism. 'Conviviality' must be understood not as a great collective coming together around a single, emancipatory vision that will thereby supposedly conclusively dispel disagreement and disharmonious living. The CP/KS approach and the Great Challenge of cosmopolitized globalism in particular demands that we accept that *there is no such vision*; at least as one that is specifiable *ex ante*. Rather, con-viviality connotes both the strategic-ethical imperative towards and mode for learning and building up new ethically-enlightened ways of dealing with – while optimally accommodating – (possibly profound, ontopolitical (Tyfield 2017)) differences and disagreements as and when they arise, *as they inevitably must* in such imperfectible, dynamic, complex systems as those of humans, living non-humans and socio-technologies.

The test of neo-socialist (and/or phronetic) con-viviality is thus not explicit and avowed adherence to the ultimate good of 'living together' (and always 'thusly', according to a specific, formulated and prematurely universal vision). Rather it is how skilfully, ethically and creatively we enact the *learning* of *how* to live together, building shared cosmopolitical 'homes in the world' (cf. Massey 1994) *without* adherence demanded to *any* such 'ultimate good' but simply by *ourselves* practising virtues of ethical complex system government (and, in time, Government). This neo-socialism would thus have to *learn from*, not just dictate to, each and every one of the multiple ethically-hopeful initiatives and experiments – of innovation-as-politics

– around the world today (e.g. Duara 2014). And a 'neo-socialism' will only be possible to the extent phronesis and the associated virtues are deliberately cultivated, *both* as means and resources in relevant political 'struggle' *and* as ends, over time profoundly shaping and situating the very goals of all such concrete agonistic agency too.

In this way, therefore, we – who else? – may yet dodge (or at least dance with) the false dichotomy of 'reform' vs 'revolution' and instead move ever more closely towards harnessing the power momentum of liberalism 2.0 in turn for the complex ethical *self*-government of complex systems; forging with virtuous innovation a phronetic civilization that does not just address the Four Challenges and defeats a threatening new barbarism, but does so in ways that are equitable, positively ethically enriching and enlivening (Keane 2015).

Notes

1 By 'literalist' I mean the specifically European approach to knowledge that treats knowledge claims as primarily representationally objective truths to be debated on that basis, regarding what knowledge *says*, as opposed to a more 'pragmatist' common-sense characteristic of, for instance, Eastern and South Asian cultures and traditions of philosophy, concerned with what knowledge claims *do* (see e.g. Duara 2014; Nisbet 2005). As Duara (2014) has set out in compelling detail, for instance, the resurgence of East and South on the global stage in the early twenty-first century suggests a – much-needed – rebalancing towards the latter approaches. Conversely, van der Pijl (2008) notes the irony of how the emergence of non-Western global powers for the first time in the modern, globalizing age is occurring alongside unprecedented heights in Western/Northern cultural dominance, not least in the flow of many of the best students from China and India to study highly positivistic STEM, economics and business degrees in the citadels of these paradigms at leading American and British universities.

2 Though there are also intra-doctrinal disagreements about just what role the state has to play (Mirowski and Plehwe 2009).

3 Notice how it also thereby reframes and affords a more 'progressive' but not unlimited and totally fluid politics of 'natural' boundaries of human identity and behaviour, including importantly gender – a crucial battleground of the transition to liberal 2.0 system government, as Mason (2015) rightly notices (see also Rose 2016).

4 Note also how the particularly *American* catastrophe of the Great Depression of the 1930s catalyzed the system-necessary inclusion of the (Euro-)American working classes *into* the bargain of global capitalism, albeit on tough terms (De Angelis 2000), in order to defeat the system threat of socialism, and thereby *drove* the US to global hegemony, rather than stalling its ascendancy. In similar vein, it seems equally plausible that a looming political economic crisis in China – whether from a real estate or debt crash or … – would simply catapult the historic bloc of China's middle-risk-class to greater heights of system dominance, both within China and globally, taking China with it to global hegemony. But in this case, any overthrow of the CCP would be more like a 1688 Glorious Revolution: an essentially *conservative* and bourgeois takeover aiming to *preserve*, not upend, the political economic status quo that the CCP could no longer provide or guarantee or that it directly frustrates.

5 And as such, it will likely continue to pose China enormous political problems, even as China rises to global hegemony, further compounding the non-linear processual nature of Chinese ascendancy. Indeed, if we compare the conditions outlined in Chapter 4 of China's long-term 'harmonious' regime of government with those described here as likely to prevail through much of the twenty-first century, it also

seems clear that far from restoring China to a durable position of global centrality and cultural supremacy, the twenty-first century as China's '*century*' will be as brief an interlude in the *longue durée* of global (capitalist) system dynamics as was the American twentieth century (actually more like 60 years (Arrighi 2007, 1994)) – and probably even briefer.

6 As elaborated in Table 7.1, these insights point to the essentially *productive*, rather than *re*-productive, aspect of risk-class, just as for 'class' with the emergence of industrial capitalism in the eighteenth/nineteenth centuries, with the emergent *middle* class – *not* the working class – thus the constituency *opposing* the incumbent, dying capitalist regime and forcing a renewal of capitalism (cf. Mason 2015). Against a familiar critical tradition of sociology, therefore, which is focused on critique of the reality of (the moral and political scandal of) class and its inequalities, the analysis here is ironically more in the spirit of Marx and Engels, whose seminal discussions of class at the dawn of industrial capitalism were precisely to identify a *new, emerging and hugely system-productive* form of social inequality, even as it seeks to go beyond that analysis to illuminate the renewal of class in an age of post-industrial global-risk capitalism. The corollary of this is, in turn, that we must think beyond the industrial working class (even including that of the global South and women) as the sole and necessary agents of the transcendence of capitalism (as Mason (2015) rightly notes) (see below).

7 This dynamic of Chinese economic and Western political leadership regarding learning to work with global complexity also then adds another way in which Chinese hegemony is to noone's clear and sole advantage. And this, in turn, affords continuing and profound scepticism about Chinese global leadership that will themselves feed liberty-security dynamics at the geopolitical level, driving the epochal profundity and cosmopolitized compulsion to deeper mutual understanding of China and the West (and the world more broadly) as a medium/long-term process (cf. Han and Park 2014; Tyfield 2016).

8 The latter is also evidenced by the contemporary popularity (including in China!) of American and British television dramas and comedies about political intrigue, spin and manoeuvring, e.g. 'House of Cards', 'Veep', 'Thick of It' etc....

9 By 'power majority' we mean a group that is systemically located within the complex power/knowledge system of the day such that they have disproportionate influence on system government and political trajectories. They are not, therefore, the demographic nor even the democratic majority, though they may be both.

10 This is further complicated by the sloppy, confusing and out-dated identification of the progressive Left *as* 'liberalism', particularly in the Anglosphere. The irony here is that a redefined and self-confident 'liberalism' is itself a necessary (but insufficient) step for the renaissance of the former, not least in its subsequent clear differentiation.

11 Compare again the nineteenth century, when key figures of an emergent socialism included such enlightened bourgeois figures as Marx, Engels, Owen, Morris and the Webbs.

References

Arrighi, G. (1994) *The Long Twentieth Century*, London: Verso.

Arrighi, G. (2007) *Adam Smith in Beijing: Lineages of the Twenty-First Century*, London: Verso.

Austin, G. (2014) *Cyber Policy in China*, Cambridge: Polity.

Barber, B. (2014) *If Mayors Ruled the World: Dysfunctional Nations, Rising Cities*, New Haven, CT: Yale University Press.

Beck, U. and E. Grande (2007) *Cosmopolitan Europe*, Cambridge: Polity.

Biel, R. (2012) *The Entropy of Capitalism*, Boston and Leiden: Brill.

Birch, K. (2016) 'Rethinking value in the bio-economy: Finance, assetization, and the management of value', *Science, Technology & Human Values*, 0162243916661633.

Bonneuil, C. and J.-B. Fressoz (2016) *The Shock of the Anthropocene: The Earth, History and Us.* London: Verso Books.

Bregman, R. (2016) *Utopia for Realists*, Amsterdam: The Correspondent.

Bromwich, D. (2016) 'What are we allowed to say?' *London Review of Books* 38(18): 3–10.

Brown, K. (2016) *CEO, China: The Rise of Xi Jinping*, London: I.B. Tauris & Co.

Cabrera, L. (ed.) (2012) *Global Governance, Global Government: Institutional Visions for an Evolving World System*, New York: SUNY.

Calder, G. (2016) *How Inequality Runs in Families*, London: Policy Press.

Callon, M., Lascoumes, P. and Barthe, Y. (2009) *Acting in an Uncertain World: An Essay on Technical Democracy*, translated G. Burchell, Cambridge, MA: MIT Press.

Chandler, D. (2014) *Resilience: The Governance of Complexity*, Abingdon and New York: Routledge.

Christian, D. and McNeill, W.H. (2011) *Maps of Time: An Introduction to Big History*, Berkeley: University of California Press.

Clark, N. (2014) 'Geo-politics and the disaster of the Anthropocene', *The Sociological Review*, 62(S1): 19–37.

De Angelis, M. (2000) *Keynesianism, Social Conflict and Political Economy*, Houndsmill, Basingstoke: Macmillan.

Dean, K. (2003) *Capitalism and Citizenship: The Impossible Partnership*, London: Routledge.

Duara, P. (2014) *The Crisis of Global Modernity*. Cambridge: Cambridge University Press.

The Economist (2016a) 'A new age of discovery', 2 January.

The Economist (2016b) 'Art of the lie', 10 September.

The Economist (2016c) 'Rose thou art sick', 2 April.

Elvin, M. (1993) 'Three thousand years of unsustainable growth: China's environment from archaic times to the present', *East Asian History* 6: 7–50.

Esbjörn-Hagens, S. (ed.) (2010) *Integral Theory in Action*, New York: SUNY Press.

Flinders, M. (2013) *Defending Politics*, Oxford: Oxford University Press.

Foucault, M. (1991) *Discipline and Punish*, translated by A. Sheridan, London: Penguin.

Foucault, M. (2004) *Society Must be Defended*: Lectures at the Collège de France 1975–1976. Translated by David Macey. London: Penguin.

Foucault, M. (2009) *Security, Territory, Population: Lectures at the Collège de France 1977–1978*. Translated by Graham Burchell. Basingstoke: Palgrave Macmillan.

Foucault, M. (2010) *The Birth of Biopolitics: Lectures at the Collège de France 1978–1979*. Translated by Graham Burchell. Basingstoke: Palgrave Macmillan.

Frank, T. (2005) *What's the Matter with Kansas?*, New York: Holt McDougal.

Freese, B. (2003) *Coal: A Human History*, Cambridge, MA: Perseus.

Fukuyama, F. (2006) *The End of History and the Last Man*, New York: Simon and Schuster.

Gay, R. (2013) 'Mainstreaming wellbeing: An impact assessment for the right to health', in M. Grodin, D. Tarantola, G. Annas and S. Gruskin (eds), *Health and Human Rights in a Changing World*, Abingdon and New York: Routledge: 212–232.

Glaeser, E. (2011) *Triumph of the City*. London: Pan Macmillan.

Goldstein, J. and D. Tyfield ([2017]) 'Green Keynesianism: Bringing the state back in(to question)?, *Science as Culture* (forthcoming).

Gramsci, A. (1971) *Selections from the Prison Notebooks*, London: Lawrence and Wishart.

Guardian (2016) 'The Chilcot Enquiry: Editorial', 6 July, www.theguardian.com/commentisfree/2016/jul/06/the-guardian-view-on-the-chilcot-report-a-country-ruined-trust-shattered-a-reputation-trashed.

Habermas, Jürgen (1989) *The Structural Transformation of the Public Sphere: An Inquiry into a Category of Bourgeois Society*, Oxford: Polity Press.

Haidt, J. (2012) *The Righteous Mind*, London: Penguin.

Hamilton, C., F. Gemenne and C. Bonneuil (eds) (2015) *The Anthropocene and the Global Environmental Crisis: Rethinking Modernity in a New Epoch*, London: Routledge.

Han, S.J. and Y.D. Park (2014) 'Another cosmopolitanism: a critical reconstruction of the neo-Confucian conception of *Tianxiaweigong* in the age of global risks', *Development and Society* 43(2): 185–206.

Harari, Y.N. (2015) *Sapiens: A Brief History of Humankind*, London: Vintage.

Harris, P.G. (2011) *China's Responsibility for Climate change: Ethics, Fairness and Environmental Policy*, London: Policy Press.

Harvey, D. (2008) 'The right to the city', *New Left Review* 53: 23–40.

Hobsbawm, E. (1962) *The Age of Revolution*, London: Abacus.

Hofmann, M. and Y. Sun (2016) 'Rebalancing what and upgrading whom? Compensation and education in China as a core driver to further pave the way of steady social development', paper presented at the *International Symposium on Innovation-Driven Development*, Sun Yat Sen University, Guangzhou, 13–15 June.

Huber, M. (2013) *Lifeblood: Oil, Freedom, and the Forces of Capital*. Minneapolis: University of Minnesota Press.

Hung, H.F. (2016) *The China Boom*, New York: Columbia University Press.

Jessop, B. (2006) 'Spatial fixes, temporal fixes, and spatio-temporal fixes', in N. Castree and D. Gregory (eds), *David Harvey: a Critical Reader*, Oxford: Blackwell: 142–166.

Johnson, C. (2006) *Nemesis: The Last Days of the American Republic*, New York: Holt Paperbacks.

Jones, O. (2012) *Chavs: The Demonization of the Working Class*, London: Verso.

Keane, W. (2015) *Ethical Life: Its Natural and Social Histories*, Princeton, NJ: Princeton University Press.

Latour, B. (2009) *Politics of Nature*. Cambridge, MA: Harvard University Press.

Losurdo, D. (2010) *Liberalism: A Counter-History*, London: Verso.

MacIntyre, A. (1997) *After Virtue*, London: Duckworth.

Markusson, N., M.D. Gjefsen, J. Stephens and D. Tyfield (2017) 'The Political Economy of Technical Fixes: The (Mis)Alignment of 'Clean Fossil' and Political Regimes', *Energy Research & Social Science* 23: 1–10.

Mason, P. (2015) *PostCapitalism*, London: Allen Lane.

Massey, D. (1994) *Space, Place and Gender*, Minneapolis: University of Minnesota Press.

Mazzucato, M. (2011) *The Entrepreneurial State*. London: Demos.

McDougall, J. (2016) 'No, this isn't the 1930s – but, yes, this is fascism', *The Conversation*, 16 November, available at: https://theconversation.com/no-this-isnt-the-1930s-but-yes-this-is-fascism-68867

McGilchrist, I. (2009) *The Master and His Emissary: The Divided Brain and the Making of the Western World*. New Haven, CT: Yale University Press.

McGranahan, G., D. Schensul and G. Singh (2016) 'Inclusive urbanization: Can the 2030 Agenda be delivered without it?', *Environment and Urbanization* 28(1): 13–34.

McNally, C.A. (2012) 'Sino-Capitalism: China's re-emergence and the international political economy', *World Politics* 64(4): 741–776.

Micklethwait, J. and Wooldridge, A. (2015) *The Fourth Revolution: The Global Race to Reinvent the State*, London: Penguin.

Mirowski, P. and Plehwe, D. (eds) (2009) *The Road From Mont Pélerin: The Making of the Neo-Liberal Thought Collective*, Cambridge, MA: Harvard University Press.

Mitchell, T. (2011) *Carbon Democracy*, London: Verso.

Morozov, E. (2014) *To Save Everything, Click Here*, London: Allen Lane.

Morris, I. (2010) *Why the West Rules – For Now*, London: Profile Books.

Müller, J.-W. (2016) 'Capitalism in one family', *London Review of Books* 38(23): 10–14.

Nisbet, R.E. (2005) *The Geography of Thought*, London and Boston: Nicholas Brealey Publishing.

Nordmann, A. (2014) 'Responsible innovation, the art and craft of anticipation', *Journal of Responsible Innovation* 1(1): 87–98.

Nye, D. (1998) *Consuming Power*, Cambridge, MA: MIT Press.

Organisation for Economic Cooperation and Development (OECD) (2016) *Making Cities Work for All*, Paris: OECD.

Painter, A. and C. Thoung (2015) *Creative Citizen, Creative State*, London: RSA.

Palsson, G., B. Szerszynski, *et al.* (2013) 'Reconceptualizing the "Anthropos" in the Anthropocene: Integrating the social sciences and humanities in global environmental change research', *Environmental Science & Policy* 28: 3–13.

Peck, J. and A. Tickell (2002) 'Neoliberalizing space', *Antipode* 34(3): 380–404.

Pei, M.X. (2016) *China's Crony Capitalism: Dynamics of Regime Decay*, Cambridge, MA: Harvard University Press.

Perez, C. (2016) 'Capitalism, technology and a green global Golden Age', in M. Jacobs and M. Mazzucato (eds), *Rethinking Capitalism*, Chichester: Wiley Blackwell: 191–217.

Pinto, M.F. (2017) 'Agnotology and the new politicization of science and scientization of politics', in D. Tyfield, R. Lave, S. Randalls and C. Thorpe (eds), *The Routledge Handbook of the Political Economy of Science*, London: Routledge: 341–350.

Polanyi, K. (1944/1957) *The Great Transformation*, Boston: Beacon Press.

Rainie, L. and B. Wellman (2014) *Networked*, Cambridge, MA: MIT Press.

Reed, H. and S. Lansley (2016) *Universal Basic Income: An Idea Whose Time Has Come?*, London: Compass.

Rein, S. (2015) *The End of Copycat China*, Hoboken, NJ: John Wiley & Sons, Inc.

Rifkin, J. (2011) *The Third Industrial Revolution: How Lateral Power is Transforming energy, the Economy, and the World*, Basingstoke: Macmillan.

Rose, J. (2016) 'Who do you think you are?', *London Review of Books* 38(9): 3–13.

Rupert, M. (1993) 'Alienation, capitalism and the inter-state system: toward a Marxian/Gramscian critique', in S. Gill (ed.), *Gramsci, Historical Materialism and International Relations*, Cambridge: Cambridge University Press: 67–87.

Sachsenmaier, D. (2011) Global Perspectives on Global History: Theories and Approaches in a Connected World, Cambridge: Cambridge University Press.

Shambaugh, D. (2016) *China's Future*, Cambridge: Polity.

Shapin, Stephen (1995) *A Social History of Truth: Civility and Science in Seventeenth-Century*, Chicago: University of Chicago Press.

Spier, F. (2015) *Big History and the Future of Humanity*, Chichester: Wiley Blackwell.

Starrs, S. (2014) 'The chimera of global convergence', *New Left Review* 87: 81–96.

Steffen, A. (2016) 'Trump, Putin and the pipelines to nowhere', *Medium*, 16 December https://medium.com/@AlexSteffen/trump-putin-and-the-pipelines-to-nowhere-742d745ce8fd#.wwo5n3yuy.

Steffen, W., W. Broadgate, L. Deutsch, O. Gaffney and C. Ludwig (2015) 'The trajectory of the Anthropocene: the Great Acceleration', *The Anthropocene Review* 2(1): 81–98.

Stern, A. with L. Kravitz (2016) *Raising the Floor*, New York: PublicAffairs.

Stilgoe, J., R. Owen, and P. Macnaghten (2013) 'Developing a framework of Responsible Innovation', *Research Policy* 42(9): 1568–1580.

Szerszynski, B. (2015) 'Gods of the Anthropocene: geo-spiritual formations in the Earth's new epoch', *Theory, Culture & Society* 34(2–3): 253–275.

Therborn, G. (2016) 'An age of progress?' *New Left Review* 99: 27–37.

Thompson, E.H. (1980) 'Notes on exterminism, the last stage of civilization', *New Left Review* I/121, May-June.

Tse, E. (2016) *China's Disruptors*, London: Portfolio Penguin.

Tyfield, D. (2014) 'Not just complex, but dangerous – Response to Levins', *International Critical Thought* 4(2): 241–254.

Tyfield, D. (2016) 'Realizing the Beckian vision: Cosmopolitan cosmopolitanism and low-carbon China as political education. *In memoriam* Ulrich Beck', *Theory, Culture & Society* 33(7–8): 301–309.

Tyfield, D. (2017, forthcoming) 'Mobilizing the emergence of phronetic TechnoScienceSocieties: Low-carbon e-mobility in China', *Sociology of Sciences Yearbook 2017.*

Tyfield, D., R. Lave, S. Randalls and C. Thorpe (2016) 'Introduction', in D. Tyfield, R. Lave, S. Randalls and C. Thorpe (eds), *The Routledge Handbook of the Political Economy of Science*, London: Routledge.

Tyfield, D., D. Zuev, P. Li and J. Urry (2016) *The Politics and Practices of Low-Carbon Urban Mobility in China:4 Future Scenarios*, CeMoRe report, Lancaster University.

Urry, J. (2013) *Societies Beyond Oil*. London: Zed.

Van der Pijl, K. (2008) 'China's challenge to the West in the 21st century', Centre for Global Political Economy, University of Sussex.

Van Reybrouck, D. (2016) *Against Elections: The Case for Democracy*, Oxford: Bodley Head.

Vermeylen, S. (2014) 'Who owns the moon?', *The Conversation* 17 October. https://theconversation.com/who-owns-the-moon-32721.

Von Schomberg, R. (2013) 'A vision of Responsible Research and Innovation', in R. Owen, J. Besant and M Heintz (eds), *Responsible Innovation: Managing the Responsible Emergence of Science and Innovation in Society*, Chichester: Wiley: 51–74.

Walker, G. (2012) *Environmental Justice: Concepts, Evidence and Politics*, London: Routledge.

Wilkinson, A., R. Kupers and D. Mangalagiu (2013) 'How plausibility-based scenario practices are grappling with complexity to appreciate and address 21st century challenges', *Technological Forecasting and Social Change* 80: 699–710.

Wright, D. (2011) *The Six Perfections*, New York: Oxford University Press.

Wright, E.O. (2010) *Envisioning Real Utopias*, London: Verso.

Yusoff, K. (2013) 'Geologic life: prehistory, climate, futures in the Anthropocene', *Environment and Planning D: Society and Space* 31(5): 779–795.

Index

Page numbers in *italics* denote tables, those in **bold** denote figures.

Taylor & Francis eBooks

Helping you to choose the right eBooks for your Library

Add Routledge titles to your library's digital collection today. Taylor and Francis ebooks contains over 50,000 titles in the Humanities, Social Sciences, Behavioural Sciences, Built Environment and Law.

Choose from a range of subject packages or create your own!

Benefits for you

» Free MARC records
» COUNTER-compliant usage statistics
» Flexible purchase and pricing options
» All titles DRM-free.

Benefits for your user

» Off-site, anytime access via Athens or referring URL
» Print or copy pages or chapters
» Full content search
» Bookmark, highlight and annotate text
» Access to thousands of pages of quality research at the click of a button.

| REQUEST YOUR **FREE** INSTITUTIONAL TRIAL TODAY | **Free Trials Available** We offer free trials to qualifying academic, corporate and government customers. |

eCollections – Choose from over 30 subject eCollections, including:

Archaeology	Language Learning
Architecture	Law
Asian Studies	Literature
Business & Management	Media & Communication
Classical Studies	Middle East Studies
Construction	Music
Creative & Media Arts	Philosophy
Criminology & Criminal Justice	Planning
Economics	Politics
Education	Psychology & Mental Health
Energy	Religion
Engineering	Security
English Language & Linguistics	Social Work
Environment & Sustainability	Sociology
Geography	Sport
Health Studies	Theatre & Performance
History	Tourism, Hospitality & Events

For more information, pricing enquiries or to order a free trial, please contact your local sales team:
www.tandfebooks.com/page/sales